Campbel
Fall 1999

Electrochemistry at Metal and Semiconductor Electrodes

Electrochemistry at Metal and Semiconductor Electrodes

by
Norio Sato
Emeritus Professor,
Graduate School of Engineering,
Hokkaido University,
Sapporo, Japan

1998
ELSEVIER
Amsterdam – Lausanne – New York – Oxford – Shannon – Singapore – Tokyo

ELSEVIER SCIENCE B.V.
Sara Burgerhartstraat 25
P.O. Box 211, 1000 AE Amsterdam, The Netherlands

Library of Congress Cataloging-in-Publication Data

Sato, Norio.
    Electrochemistry at metal and semiconductor electrodes / by Norio
  Sato.
       p.   cm.
    Includes bibliographical references (p.   -   ) and index.
    ISBN 0-444-82806-0 (alk. paper)
    1. Electrodes, Oxide. 2. Semiconductors.   I. Title.
  QD572.O85S37 1998
  541.3'724--dc21   98-36139
                                                                 CIP

ISBN: 0.444.82806.0

© 1998 Elsevier Science B.V. All rights reserved.

No part of this publication may be reproduced, stored in a retrieval system or transmitted in any form or by any means, electronic, mechanical, photocopying, recording or otherwise, without the prior written permission of the publisher, Elsevier Science B.V., Copyright & Permissions Department, P.O. Box 521, 1000 AM Amsterdam, The Netherlands.

Special regulations for readers in the U.S.A. – This publication has been registered with the Copyright Clearance Center Inc. (CCC), 222 Rosewood Drive, Danvers, MA, 01923. Information can be obtained from the CCC about conditions under which photocopies of parts of this publication may be made in the U.S.A. All other copyright questions, including photocopying outside of the U.S.A., should be referred to the publisher.

No responsibility is assumed by the publisher for any injury and/or damage to persons or property as a matter of products liability, negligence or otherwise, or from any use or operation of any methods, products, instructions or ideas contained in the material herein.

⊚ The paper used in this publication meets the requirements of ANSI/NISO Z39.48-1992 (Permanence of Paper).

Printed in The Netherlands.

PREFACE

Electrochemistry at Electrodes is concerned with the structure of the electrical double layer and charge transfer reactions across the electrode/electrolyte interface. The purpose of this text is to integrate modern electrochemistry with semiconductor physics; this approach provides a quantitative basis for understanding electrochemistry at metal and semiconductor electrodes.

Electrons and ions are the principal particles that play the main role in electrochemistry. This text, therefore, emphasizes the *energy level concepts* of electrons and ions rather than the phenomenological thermodynamic and kinetic concepts on which most of the classical electrochemistry texts are based. This rationalization of the phenomenological concepts in terms of the physics of semiconductors should enable readers to develop more *atomistic* and *quantitative* insights into processes that occur at electrodes.

This book incorporates from many traditional disciplines of science and engineering such as interfacial chemistry, biochemistry, enzyme chemistry, membrane chemistry, metallurgy, modification of solid interfaces, and materials corrosion.

This text is intended to serve as an introduction for the study of advanced electrochemistry at electrodes and is aimed towards *graduates* and *senior undergraduates* studying materials and interfacial chemistry or those beginning research work in the field of electrochemistry.

Chapter 1 introduces a concept of energy levels of particles in physicochemical ensembles. Electrons are the *Fermi particles* of which the energy levels are given by the Fermi levels, while ions are the *Boltzmann particles* of which the energy is distributed in an exponential Boltzmann function. In Chapter 2 the energy levels of electrons in solid metals, solid semiconductors, and aqueous solutions are discussed. Electrons in metals are in *delocalized energy bands*; electrons in semiconductors are in delocalized energy bands as well as in localized levels; and *redox electrons* associated with redox particles in aqueous solutions are in *localized levels* which are split into the occupied (reductant) and the vacant (oxidant) electron levels due to the Franck-Condon principle. Chapter 3 introduces the energy levels of ions in gas, liquid, and solid phases. In aqueous solution, the acidic and basic proton levels in water molecules interrelate with proton levels in solute particles such as acetic acid.

In Chapter 4 the physical basis for the *electrode potential* is presented based on electron and ion levels in the electrodes, and the electronic and ionic electrode potentials are defined. Chapter 5 deals with the structure of the interfacial electrical double layer. The *potential of zero charge* of metal electrodes and the *flat band potential* of semiconductor electrodes are shown to be characteristic of individual electrodes. The interface of the semiconductor electrode is described as either in the state of *band edge level pinning* or in the state of *Fermi level pinning*. Chapter 6 introduces electrochemical cells for producing electric energy (chemical cells) and for producing chemical substances (electrolytic cells).

In Chapter 7 the general kinetics of electrode reactions are presented with the kinetic parameters such as the stoichiometric number, reaction order, and activation energy. In most cases the reaction affinity is distributed in multiple steps rather than in a single particular rate step. Chapter 8 discusses the kinetics of transfer reactions of electrons across the electrode interfaces. Electron transfer proceeds via the *quantum mechanical tunneling* from the occupied electron level to the vacant electron level. The complexation and adsorption of redox particles influence the rate of electron transfer by shifting the electron level of redox particles. Chapter 9 discusses the kinetics of ion transfer reactions which are based upon *activation processes* of Boltzmann particles.

Chapter 10 deals with photoelectrode reactions at semiconductor electrodes in which the concentration of minority carriers is increased by photoexcitation thereby enabling the transfer of electrons to occur that can not proceed in the dark. The concept of *quasi-Fermi level* is introduced to account for the photoenergy gain in semiconductor electrodes. Chapter 11 discusses the coupled electrode (*mixed electrode*) at which anodic and cathodic reactions occur at the same rate on a single electrode; this concept is illustrated by the example of corroding metal electrodes in aqueous solutions.

I wish to thank the Japan Technical Information Service for approval to reproduce diagrams from a book "Electrode Chemistry" which I authored. Special acknowledgment is due to Professor Dr. Roger W. Staehle who has examined and edited the manuscript. I am also grateful to Dr. Takeji Takeuchi for his help in preparing the camera-ready manuscript. Finally I am grateful to my wife, Yuko, for her constant love and support throughout my career.

Norio Sato.

Sapporo, Japan
April, 1998

CONTENTS

CHAPTER 1
THE ENERGY LEVEL OF PARTICLES 1
   1.1 Particles and Particle Ensembles 1
   1.2 Chemical Potential and Electrochemical Potential 4
   1.3 Electrochemical Potential of Electrons 5
   1.4 The Reference Level of Particle Energy 8
   1.5 Electrostatic Potential of Condensed Phases 9
   1.6 Energy Levels of Charged Particles in Condensed Phases 11
   References 13

CHAPTER 2
THE ENERGY LEVEL OF ELECTRONS 15
   2.1 Energy Levels of Electrons in Condensed Phases 15
   2.2 Electrons in Metals 19
      2.2.1 Energy band and the Fermi level 19
      2.2.2 The real potential and the chemical potential of electrons in metals 21
   2.3 Electron Energy Bands of Semiconductors 24
   2.4 Electrons and Holes in Semiconductors 27
      2.4.1 Intrinsic semiconductors 27
      2.4.2 n-type and p-type semiconductors 29
   2.5 Energy Levels of Electrons in Semiconductors 32
   2.6 Metal Oxides 35
      2.6.1 Formation of electron energy bands 35
      2.6.2 Localized electron levels 38
   2.7 The Surface of Semiconductors 39
      2.7.1 The surface state 39
      2.7.2 The space charge layer 42
      2.7.3 Surface degeneracy (Quasi-metallization of surfaces) 44
   2.8 Amorphous Semiconductors 44
   2.9 Electron Energy Bands of Liquid Water 45
   2.10 Redox Electrons in Aqueous Solution 47
      2.10.1 Electron levels of gaseous redox particles 47
      2.10.2 Electron levels of hydrated redox particles 48
      2.10.3 Fluctuation of electron energy levels 51
      2.10.4 The Fermi level of hydrated redox electrons 53
   2.11 The Electron Level of Normal Hydrogen Electrode 55
   References 58

CHAPTER 3
THE ENERGY LEVEL OF IONS 61
   3.1 Ionic Dissociation of Gaseous Molecules 61
   3.2 Metal Ion Levels in Solid Metals 63

    3.2.1 The unitary energy level of surface metal ions 63
    3.2.2 Metal ion levels at the surface and in the interior 65
 3.3 Ion Levels of Covalent Semiconductors 67
    3.3.1 The unitary level of surface ions 67
    3.3.2 Ion levels at the surface and in the interior 69
 3.4 Ion Levels of Compound Semiconductors 71
    3.4.1 The unitary level of surface ions 71
    3.4.2 Ion levels at the surface and in the interior 74
 3.5 Ion Levels in Aqueous Solution 76
    3.5.1 Levels of hydrated ions 76
    3.5.2 Proton levels in aqueous solution 78
 3.6 Thermodynamic Reference Level for Ions 85
 References 86

CHAPTER 4
ELECTRODE POTENTIAL 87
 4.1 Electrode 87
    4.1.1 Electrode 87
    4.1.2 Anode and cathode 88
    4.1.3 Electronic electrode and ionic electrode 88
    4.1.4 Polarizable and nonpolarizable electrodes 89
 4.2 The Interface of Two Condensed Phases 90
    4.2.1 Potential difference between two contacting phases 90
    4.2.2 The interface of zero charge 93
    4.2.3 Interfaces in charge transfer equilibrium 94
 4.3 Electrode Potential 96
    4.3.1 Electrode potential defined by electron energy levels 96
    4.3.2 Electrode potential and ion energy levels in electrodes 101
 4.4 Electrode Potential in Charge Transfer Equilibrium 103
    4.4.1 Electrode potential in electron transfer equilibrium 103
    4.4.2 Electrode potential in ion transfer equilibrium 105
    4.4.3 Potential of film-covered ionic electrodes in equilibrium 107
    4.4.4 Potential of gas electrodes in equilibrium 108
 4.5 Measurement of Electrode Potentials 110
 4.6 Potential of the Emersed Electrode 112
    4.6.1 Potential of emersed electrodes in vacuum 113
    4.6.2 Potential of emersed electrodes in inactive gas 114
 References 117

CHAPTER 5
ELECTRIC DOUBLE LAYER AT ELECTRODE INTERFACES 119
 5.1 Solid Surface and Adsorption 119
    5.1.1 Clean surface of solids 119
    5.1.2 Adsorption 121
    5.1.3 Electron level of adsorbed particles 122
 5.2 Electric Double Layer at Solid/Aqueous Solution Interfaces 127
    5.2.1 Electric double layer model 127
    5.2.2 Diffuse charge layer (Space charge layer) 129
 5.3 The Potential of Zero Charge on Metal Electrodes 132
    5.3.1 Classical model of the compact double layer at interfaces 132
    5.3.2 The potential of zero charge 135
 5.4 Thermodynamics of Adsorption on Metal Electrodes 138
    5.4.1 Gibbs' adsorption equation 138
    5.4.2 Ion adsorption on mercury electrodes 139

    5.4.3 Contact adsorption of ions 142
  5.5 Electric Double Layer at Metal Electrodes 143
    5.5.1 Interfacial electric capacity (Electrode capacity) 143
    5.5.2 The effective image plane on metal surfaces 144
    5.5.3 The closest approach of water molecules to electrode interfaces 146
    5.5.4 Electric capacity of the compact layer 148
    5.5.5 Potential difference across the compact double layer 150
  5.6 Contact Adsorption and Electric Double Layer 151
    5.6.1 Contact adsorption and work function 151
    5.6.2 Interfacial dipole moment induced by contact adsorption 153
    5.6.3 Interfacial potential difference affected by contact adsorption 155
  5.7 Particle Adsorption on Metal Electrodes 158
    5.7.1 Adsorption of water molecules 158
    5.7.2 Coadsorption of water molecules and third-particles 161
    5.7.3 Surface lattice transformation due to contact adsorption 162
    5.7.4 Electron energy levels of adsorbed particles 165
  5.8 Electric Double Layer at Semiconductor Electrodes 168
    5.8.1 Electric double layer model 168
    5.8.2 Potential distribution across the electrode interface 169
  5.9 Band Edge Level Pinning and Fermi Level Pinning 171
  5.10 The Space Charge Layer of Semiconductor Electrodes 174
    5.10.1 Space charge layers 174
    5.10.2 Differential electric capacity of space charge layers 176
    5.10.3 Schottky barrier 181
  5.11 The Compact Layer at Semiconductor Electrodes 181
    5.11.1 Hydroxylation of electrode interfaces 181
    5.11.2 The compact layer 184
    5.11.3 Differential electric capacity of electrode interfaces 187
  5.12 The Surface State of Semiconductor Electrodes 188
    5.12.1 Surface states 188
    5.12.2 Differential electric capacity of surface states 190
  5.13 The Flat Band Potential of Semiconductor Electrodes 192
    5.13.1 Flat band potential 192
    5.13.2 Band edge potential 195
  References 196

CHAPTER 6
ELECTROCHEMICAL CELLS 201
  6.1 Electrochemical Cells 201
  6.2 Electromotive Force of Electrochemical Cells 204
  6.3 Equilibrium Potential of Electrode Reactions 206
    6.3.1 Equilibrium potential of electron transfer reactions 206
    6.3.2 Equilibrium potential of ion transfer reactions 208
  6.4 Electrochemical Reference Level for Hydrated Ions 210
  References 211

CHAPTER 7
ELECTRODE REACTIONS 213
  7.1 Electrode Reactions 213
    7.1.1 Electron transfer and ion transfer reactions 213
    7.1.2 Cathodic and anodic reactions 213
    7.1.3 Electron transfer of hydrated particles and adsorbed particles 214
  7.2 Reaction Rate 216
    7.2.1 Forward and backward reaction affinities 216

    7.2.2 Reaction rate 217
    7.2.3 Polarization curve of electrode reactions 218
  7.3 Reaction Mechanism 220
    7.3.1 The stoichiometric number of reactions 220
    7.3.2 The activation energy 221
    7.3.3 Quantum tunneling and activated flow of particles 223
    7.3.4 The reaction order..225
  7.4 Rate-Determining Steps of Reactions 226
    7.4.1 Reaction of elementary steps in series...226
    7.4.2 Reaction rate determined by a single step 228
    7.4.3 Reaction rate determined by multiple steps...229
    7.4.4 Affinity distributed to elementary steps...230
    7.4.5 Rate of multistep reactions...232
  References..233

CHAPTER 8
ELECTRODE REACTIONS IN ELECTRON TRANSFER..235
  8.1 Electron Transfer at Metal Electrodes..235
    8.1.1 Kinetics of electron transfer 235
    8.1.2 The state density of redox electrons 238
    8.1.3 Exchange reaction current at the equilibrium potential 240
    8.1.4 Reaction current under polarization...242
    8.1.5 Diffusion and reaction rate...245
  8.2 Electron Transfer at Semiconductor Electrodes 249
    8.2.1 Semiconductor electrodes compared with metal electrodes 249
    8.2.2 The conduction band and the valence band mechanisms 250
    8.2.3 Electron state density in redox electrode reactions 252
    8.2.4 Exchange reaction current at the equilibrium potential 254
  8.3 Reaction Current at Semiconductor Electrodes..258
    8.3.1 Reaction current under polarization...258
    8.3.2 Reaction current versus potential curve...262
    8.3.3 The transport overvoltage of minority carriers 266
    8.3.4 Recombination of minority carriers 267
    8.3.5 Polarization curves of redox electron transfers..268
    8.3.6 Redox Fermi level and band edge level 270
    8.3.7 Electron transfer via the surface state 272
    8.3.8 Electron tunneling through the space charge layer 274
  8.4 Complexation and Adsorption in Electron Transfer Reactions 274
    8.4.1 Complexation shifts the redox electron level 275
    8.4.2 Contact adsorption shifts the redox electron level..278
  8.5 Electron Transfer at Film-Covered Metal Electrodes 281
    8.5.1 Electron transfer between the electrode metal and the redox particles.....................282
    8.5.2 Electron transfer between the film and the redox particles 284
    8.5.3 Polarization curves observed..286
  References..287

CHAPTER 9
ELECTRODE REACTIONS IN ION TRANSFER 289
  9.1 Metal Ion Transfer at Metal Electrodes...289
    9.1.1 Metal ion transfer in a single elemental step...289
    9.1.2 Metal ion transfer in a series of two elemental steps 294
  9.2 Ion Transfer at Semiconductor Electrodes..298
    9.2.1 Surface atom ionization of covalent semiconductor electrodes 298
    9.2.2 Dissolution of covalent semiconductors 302

    9.2.3 Dissolution of ionic semiconductors 305
    9.2.4 Oxidative and reductive dissolution of ionic semiconductors 309
  9.3 Ion Adsorption on Metal Electrodes 314
    9.3.1 Ion adsorption equilibrium 314
    9.3.2 Electron levels of adsorbed ions 315
  9.4 Ion Adsorption on Semiconductor Electrodes 317
    9.4.1 Ion adsorption equilibrium 317
    9.4.2 Electron levels of adsorbed ions 317
    9.4.3 Proton levels on electrode surfaces 319
  References 322

CHAPTER 10
SEMICONDUCTOR PHOTOELECTRODES 325
  10.1 Quasi-Fermi Level of Excited Electrons and Holes 325
    10.1.1 Quasi-Fermi level 325
    10.1.2 Quasi-Fermi levels and electrode reactions 328
  10.2 Photopotential 330
  10.3 Photoexcited Electrode Reactions 334
    10.3.1 Photoexcited electrode reaction current (Photocurrent) 334
    10.3.2 The range of electrode potential for photoelectrode reactions 338
    10.3.3 The flat band potential of photoexcited electrodes 344
  10.4 The Rate of Photoelectrode Reactions 347
    10.4.1 Anodic transfer reactions of photoexcited holes 347
    10.4.2 Generation and transport of holes 349
    10.4.3 Interfacial overvoltage of hole transfer 350
    10.4.4 Recombination of photoexcited holes in anodic reactions 352
    10.4.5 Cathodic hole injection reactions 354
  10.5 Photoelectrochemical Cells 357
  10.6 Photoelectrolytic Cells 357
    10.6.1 Photoelectrolytic cells of metal and semiconductor electrodes 357
    10.6.2 Photoelectrolytic cells of two semiconductor electrodes 364
  10.7 Photovoltaic Cells 367
  References 371

CHAPTER 11
MIXED ELECTRODES 373
  11.1 The Single Electrode and The Mixed Electrode 373
  11.2 Catalytic Reactions on Mixed Electrodes 375
  11.3 Mixed Electrode Potential 377
  11.4 Passivation of Metal Electrodes 381
    11.4.1 Polarization curve of anodic metal dissolution 381
    11.4.2 Metal dissolution in the passive and transpassive states 383
    11.4.3 Spontaneous passivation of metal electrodes 387
  References 389

LIST OF SYMBOLS 391

INDEX 397

CHAPTER 1

THE ENERGY LEVEL OF PARTICLES

1.1 Particles and Particle Ensembles

Materials and substances are composed of particles such as molecules, atoms and ions, which in turn consist of much smaller particles of electrons, positrons and neutrons. In electrochemistry, we deal primarily with charged particles of ions and electrons in addition to neutral particles. The sizes and masses of ions are the same as those of atoms: for relatively light lithium ions the radius is 6×10^{-11} m and the mass is 1.1×10^{-26} kg. In contrast, electrons are much smaller and much lighter than ions, being 1/1,000 to 1/10,000 times smaller (classical electron radius = 2.8×10^{-15} m, electron mass = 9.1×10^{-31} kg). Due to the extremely small size and low mass of electrons, the quantization of electrons is more pronounced than that of ions. Note that the electric charge carried by an electron ($e = -1.602 \times 10^{-19}$ C) is conventionally used to define the elemental unit of electric charge.

In general, a single particle has unitary properties of its own. In addition, a large number of particles constitutes a statistical ensemble that obeys ensemble properties based on the statistics that apply to the particles. According to the quantum statistical mechanics, the particles with half an odd integer spin such as electron and positron follow the *Fermi statistics*, and the particles with an even integer spin such as photon and phonon follow the *Bose-Einstein statistics*. For heavy particles of ions and atoms, which also follow either the Fermi or the Bose-Einstein statistics, both Fermi and Bose-Einstein statistics become indistinguishable from each other and may be represented approximately by the *Boltzmann statistics* in the temperature range of general interest.

The particles that obey Fermi statistics are called *Fermi particles* or *fermions*. The probability density of Fermi particles in their energy levels is represented by the Fermi function, $f(\varepsilon)$, that gives the probability of fermion occupation in an energy level, ε, as shown in Eqn. 1–1:

$$f(\varepsilon) = \cfrac{1}{\exp\left(\cfrac{\varepsilon - \varepsilon_F}{kT}\right) + 1} \,, \tag{1-1}$$

where k is the Boltzmann constant, T is the absolute temperature, and ε_F is the thermodynamic potential of Fermi particle called the *Fermi level* or *Fermi energy*. Fermi statistics permits only one energy eigenstate to be occupied by one particle.

Particles that obey Bose-Einstein statistics are called *Bose particles* or *bosons*. The probability density of bosons in their energy levels is represented by the Bose-Einstein function as shown in Eqn. 1–2:

$$f(\varepsilon) = \cfrac{1}{\exp\left(\cfrac{\varepsilon - \varepsilon_B}{kT}\right) - 1} \,, \tag{1-2}$$

where ε_B is the thermodynamic potential of Bose particles, called the *Bose-Einstein level* or *Bose-Einstein condensation level*. In Bose-Einstein statistics one energy eigenstate may be occupied by more than one particle.

Fig. 1–1 shows the two probability density functions. In Fermi statistics, the particle occupation probability (Fermi function) becomes equal to unity at energy levels slightly lower than the Fermi level ($f(\varepsilon) \doteq 1$ at $\varepsilon < \varepsilon_F$) and to zero at energy levels slightly higher than the Fermi level ($f(\varepsilon) \doteq 0$ at $\varepsilon > \varepsilon_F$), apparently decreasing from one to zero in a narrow energy range around the Fermi level, ε_F, with increasing

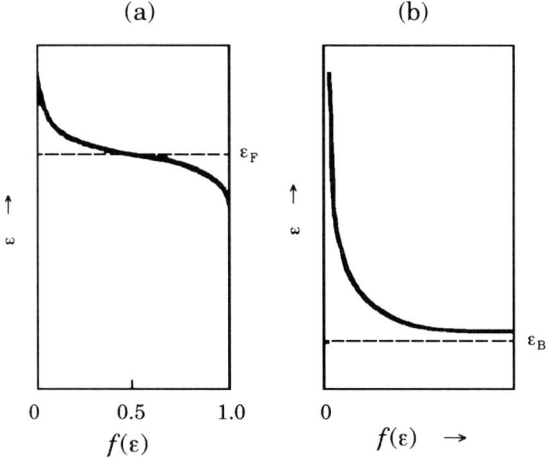

Fig. 1–1. Probability density functions of particle energy distribution: (a) Fermi function, (b) Bose-Einstein function. ε = particle energy; $f(\varepsilon)$ = probability density function; ε_F = Fermi level; ε_B = Bose-Einstein condensation level.

particle energy. On the other hand, in Bose-Einstein statistics the particle occupation probability decreases nearly exponentially with increasing particle energy above the Bose-Einstein level, ε_B. At high energy levels ($\varepsilon \gg \varepsilon_F$, $\varepsilon \gg \varepsilon_B$), both Fermi and Bose statistics may be approximated by the classical Boltzmann distribution function shown in Eqn. 1–3:

$$f(\varepsilon) = C \exp\left(\frac{-\varepsilon}{kT}\right), \tag{1-3}$$

where C is a normalization constant, and the exponential factor of $\exp(-\varepsilon/kT)$ is called the *Boltzmann factor*. The Boltzmann function is valid for particle ensembles of low density at relatively high temperature.

According to quantum statistics, a particle would be in a state of *degeneracy* if the particle ensemble follows either the Fermi or the Bose-Einstein statistics. We may assume that a particle is in the state of degeneracy at low temperatures and in the state of nondegeneracy at high temperatures. The transition temperature, T_c, (*degeneracy temperature*) between the two states is proportional to the 2/3 power of particle density, n, and inversely proportional to the particle mass, m. The degeneracy temperature for Fermi particles, that is called the *Fermi temperature*, is given by $T_c = \varepsilon_F / k = (h^2 / 8mk) \times (3n/\pi)^{2/3}$, where h is the Planck constant. The transition temperature from degeneracy to nondegeneracy is estimated to be about 10,000 K for free electrons in metals and about 1 K for ions and atoms in condensed phases. Therefore, the electrons in metal crystals are degenerated Fermi particles, while the ions and atoms in condensed phases are nondegenerated Boltzmann particles in the temperature range of general interest.

In quantum mechanics, the energy of particles is quantized into a series of allowed energy levels, $\varepsilon_n = n^2 h^2 / (8 m a^2)$; where a is the space size for a particle, m is the particle mass, and n ($n = 1, 2, 3, \cdots$) is the quantum number. The interval of allowed energy levels is then given by $\Delta\varepsilon = \varepsilon_{n+1} - \varepsilon_n = (2n + 1) h^2 / (8 m a^2)$, indicating that the greater the particle mass and the greater the particle space size, the smaller are the energy level intervals and, hence, the less are the quantization effects. The transition from the quantized energy levels to the continuous energy levels corresponds to the degeneracy–nondegeneracy transition of particle ensembles.

The particles we will deal with in this textbook are mainly electrons and ions in condensed solid and liquid phases. In condensed phases the ions are the classical Boltzmann particles and the electrons are the degenerated Fermi particles.

1.2 Chemical Potential and Electrochemical Potential

According to classical thermodynamics, the energy of particles may be represented by the entropy, the internal energy, the enthalpy, the free energy, and the free enthalpy, depending on the independent variables we choose to describe the particle ensemble system. We use in this textbook the free enthalpy, G, (also called the Gibbs free energy or Gibbs energy) with independent variables of temperature, T, and pressure, p; and the free energy, F, (also called the Helmholtz free energy) with independent variables of temperature, T, and volume, V.

The differential energy of a substance particle, i, in a particle ensemble is called the *chemical potential*, μ_i, when the particle is electrically neutral (atoms and molecules),

$$\mu_i = \left(\frac{\partial G}{\partial x_i} \right)_{p,T,x} = \left(\frac{\partial F}{\partial x_i} \right)_{V,T,x} ; \tag{1-4}$$

and the differential energy is called the *electrochemical potential*, $\bar{\mu}_i$, when the particle is electrically charged (ions and electrons),

$$\bar{\mu}_i = \left(\frac{\partial G}{\partial x_i} \right)_{p,T,x,\phi} = \left(\frac{\partial F}{\partial x_i} \right)_{V,T,x,\phi} ; \tag{1-5}$$

where x_i is the *molar fraction* of particle i and ϕ is the *inner potential* (electrostatic potential) of the particle ensemble. In Eqns. 1–4 and 1–5 we may use, instead of the molar fraction, x_i, the particle concentration, n_i, in terms of the number of particles in unit volume of the particle ensemble. For an ensemble comprising only the same particles of pure substance, the chemical potential becomes equal to the free enthalpy or free energy divided by the total number of particles in the ensemble ($\mu_i = G/N_i = F/N_i$), and so does the electrochemical potential ($\bar{\mu}_i = G/N_i = F/N_i$). The chemical potential may be defined not only for noncharged neutral particles but it can also be defined for charged particles by subtracting the electrostatic energy from the electrochemical potential of a charged particle, as is shown in Eqn. 1–9.

For an ensemble comprising a number of different kinds of substance particles, chemical thermodynamics introduces the *absolute activity*, λ_i, to represent the chemical potential, μ_i, of component i as shown in Eqn. 1–6:

$$\mu_i = k\,T \ln \lambda_i . \tag{1-6}$$

Further, introducing a standard state (reference state) where the chemical potential of component i is μ_i^* and the absolute activity is λ_i^*, we obtain from Eqn. 1–6 the following equation:

$$\mu_i - \mu_i^* = k\,T \ln \frac{\lambda_i}{\lambda_i^*} \;. \tag{1-7}$$

The ratio $\lambda_i / \lambda_i^* = a_i$ is called the *relative activity* or simply the *activity*, which of course depends on the standard state chosen. In general, the standard state of substances is chosen either in the state of pure substance ($x_i \to 1$) based on the Raoult's law [$\mu_i^* = (\partial G / \partial x)_{x \to 1}$] or in the state of infinite dilution ($x_i \to 0$) based on the Henry's law [$\mu_i^* = (\partial G / \partial x)_{x \to 0}$].

The ratio of the activity, a_i, to the molar fraction, x_i, or to the concentration, n_i, is the *activity coefficient*, $\gamma_i = a_i / x_i$ or $\gamma_i = a_i / n_i$. Then, Eqn. 1–7 yields Eqn. 1–8:

$$\mu_i = \mu_i^* + k\,T \ln a_i = \mu_i^* + k\,T \ln \gamma_i + k\,T \ln x_i \;. \tag{1-8}$$

The chemical potential, μ_i^*, in the standard state is the *unitary energy level* of component i in a particle ensemble, and the term $k\,T \ln(\gamma_i x_i)$ is the *communal energy*, in which the term $k\,T \ln x_i$ is called the *cratic energy* representing the mixing energy due to the indistinguishability of identical particles in an ensemble of particles [Gurney, 1953].

For charged particles the electrostatic energy of $z_i e \phi$ has to be added to the chemical potential, μ_i, to obtain the electrochemical potential, $\bar{\mu}_i$, as shown in Eqn. 1–9:

$$\bar{\mu}_i = \mu_i + z_i e \phi = \mu_i^* + k\,T \ln a_i + z_i e \phi \;, \tag{1-9}$$

where z_i is the charge number of component i, e is the elemental charge, and ϕ is the electrostatic inner potential of the ensemble.

1.3 Electrochemical Potential of Electrons

For high density electron ensembles such as free valence electrons in solid metals where electrons are in the state of degeneracy, the distribution of electron energy follows the Fermi function of Eqn. 1–1. According to quantum statistical dynamics [Davidson, 1962], the *electrochemical potential*, $\bar{\mu}_e$, of electrons is represented by the Fermi level, ε_F, as shown in Eqn. 1–10:

$$\bar{\mu}_e = \left(\frac{\partial G}{\partial n_e} \right)_{p,T,x,\phi} = \left(\frac{\partial F}{\partial n_e} \right)_{V,T,x,\phi} = \varepsilon_F \;, \tag{1-10}$$

where n_e is the electron concentration in the electron ensembles.

The "state density", $D(\varepsilon)$, of electrons is defined as the number of energy eigenstates, each capable of containing one electron, for unit energy interval (energy differential) for unit volume of the electron ensemble. According to the electron theory of metals [Blakemore, 1985], the state density of free electrons in metals is given by a parabolic function of electron energy ε as shown in Eqn. 1–11:

$$D(\varepsilon) = \frac{1}{2\pi^2}\left(\frac{2m_e}{\hbar^2}\right)^{\frac{2}{3}}(\varepsilon - \varepsilon_0)^{\frac{1}{2}}, \tag{1-11}$$

where ε_0 is the potential energy of electrons (the Hartree potential) in metals. The concentration, $n_e(\varepsilon)$, of electrons that occupy the eigenstates at an energy level of ε is given by the product of the state density and the probability density of Fermi function as in Eqn. 1–12:

$$n_e(\varepsilon) = D(\varepsilon) \cdot f(\varepsilon) = \frac{D(\varepsilon)}{\exp\left(\dfrac{\varepsilon - \varepsilon_F}{kT}\right) + 1}. \tag{1-12}$$

Similarly, the concentration of eigenstates vacant of electrons is given by Eqn. 1–13:

$$D(\varepsilon) - n_e(\varepsilon) = D(\varepsilon)\{1 - f(\varepsilon)\} = \frac{D(\varepsilon)}{\exp\left(\dfrac{\varepsilon_F - \varepsilon}{kT}\right) + 1}. \tag{1-13}$$

It follows from Eqns. 1–12 and 1–13 that the state density is *half occupied* by electrons with the remaining half vacant for electrons at the Fermi level, ε_F, as shown in Fig. 1–2. Since the Fermi temperature of electrons ($T_c = \varepsilon_F/k$) in electron

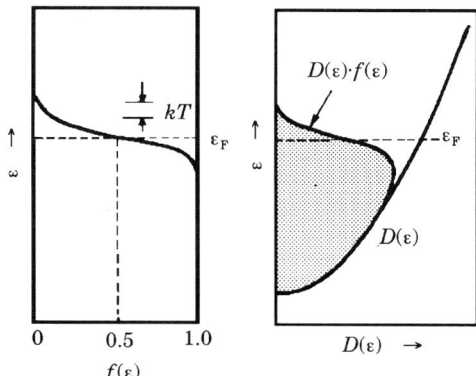

Fig. 1–2. Energy distribution of electrons near the Fermi level, ε_F, in metal crystals: ε = electron energy; $f(\varepsilon)$ = distribution function (probability density); $D(\varepsilon)$ = electron state density; $D(\varepsilon) \cdot f(\varepsilon)$ = occupied electron state density.

ensembles of high electron density (electrons in metals) is very high ($T_c \doteq 10{,}000$ K), the density of the occupied electron states (eigenstates) changes appreciably only within an energy range of several kT around the Fermi level in the temperature range of general interest as shown in Fig. 1–2.

The total concentration, n_e, of electrons that occupy the eigenstates as a whole is obtained by integrating Eqn. 1–12 with respect to the particle energy, ε, as shown in Eqn. 1–14:

$$n_e = \int_{\varepsilon_0}^{+\infty} D(\varepsilon) \cdot f(\varepsilon)\, d\varepsilon = \int_{\varepsilon_0}^{+\infty} \frac{D(\varepsilon)}{\exp\left(\dfrac{\varepsilon - \varepsilon_F}{kT}\right) + 1}\, d\varepsilon \,. \tag{1-14}$$

Equating Eqn. 1–14 with the electron concentration in the electron ensemble, we obtain the *Fermi level*, ε_F, as a function of the electron concentration, n_e, as shown in Eqn. 1–15:

$$\varepsilon_F = \varepsilon_0 + \frac{(3\pi^2 n_e)^{\frac{2}{3}} \hbar^2}{2 m_e}, \tag{1-15}$$

where ε_0 is the lowest level of the allowed energy band for electrons, m_e is the electron mass, and $\hbar$ denotes $\hbar = h/2\pi$.

For low density electron ensembles such as electrons in semiconductors, where electrons are usually allowed to occupy energy bands much higher and much lower than the Fermi level, the probability density of electron energy distribution may be approximated by the Boltzmann function of Eqn. 1–3, as shown in Fig. 1–3. Therefore, the total concentration, n_e, of electrons that occupy the allowed

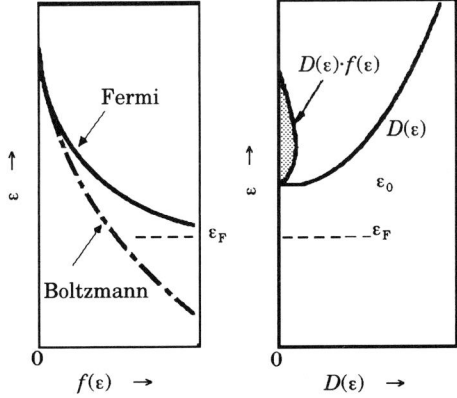

Fig. 1–3. Probability density of electron energy distribution, $f(\varepsilon)$, state density, $D(\varepsilon)$, and occupied electron density, $D(\varepsilon) \cdot f(\varepsilon)$, in an allowed energy band much higher than the Fermi level in solid semiconductors, where the Boltzmann function is applicable.

electron levels may be obtained in the form of Boltzmann function as given by Eqn. 1–16:

$$n_e = \int_{\varepsilon_0}^{\infty} D(\varepsilon) \cdot f(\varepsilon)\, d\varepsilon \doteq N_0 \exp\left(-\frac{\varepsilon_0 - \varepsilon_F}{kT}\right), \quad (1\text{–}16)$$

where N_0 is the effective state density of electrons in the allowed energy band which, according to the semiconductor physics, is given by Eqn. 1–17:

$$N_0 = 2 \left(\frac{m_e kT}{2\pi h^2}\right)^{\frac{3}{2}}. \quad (1\text{–}17)$$

From Eqn. 1–16 we obtain the *Fermi level*, ε_F, and the *electrochemical potential*, $\bar{\mu}_e$, of electrons as shown in Eqn. 1–18:

$$\bar{\mu}_e = \varepsilon_F = \varepsilon_0 - kT \ln \frac{N_0}{n_e}. \quad (1\text{–}18)$$

Since electrons are charged particles, the electrochemical potential of electrons (Fermi level, ε_F) depends on the inner potential, ϕ, of the electron ensemble as in Eqn. 1–19:

$$\bar{\mu}_e = \varepsilon_F = \mu_e - e\phi = \varepsilon_{F(\phi=0)} - e\phi \quad . \quad (1\text{–}19)$$

In general, the chemical potential of electrons, μ_e, is characteristic of individual electron ensembles, but the electrostatic energy of $-e\phi$ varies with the choice of zero electrostatic potential. In electrochemistry, as is described in Sec. 1.5, the reference level of electrostatic potential is set at the outer potential of the electron ensemble.

1.4 The Reference Level of Particle Energy

Units of the energy scale are usually expressed in counts of kJ or eV, and the numerical value of energy levels depends on the reference level chosen. It is the relative energy level that is important in physical chemistry, and the choice of the reference zero level is a matter of convention. Followings are different reference levels which are used in different fields of science:

(1) *The isolated rest state of a given particle at infinity in vacuum (temperature T)*: This zero energy level is used in physics. The rest state of particles is hypothetical having the energy only due to the internal freedom of particles. We call the rest electron the vacuum electron, $e_{(vac)}$, and its energy the *vacuum electron level*, $\varepsilon_{e(vac)} = 0$.

(2) *The ideal gaseous state of a given particle in the standard state of pressure and temperature chosen (e.g. pressure p = 1 atm., temperature T)*: The energy of an ideal gaseous particle ensemble consists of the internal energy and the translational energy of the particle. We call the ideal electron gas in the standard state the standard gaseous electron, $e_{(STD)}$, and its energy the *standard gaseous electron level*, $\varepsilon_{e(STD)}$. According to the statistical dynamics, the standard gaseous electron level referred to the vacuum electron level is given by $kT \ln\{\pi \hbar n_e)/(m_e kT)^{3/2}\}$, which is about 0.02 eV at room temperature and may be negligible compared with the energy of chemical reactions of the order of 1 eV; where n_e is the electron concentration and m_e is the electron mass. Therefore, the standard gaseous electron level, $\varepsilon_{e(STD)}$, may be approximated by the vacuum electron level, $\varepsilon_{e(vac)}$. The ideal standard gaseous state is not always realizable with all kinds of particles and, thus, it is frequently hypothetical with some substance particles (such as iron which is solid in the standard state). Further, for charged particles the electrostatic energy has also to be taken into account, which depends on the electrostatic potential. We may place the reference level of electrostatic energy at infinity in vacuum or at the outer potential just outside the particle ensemble. In electrochemistry the standard gaseous state at the outer potential is frequently taken to be the reference zero level of particle energy.

(3) *The stable state of atoms at the standard temperature 25 °C and pressure 1 atm.*: Atoms are stable at room temperature and pressure either in the state of gas (e.g. molecular oxygen), liquid (e.g. mercury), or solid (e.g. iron). In chemical thermodynamics, the stable state of atoms at the standard state is conventionally assumed to be the reference zero level of particle energy to derive the chemical potential of various particles. The relation between the reference level of the standard gaseous state and that of the standard stable state can be derived thermodynamically.

(4) *The state of unit activity of hydrated proton at the standard temperature 25°C and pressure 1 atm.*: In the electrochemistry of aqueous solution, the scale of chemical potential for hydrated ions takes as the reference zero the standard chemical potential of hydrated protons at unit activity; in addition the standard stable state energy of atoms is set equal to zero.

1.5 Electrostatic Potential of Condensed Phases

The electrostatic *inner potential*, ϕ, of a condensed phase (liquid or solid) is defined as the differential work done for a unit positive charge to transfer from the zero level at infinity into the condensed phase. In cases in which the condensed

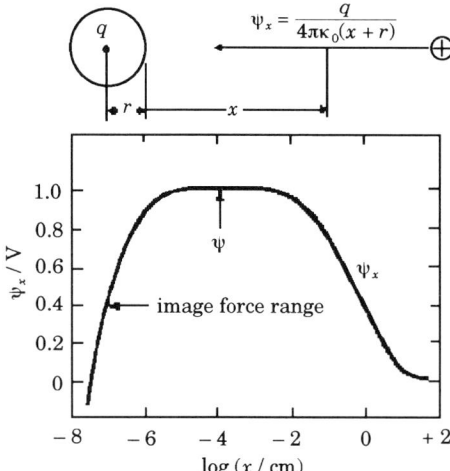

Fig. 1–4. Electrostatic potential profile near a charged metal sphere: x = distance from the metal surface; ψ = outer potential; ψ_x = electrostatic potential as a function of x. [From Parsons, 1954.]

phase is charged, an approaching unit charge is affected by the electric field of the charged phase before it enters into the phase interior. The electrostatic potential at the position just outside the charged phase (the position of the closest approach but beyond the influence of image force) is called the *outer potential*, ψ. Fig. 1–4 shows the electrostatic potential profile outside a charged metal sphere.

The *surface potential difference*, χ, is defined as the differential work done for a unit positive charge to transfer from the position of the outer potential into the condensed phase. This potential difference arises from the surface electric dipoles, such as the dipole of water molecules at the surface of liquid water and the surface dipole due to the spread-out of electrons at the metal surface. The magnitude of χ appears to remain constant whether the condensed phase is charged or uncharged.

The *inner potential*, therefore, consists of the outer potential and the surface potential difference as shown in Eqn. 1–20 and in Fig. 1–5:

$$\phi = \psi + \chi \ . \tag{1-20}$$

The outer potential, ψ, depends on the electric charge on the condensed phase, but the surface potential difference, χ, is usually assumed to be characteristic of individual condensed phases. For noncharged condensed phases, the outer potential is zero ($\psi = 0$) and the inner potential becomes equal to the surface potential

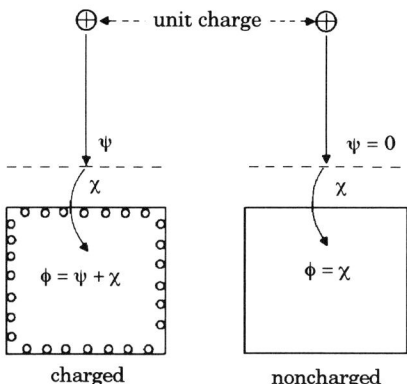

Fig. 1-5. Electrostatic potential of charged and noncharged condensed phases: ϕ = inner potential; ψ = outer potential; χ = surface potential difference.

difference. The magnitude of χ is + 0.13 V for liquid water [Trasatti, 1980] and is in the range of + 0.1 to + 5.0 V for solid metal crystals [Trasatti, 1974].

The outer potential, ψ, can be measured physically as a difference of electrostatic potential between two points in the same gas or vacuum phase. The surface potential difference, χ, which is a difference of electrostatic potential between two different phases, cannot be measured so that the inner potential, ϕ, also cannot be measured in a straightforward way.

1.6 Energy Levels of Charged Particles in Condensed Phases

In electrochemistry, we deal with the energy level of charged particles such as electrons and ions in condensed phases. The electrochemical potential, $\bar{\mu}_i$, of a charged particle i in a condensed phase is defined by the differential work done for the charged particle to transfer from the standard reference level (e.g. the standard gaseous state) at infinity (ϕ = 0) to the interior of the condensed phase. The electrochemical potential may be conventionally divided into two terms; the chemical potential μ_i and the electrostatic energy $z_i e \phi$ as shown in Eqn. 1–21:

$$\bar{\mu}_i = \mu_i + z_i e \phi \quad . \tag{1-21}$$

Eqns. 1–20 and 1–21 yield Eqn. 1–22:

$$\bar{\mu}_i = \mu_i + z_i e \chi + z_i e \psi = \alpha_i + z_i e \chi \, , \tag{1-22}$$

where α_i is the differential energy required for a charged particle i to transfer from the standard gaseous state at the outer potential to the interior of the condensed phase. This energy level α_i is defined as the "*real potential*" of charged particles [Lange, 1933]:

$$\alpha_i = \mu_i + z_i e \chi \, . \tag{1-23}$$

Fig. 1–6 shows schematically the relationship between $\bar{\mu}_i$, μ_i, and α_i. In the case of electrons in solids, the real potential α_e corresponds to the *negative work function* $-\Phi$ (= α_e); work function Φ is the differential energy required for emission of electrons from solids.

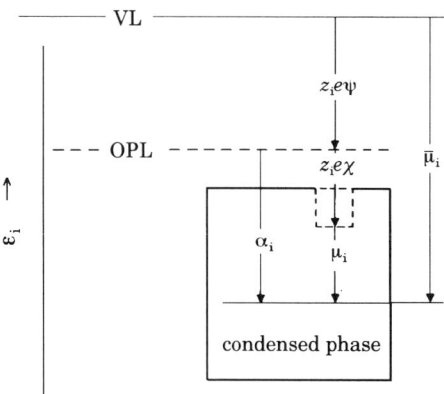

Fig. 1–6. Energy level of a charged particle i in a condensed phase: ε_i = energy of particle i; $\bar{\mu}_i$ = electrochemical potential; α_i = real potential; μ_i = chemical potential; z_i = charge number of particle i; VL = vacuum infinity level; OPL = outer potential level.

The real potential of a charged particle represents the energy level of the particle in condensed phases, referred to the energy level of the particle in the standard gaseous state at the outer potential of the condensed phases. In contrast to the electrochemical potential that depends on the electrostatic charge of the condensed phases, the real potential gives the energy level characteristic of individual particles in individual condensed phases, irrespective of the amount of electrostatic charge and the outer potential of condensed phases. For noncharged condensed phases of which the outer potential is zero ($\psi = 0$), the real potential becomes equal to the electrochemical potential ($\alpha_i = \bar{\mu}_i$).

In this textbook, we use the *real potential* α_i rather than the electrochemical potential $\bar{\mu}_i$ to represent the energy level of charged particles in condensed phases.

References

[Blakemore, 1985]: J. S. Blakemore, *Solid State Physics*, Cambridge University Press, London, (1985).

[Davidson, 1962]: N. Davidson, *Statistical Mechanics*, Sec. 6-16, McGraw-Hill Inc., New York, (1962).

[Gurney, 1953]: R. W. Gurney, *Ionic Processes in Solution*, p. 90, McGraw-Hill Book Co. Inc., New York, (1953).

[Parsons, 1954]: R. Parsons, *Modern Aspects of Electrochemistry*, (Edited by J. O'M. Bockris, B. E. Conway), p. 103, Butterworth Sci. Publ., London, (1954).

[Trasatti, 1974]: S. Trasatti, *J. Electroanal. Chem.*, **52**, 313(1974).

[Trasatti, 1980]: S. Trasatti, *Comprehensive Treatise of Electrochemistry*, **Vol. 1**, (Edited by J. O'M. Bockris, B. E. Conway, E. Yeager), p. 45, Plenum Press, New York, (1980).

CHAPTER 2

THE ENERGY LEVEL OF ELECTRONS

2.1 Energy Levels of Electrons in Condensed Phases

According to quantum mechanics, electrons in atoms occupy the allowed energy levels of atomic orbitals that are described by four quantum numbers: the principal, the azimuthal, the magnetic, and the spin quantum numbers. The orbitals are usually expressed by the principal quantum numbers 1, 2, 3, ····, increasing from the lowest level, and the azimuthal quantum numbers conventionally expressed by s (sharp), p (principal), d (diffuse), f (fundamental), ···· in order. For instance, the atom of oxygen with 8 electrons is described by $(1s)^2 (2s)^2 (2p)^4$, where the superscript indicates the number of electrons occupying the orbitals, as shown in Fig. 2–1.

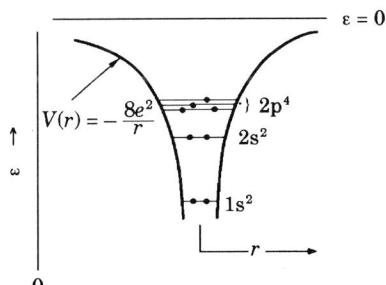

Fig. 2–1. Atomic orbital levels of oxygen occupied by electrons: ε = electron energy; $V(r)$ = potential energy of atomic oxygen nucleus; r = distance from atomic nucleus.

As two atoms X and Y form a molecule XY, the atom–atom interaction splits each atomic frontier orbital into two molecular orbitals: a *bonding* molecular orbital at low energy level and an *antibonding* molecular orbital at high energy level as shown in Fig. 2–2. Similarly, a molecule composed of many atoms has

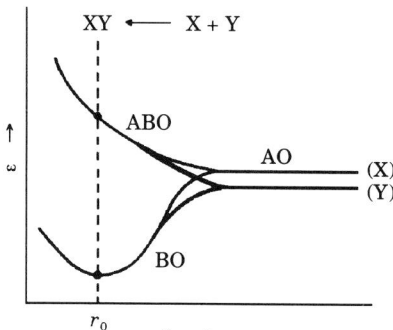

Fig. 2–2. Formation of molecular orbital levels from atomic orbital levels: r = distance between X and Y; r_0 = stable atom-atom distance in molecule XY; AO = atomic orbital; BO = bonding orbital; ABO = antibonding orbital.

the bonding, nonbonding, and antibonding molecular orbitals, the number of which equals the total number of frontier atomic orbitals of the constituent atoms. These molecular orbitals are filled with electrons successively from the lowest level to the highest occupied level.

In the case of condensed phases such as solid crystals, the molecular orbital levels are so dense (the energy state density is so great) that they form, instead of narrow orbital energy levels, the relatively wide orbital *energy bands* of bonding, nonbonding, and antibonding characters which electrons are allowed to occupy, as shown in Fig. 2–3. The inner orbital bands at low energy levels are located

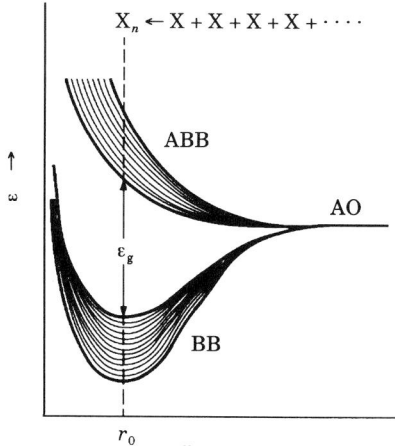

Fig. 2–3. Formation of electron energy bands in constructing a solid crystal X_n from atoms of X: r_0 = stable atom-atom distance in crystal; BB = bonding band; ABB = antibonding band; ε_g = band gap.

deep into the potential barrier of lattice atoms: hence, the localized bands attached to the lattice atoms. The frontier orbital bands, on the other hand, are at energy levels comparable to or higher than the potential barrier of lattice atoms: hence, the *delocalized bands* in which electrons are not locally fixed at the lattice atoms but are delocalized in the whole solid crystal as shown in Fig. 2–4.

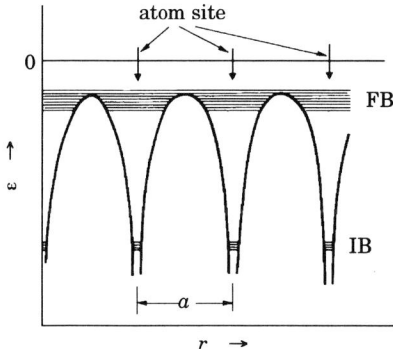

Fig. 2–4. Lattice potential energy and electron energy bands in crystals: IB = inner band; FB = frontier band.

Electrons in solids occupy the allowed energy bands successively from the inner orbital bands to the frontier bands. A solid consisting of N atoms contains in a frontier band $2N$ energy eigenstates (two electrons of different spins in an eigenstate). Therefore, a solid crystal composed of atoms having valence electrons of an odd number, such as metallic sodium and aluminum, may have a frontier band that cannot be fully occupied by electrons so that electrons are allowed to move in the band. Such a solid is called a *metal*. In contrast, in a solid composed of atoms having valence electrons of an even number, such as solid silicon and solid sodium chloride, the frontier band can be fully occupied by electrons leaving no vacant levels for electrons so that electrons may not be allowed to move in the band. Such a solid is called either an *insulator* in the case that the band gap between the filled and the vacant frontier bands is wider than several electron volts (~ 4 eV), or a *semiconductor* in the case that the band gap is narrower than several electron volts. Fig. 2–5 shows schematically the electron occupation in the frontier bands for different classes of solids.

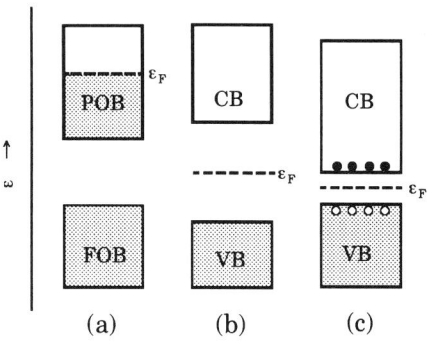

Fig. 2–5. Electron occupation in energy bands classified into (a) metals, (b) insulators, and (c) semiconductors: FOB = fully occupied band; POB = partially occupied band; CB = conduction band; VB = valence band.

In an allowed energy band for electrons with the band width of several electron volts, there are as many energy levels (electronic eigenstates) as the total number of atoms in a solid crystal. The *state density*, $D(\varepsilon)$, of electrons is defined as the number of electronic eigenstates for unit energy interval (energy differential) for unit volume of condensed phases. As is shown in Eqn. 1–11, the electron theory of solids gives the state density near the band edges as a parabolic function of electron energy, ε, as in Eqn. 2–1:

$$D(\varepsilon) = \frac{1}{2\pi^2}\left(\frac{2m_e^*}{\hbar^2}\right)^{\frac{3}{2}}(|\varepsilon - \varepsilon_{UL}|)^{\frac{1}{2}}, \qquad (2\text{–}1)$$

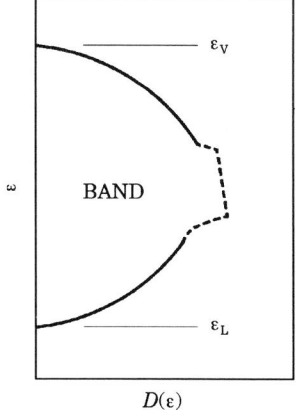

Fig. 2–6. State density distribution curve of electrons in solids: $D(\varepsilon)$ = electron state density; ε_U = upper band edge; ε_L = lower band edge.

where m_e^* is the effective mass of the electron near the lower band edge, ε_L, or the effective mass of the hole near the upper band edge, ε_U: ε_{UL} represents either ε_U or ε_L. Fig. 2–6 illustrates schematically the state density distribution curve for electrons in solid crystals.

2.2 Electrons in Metals

2.2.1 Energy band and the Fermi level

Electron occupation in the frontier bands of metal crystals varies with different metals as shown in Fig. 2–7. For metallic iron the frontier bands consist of hybridized 4s-3d-4p orbitals, in which 4s and 3d are partially occupied by electrons but 4p is vacant for electrons. Fig. 2–8 shows the electron state density curve of metallic iron, where the 3d and 4s bands are partially filled with electrons. Electrons in metals occupy the energy states in a frontier band successively from the lower band edge level to the Fermi level, leaving the higher levels vacant.

The *Fermi level* of electrons in metals, as shown in Eqn. 1–15, is given by Eqn. 2–2:

$$\varepsilon_F = \varepsilon_0 + \frac{(3\pi^2 n_e)^{\frac{2}{3}} \hbar^2}{2 m_e}, \qquad (2\text{–}2)$$

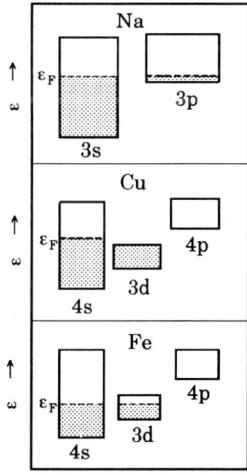

Fig. 2–7. Frontier energy bands partially occupied by electrons in metallic sodium, copper, and iron.

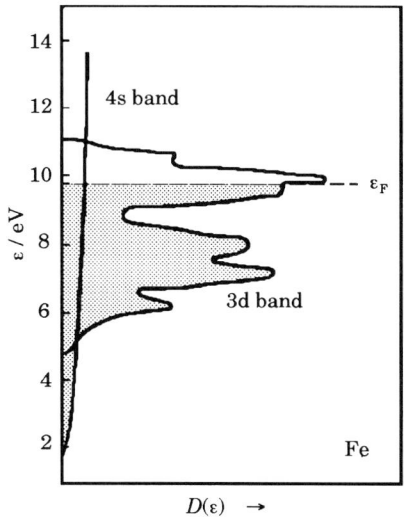

Fig. 2–8. State density distribution curve of 3d and 4s frontier bands partially occupied by electrons in metallic iron: [From Fujita, 1996.]

which indicates that the greater the valence electron density n_e in metals the higher the Fermi level ε_F. As is described in Sec. 1.3, the Fermi level represents the electrochemical potential of electrons, that is the electron energy level in metals referred to the vacuum electron level at infinity. The electrochemical potential, $\bar{\mu}_e$, of electrons in metals is expressed as shown in Eqn. 2–3:

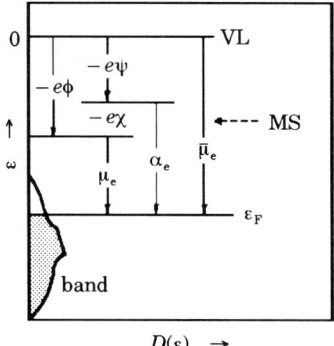

Fig. 2–9. The electrochemical potential, $\bar{\mu}_e$, the real potential, α_e, and the chemical potential, μ_e, of electrons in metals: ϕ = inner potential; χ = surface potential difference; ψ = outer potential; MS = metal surface; VL = vacuum infinity level.

$$\bar{\mu}_e = \varepsilon_F = \left(\frac{\partial F}{\partial x_i} \right)_{V, T, x, \phi} = \mu_e - e\phi = \mu_e - e\chi - e\psi = \alpha_e - e\psi \ , \tag{2-3}$$

where μ_e is the chemical potential of electrons in metals, $-e\phi$ is the electrostatic energy, and α_e is the *real potential* of electrons in metals. Fig. 2–9 shows the relationship among the real potential α_e, the electrochemical potential $\bar{\mu}_e$, and the chemical potential μ_e of electrons in metals.

2.2.2 The real potential and the chemical potential of electrons in metals

In physics, the work function is used frequently to represent the energy of electrons in metals. In electrochemistry, however, we use the *real potential*, α_e, instead of the work function, Φ, to represent the energy level of electrons in metals as shown in Eqn. 2–4:

$$\alpha_e = \mu_e - e\chi = -\Phi \ . \tag{2-4}$$

We first consider the electron density distribution and the potential profile that an electron sustains in transferring across the metal surface. According to the *"jellium model"* of metals, which assumes the uniform positive charge of lattice metal ions with the same amount of negative charge carried by moving free electrons, metal electrons diffuse out of the jellium surface to a distance (~ 0.05 nm) of several tenths the Fermi wave length, λ_F, that is the wave length of electrons at the Fermi level in metals. A greater density of metal electrons gives a longer λ_F. As shown in Fig. 2–10, the electron density is distributed increasing from the electron tailing outside the surface to a constant density in the metal interior. The electron density distribution near the metal surface creates an *electric dipole* comprising an excess positive charge on the metal side and an excess negative charge of the electron tailing on the vacuum side. This spread-out of electrons creates the surface dipole of metals, generating thereby the surface potential difference, χ. The magnitude of χ is in the range of 0.1 to 5.0 V and increases with the electron tailing distance and, therefore, with the density of valence electrons, n_e, in metals.

An electron transferring across the metal surface first sustains the electrostatic coulomb potential due to the surface potential difference, χ, and then enters into the exchange and correlation potential field, V_{ex}, caused by the ion–electron and electron–electron interaction energies. The potential energy of V_{ex} is in the range from -3 eV to -15 eV, the absolute magnitude of which increases in proportion to the cube root of the electron density: hence, increasing gradually from the electron tailing outside the surface to the metal interior. As shown in Fig. 2–10, electrons in metals occupy the energy levels successively from the bottom of the

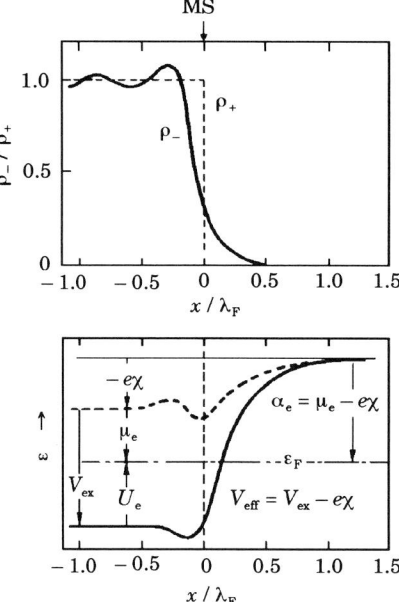

Fig. 2–10. Profile of the electron density and the electronic potential energy across the metal/vacuum interface calculated using the jellium model of metals: MS = jellium surface of metals; λ_F = Fermi wave length; ρ_+ = average positive charge density; ρ_- = negative charge density; V_{ex} = electron exchange and correlation energy; U_e = kinetic energy of electrons. [From Lange-Kohn, 1970.]

total effective potential, V_{eff} ($= V_{ex} - e\chi$), to the Fermi level, ε_F, which is higher by an amount of electron kinetic energy, U_e, than the bottom level. The kinetic energy of electrons in metals, which is given by $U_e = \hbar\, \boldsymbol{k}_F^2 / 2\, m_e$, is in the range of 1 to 2 eV for various metals, where $\boldsymbol{k}_F$ is the wave vector of electrons at the Fermi level and m_e is the effective electron mass. The chemical potential, μ_e, of electrons in metals is given by the sum of the exchange and correlation potential, V_{ex}, (a negative value) and the kinetic energy, U_e, (a positive value) as shown in Eqn. 2–5:

$$\mu_e = V_{ex} + U_e \ . \tag{2–5}$$

The *real potential*, α_e, of electrons in metals, as shown in Eqn. 2–4, comprises the electrostatic surface term, $-e\chi$, due to the surface dipole and the chemical potential term, μ_e, determined by the bulk property of metal crystals. In general, the electrostatic surface term is greater the greater the valence electron density in metals; whereas, the chemical potential term becomes greater the lower the valence electron density in metals.

Fig. 2–11 compares the work function, Φ, observed with that calculated based on the jellium model as a function of the electron density, n_e, in metals: n_e is represented in terms of the Wigner-Seitz radius which is inversely proportional to the cube root of n_e. The chemical potential term (μ_e = – 1.5 to – 2.5 eV) predominates in the work function of metals of low valence electron density, while on the contrary the surface term ($-e\chi$ = – 0.1 to –5.0 eV) predominates for metals of high valence electron density. The group of d-metals, in which the frontier band is d-orbitals, belongs to the metal of high valence electron density and their work function is determined mainly by the surface dipole of the metals. The work function of d-metals, therefore, depends on the surface roughness and the crystal plane at the surface; it is greater the greater is the electron density of the surface crystal plane.

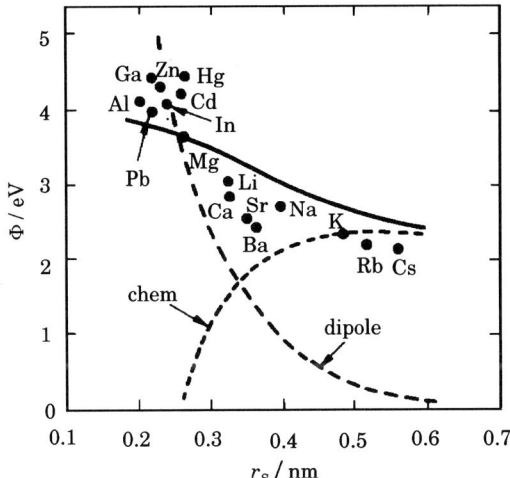

Fig. 2–11. Work function, Φ, observed and calculated using the jellium model as a function of Wigner-Seitz radius, r_S, of metals: $r_S = \{3/(4\pi n_e)\}^{1/3}$; n_e = electron density in metals; solid line = calculated work function; chem (broken line) = contribution of μ_e; dipole (broken line) = contribution of $-e\chi$; ● = measured work function. [From Lange-Kohn, 1970.]

2.3 Electron Energy Bands of Semiconductors

The energy bands of *frontier* electrons in semiconductors consist of a *valence band* (VB) fully occupied by electrons at low energy levels and a vacant *conduction band* (CB) at high energy levels; the valence and conduction bands are separated by a *forbidden* band called the *band gap*. Crystalline silicon with the diamond structure is a typical covalent bond semiconductor; its frontier bands comprise a bonding band filled with electrons and a vacant antibonding band, both arising from sp^3 hybridized orbitals, as shown in Fig. 2–12. As in the case of metals, the state density distribution in semiconductors may be represented approximately by a parabolic curve near the band edges, as shown schematically in Fig. 2–13.

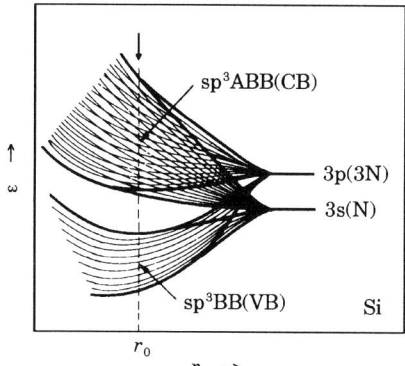

Fig. 2–12. Electron energy band formation of silicon crystals from atomic frontier orbitals: N = number of silicon atoms in crystal; r = distance between atoms; r_0 = stable atom–atom distance in crystals, sp^{3}BB = bonding band (valence band) of sp^3 hybrid orbitals; sp^{3}ABB = antibonding band (conduction band) of sp^3 hybrid orbitals.

In cases in which both the upper edge level of the valence band and the lower edge level of the conduction band are at the same wave vector of electrons (GaAs, etc.), the band gap is called the *direct band gap*; while it is called the *indirect band gap* in cases in which the two band edge levels are at different wave vectors (Si, etc.). The band gap is $\varepsilon_g = 1.1$ eV for silicon and $\varepsilon_g = 1.4$ eV for gallium arsenide. Since the band gap is relatively narrow in semiconductors, a few electrons in the fully occupied valence band are thermally excited up to the conduction band leaving positive "*holes*" (vacant electrons) in the valence band. The concentration, n_i, of thermally excited electron–hole pairs is given, to a first approximation, by the Boltzmann function as shown in Eqn. 2–6:

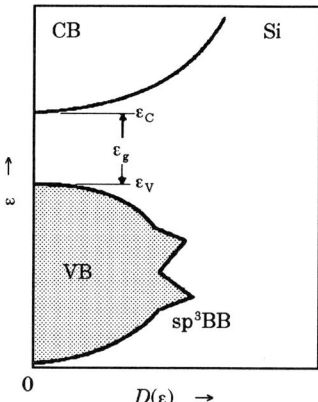

Fig. 2–13. Schematic electron state density distribution curves in the valence and conduction bands of silicon: ε_C = conduction band edge level; ε_V = valence band edge level; ε_g = band gap (1.1 eV for silicon); CB = conduction band; VB = valence band.

$$n_i = n_0 \exp \frac{-\varepsilon_g}{2kT}, \tag{2-6}$$

where n_0 is the concentration of electrons at the upper edge of the valence band. The band gap of the order of one electron volt (ε_g = 1 eV), which is much greater than kT ($\doteq$ 0.03 eV at 298 K), gives rise to an extremely small concentration of electron–hole pairs ($n_i \doteq n_0 \times 10^{-20}$) at room temperature.

Semiconductors may be classified into two groups: *intrinsic* semiconductors with no allowed electron levels in the band gap and *extrinsic* semiconductors which contain allowed electron levels localized at impurity atoms in the band gap. The addition of impurities into semiconductors is called *doping*. Phosphorous with five valence electrons doped as an impurity into semiconductor silicon, with four valence electrons for each silicon atom, produces one excess electron for each phosphorous atom. This excess electron is bound with the phosphorus atom at low temperatures but becomes free at relatively high temperatures, moving in the conduction band of the whole silicon crystal. Such an impurity that gives excess electrons is called the *donor*, and the semiconductor containing donors is called the *n-type* semiconductor. On the other hand, the doping of boron with three valence electrons into semiconductor silicon produces one *vacant electron* (one *hole*) for each boron atom, which is allowed to move in the valence band of the whole silicon crystal. The impurity giving holes is called the *acceptor*, and the semiconductor containing acceptors is called the *p-type* semiconductor. Fig. 2–14 shows the donor and acceptor impurities in semiconductor silicon.

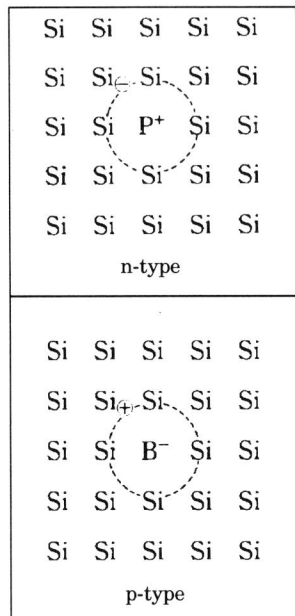

Fig. 2–14. Impurity atoms in silicon crystals: Si = silicon atom; P^+ = ionized donor of impurity phosphorus; B^- = ionized acceptor of impurity boron.

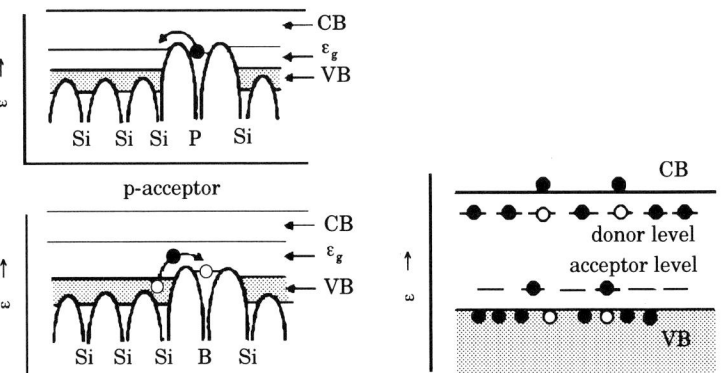

Fig. 2–15. Donor and acceptor levels in silicon crystals : ● = electron, ○ = hole.

The donor electron level, ε_D, which may be derived in the same way that the orbital electron level in atoms is derived, is usually located close to the conduction band edge level, ε_C, in the band gap ($\varepsilon_C - \varepsilon_D = 0.041$ eV for P in Si). Similarly, the acceptor level, ε_A, is located close to the valence band edge level, ε_V, in the band gap ($\varepsilon_A - \varepsilon_V = 0.057$ eV for B in Si). Fig. 2–15 shows the energy diagram for donor and acceptor levels in semiconductors. The localized electron levels close to the band edge may be called *shallow levels*, while the localized electron levels away from the band edges, associated for instance with lattice defects, are called *deep levels*. Since the donor and acceptor levels are localized at impurity atoms and lattice defects, electrons and holes captured in these levels are not allowed to move in the crystal unless they are freed from these initial levels into the conduction and valence bands.

2.4 Electrons and Holes in Semiconductors

2.4.1 Intrinsic semiconductors

The thermally excited electrons from the valence band (VB) occupy successively the levels in the conduction band (CB) in accordance with the Fermi distribution function. Since the concentration of thermally excited electrons (10^{11} to 10^{13} cm^{-3}) is much smaller than the state density of electrons (10^{19} cm^{-3}) in the conduction band, the Fermi function may be approximated by the Boltzmann distribution function. Consequently, the concentration of electrons in the conduction band is given by the following integral [Blakemore, 1985; Sato, 1993]:

$$n = \int_{\varepsilon_C}^{\infty} D_C(\varepsilon) \cdot f(\varepsilon)\, d\varepsilon \doteq N_C \exp \frac{\varepsilon_F - \varepsilon_C}{kT} , \qquad (2\text{--}7)$$

where $D_C(\varepsilon)$ is the state density in the conduction band given by Eqn. 2–1 and N_C is the *effective* state density at the lower edge of the conduction band. Semiconductor physics gives N_C as expressed in Eqn. 2–8:

$$N_C = 2 \left(\frac{m_e^* kT}{2\pi \hbar^2} \right)^{\frac{3}{2}} , \qquad (2\text{--}8)$$

where m_e^* is the effective mass of electron at the band edge. From Eqn. 2–7 the *Fermi level*, ε_F, is derived to obtain Eqn. 2–9:

$$\varepsilon_F = \varepsilon_C - kT \ln \frac{N_C}{n} . \qquad (2\text{--}9)$$

In the same way as Eqn. 2–7, we also obtain the hole concentration, p, in the valence band as shown in Eqn. 2–10:

$$p = \int_{-\infty}^{\varepsilon_V} D_V(\varepsilon) \cdot f(\varepsilon)\, d\varepsilon \doteqdot N_V \exp \frac{\varepsilon_V - \varepsilon_F}{kT}. \tag{2-10}$$

Further, Eqn. 2–10 yields the *Fermi level*, ε_F, as shown in Eqn. 2–11:

$$\varepsilon_F = \varepsilon_V + kT \ln \frac{N_V}{p}, \tag{2-11}$$

where N_V is the *effective* state density at the upper edge of the valence band given by Eqn. 2–12:

$$N_V = 2 \left(\frac{m_h^* kT}{2 \pi \hbar^2} \right)^{\frac{3}{2}}, \tag{2-12}$$

with m_h^* denoting the effective mass of hole.

The physical meaning of the effective state densities, N_C and N_V, at the band edges of the conduction and valence bands is as follows: Although the state density of electrons is distributed in the whole range of the conduction and valence bands, only a few parts of the state density near the band edge level are occupied with electrons or holes, the concentration of which is extremely small compared with the concentration of electrons in metals. Therefore, the available energy states (the integral of $D(\varepsilon)$) for electrons or holes in semiconductors may be restricted within a narrow energy range of a few kT near the band edge level, and their state density may be represented by the effective state density, N_C or N_V, at the band edge level. The magnitudes of N_C and N_V given by Eqns. 2–8 and 2–12 are of the order of 10^{19} cm^{-3} at room temperature.

In intrinsic semiconductors, the concentration of electrons, n, in the conduction band is equal to the concentration of holes, p, in the valence band as shown in Eqn. 2–13:

$$n_i = n = N_C \exp \frac{\varepsilon_F^* - \varepsilon_C}{kT} = p = N_V \exp \frac{\varepsilon_V - \varepsilon_F^*}{kT}, \tag{2-13}$$

where ε_F^* is the Fermi level of intrinsic semiconductors. From Eqn. 2–13 we obtain the concentration of electron–hole pairs, n_i, as shown in Eqn. 2–14:

$$n_i = \sqrt{N_C N_V} \, \exp \frac{-\varepsilon_g}{2kT}, \tag{2-14}$$

where ε_g is the band gap. For semiconductor germanium, the band gap is $\varepsilon_g = 0.7$ eV and the concentration of electron–hole pair is $n_i \doteqdot 2.5 \times 10^{13}$ cm^{-3}.

The *Fermi level*, ε_F^*, of intrinsic semiconductors is obtained from Eqn. 2–13 as shown in Eqn. 2–15:

$$\varepsilon_F^* = \frac{1}{2}(\varepsilon_C + \varepsilon_V) + \frac{1}{2} kT \ln \frac{N_V}{N_C} \ . \tag{2-15}$$

Since N_C is nearly equal to N_V, the Fermi level of intrinsic semiconductors is located midway in the band gap as shown in Fig. 2–16. All the equations given in the foregoing are valid under the condition that $n_i \ll N_C$ or N_V: this condition is fulfilled with usual intrinsic semiconductors.

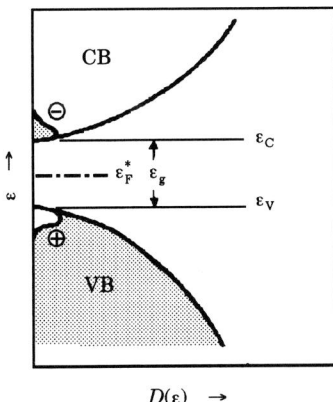

Fig. 2–16. Electron state density distribution and electron–hole pair formation in the conduction and valence bands of intrinsic semiconductors: ε_F^* = Fermi level of intrinsic semiconductors.

2.4.2 n-type and p-type semiconductors

As shown in Fig. 2–17 and Fig. 2–18, semiconductors containing impurities are classified into the following two types: *n-type* semiconductors with localized donor levels close to the conduction band, and *p-type* semiconductors with localized acceptor levels close to the valence band in the band gap. The liberation of electrons from donor levels into the conduction band and the liberation of holes from acceptor levels into the valence band are represented by the *ionization* processes of donor D and acceptor A, respectively, as shown in Eqns. 2–16 and 2–17:

$$D \rightarrow D^+ + e \ , \tag{2-16}$$

$$A \rightarrow A^- + h \ . \tag{2-17}$$

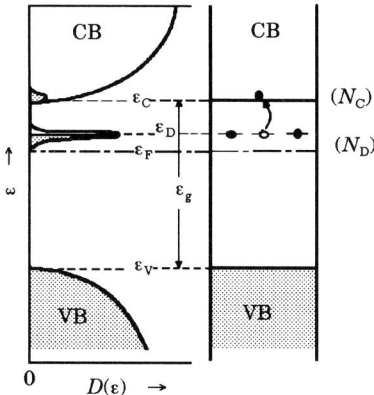

Fig. 2–17. Electron energy and state density in n-type semiconductors: ε_D = donor level; N_D = donor concentration; N_C = effective conduction band state density.

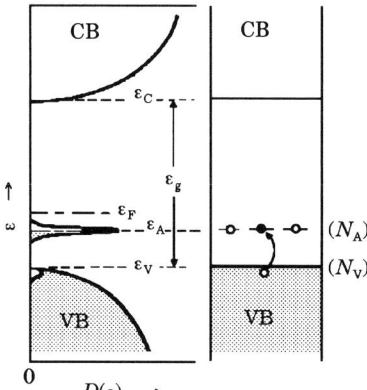

Fig. 2–18. Electron energy and state density in p-type semiconductors: ε_A = acceptor level; N_A = acceptor concentration; N_V = effective valence band state density.

The concentration of electrons, n, in the conduction band of n-type semiconductors and the hole concentration, p, in the valence band of p-type semiconductors are given by Eqn. 2–7 and Eqn. 2–10, respectively. The concentration of ionized donors, N_{D^+}, and the concentration of ionized acceptors, N_{A^-}, are derived using the Fermi function approximated by the Boltzmann function as shown in Eqns. 2–18 and 2–19, respectively:

$$N_{D^+} = \frac{N_D}{2} \exp \frac{\varepsilon_D - \varepsilon_F}{kT} \quad , \tag{2-18}$$

$$N_{A^-} = \frac{N_A}{2} \exp\frac{\varepsilon_F - \varepsilon_A}{kT}, \qquad (2\text{--}19)$$

where N_D and N_A are the concentrations of donors and acceptors, respectively. A factor 1/2 in Eqns. 2–18 and 2–19 results from the degeneracy factor 1 for the donor and acceptor levels instead of the factor 2 for the electron energy eigenstate in the conduction and valence bands.

In n-type semiconductors, the electrical neutrality is expressed as $n = N_{D^+} + p$ in the strict sense, but it may be approximated by $n \doteq N_{D^+}$ because p is much less than n. Then, Eqn. 2–18 yields Eqn. 2–20:

$$n = N_C \exp\frac{\varepsilon_F - \varepsilon_C}{kT} = N_{D^+} = \frac{N_D}{2}\exp\frac{\varepsilon_D - \varepsilon_F}{kT}. \qquad (2\text{--}20)$$

From Eqn. 2–20 we obtain the concentration of electrons, n, in the conduction band and the *Fermi level*, ε_F, of n-type semiconductors as shown in Eqns. 2–21 and 2–22:

$$n = \sqrt{\frac{N_C N_D}{2}} \exp\left(-\frac{\varepsilon_C - \varepsilon_D}{kT}\right), \qquad (2\text{--}21)$$

$$\varepsilon_F = \frac{\varepsilon_C + \varepsilon_D}{2} - \frac{kT}{2}\ln\frac{2 N_C}{N_D}, \qquad (2\text{--}22)$$

where $\varepsilon_C - \varepsilon_D$ is the energy for *donor ionization*. In Eqns. 2–15 and 2–22, it appears that the Fermi level of n-type semiconductors descends from an energy level near ε_D toward the middle of the band gap (toward the Fermi level, ε_F^*, of an intrinsic semiconductor) with decreasing concentration of donors, N_D.

The same equations as Eqns. 2–21 and 2–22 can be obtained for p-type semiconductors as shown in Eqns. 2–23 and 2–24:

$$p = \sqrt{\frac{N_V N_A}{2}} \exp\left(-\frac{\varepsilon_A - \varepsilon_V}{kT}\right), \qquad (2\text{--}23)$$

$$\varepsilon_F = \frac{\varepsilon_V + \varepsilon_A}{2} + \frac{kT}{2}\ln\frac{2 N_V}{N_A}. \qquad (2\text{--}24)$$

Thus, it appears that the Fermi level of p-type semiconductors ascends from an energy level near ε_A toward the middle of the band gap with decreasing acceptor concentration, N_A. From Eqns. 2–22 and 2–24, it follows that the Fermi level is located at levels higher for n-type semiconductors and lower for p-type semiconductors than the middle of the band gap. As described in the foregoing, the concentration of electrons, n, in the conduction band is different from the concentration of holes, p, in the valence band in extrinsic semiconductors containing impurities in contrast with intrinsic semiconductors in which n equals p.

Both electrons and holes are mobile charge carriers in semiconductors. The mobile charge carrier of which the concentration is much greater than the other is called the *majority carrier*, and the *minority carrier* is in much smaller concentrations. In n-type semiconductors, the majority carriers are electrons in the conduction band and the minority carriers are holes in the valence band. The product of the concentrations of majority and minority carriers (electrons and holes) in a semiconductor of extrinsic type (containing impurities) equals the square of the concentration of electron–hole pairs, n_i, in the same semiconductor of intrinsic type (containing no impurities):

$$n \times p = n_i^2 = (N_C N_V) \exp \frac{-\varepsilon_g}{kT} . \tag{2-25}$$

The value of this product is characteristic of individual semiconductors, amounting to $n \times p = 4.2 \times 10^{21}$ cm^{-6} for silicon and $n \times p = 6 \times 10^{26}$ cm^{-6} for germanium.

2.5 Energy Levels of Electrons in Semiconductors

As mentioned in Sec. 1.3, the electrochemical potential of electrons in condensed phases corresponds to the Fermi level of electrons in the phases. There are two possible cases of electron ensembles in condensed phases: one to which the *band model* is applicable (in the state of degeneracy where the wave functions of electrons overlap), and the other to which the band model cannot apply (in the state of nondegeneracy where no overlap of electron wave functions occurs). In the former case electrons or holes are allowed to move in the bands, while in the latter case electrons are assumed to be individual particles rather than waves and move in accord with the thermal *hopping* mechanism between the adjacent sites of localized electron levels.

In the *hopping model* the electrochemical potential, $\bar{\mu}$, of electrons is expressed conventionally in the same way as that of ions as shown in Eqn. 2–26:

$$\bar{\mu} = \bar{\mu}^* + kT \ln x , \tag{2-26}$$

where $\bar{\mu}^*$ is the *unitary* electrochemical potential of electrons characteristic of individual electron ensembles and x is the ratio of the number of the sites occupied by electrons to the total number of sites available for electron occupation. Parameter x corresponds to the molar fraction in terms of chemistry. By use of the electron concentration, $n = xN$ (the number of moles of electrons per unit volume), instead of the molar fraction, x, we obtain Eqn. 2–27 from Eqn. 2–26:

$$\bar{\mu} = \bar{\mu}^* + k\,T \ln \frac{n}{N} = \bar{\mu}^* - k\,T \ln N + k\,T \ln n = \bar{\mu}^0 + k\,T \ln n \;, \tag{2-27}$$

where N is the total number of electron sites available for electron occupation and $\bar{\mu}^0$ is the *standard* electrochemical potential of electrons at unit concentration.

We now consider the relationship which connects the electrochemical potential of electrons in the hopping model with that in the band model. The total concentration, N, of electron sites for the hopping model may be replaced by the effective state density, N_C, for the band model. For the two models, therefore, we obtain from Eqn. 2–27 the following equation:

$$\bar{\mu} = \bar{\mu}^* + k\,T \ln \frac{n}{N^*} = \bar{\mu}^* - k\,T \ln N^* + k\,T \ln n \;, \tag{2-28}$$

where N^* represents the total electron site concentration, N, for the hopping model and the effective conduction band state density, N_C, for the band model. Introducing the *activity*, a ($a = \gamma\,n$), in place of n in Eqn. 2–28, we may write the electrochemical potential, $\bar{\mu}$, of electrons as shown in Eqn. 2–29:

$$\bar{\mu} = \bar{\mu}^* - k\,T \ln N^* + k\,T \ln n + k\,T \ln \gamma \;. \tag{2-29}$$

The concentration of electrons that occupy a part of the total concentration, N^*, of the energy states available for electrons is obtained by integrating the product of the state density, $D(\varepsilon)$, and the Fermi function, $f(\varepsilon)$, as shown in Eqn. 2–30:

$$n = \int_{\varepsilon_c}^{\infty} D(\varepsilon)\cdot f(\varepsilon)\,d\varepsilon = N^* F_{1/2}\left(\frac{\varepsilon_F - \varepsilon^*}{k\,T}\right) \;, \tag{2-30}$$

where $F_{1/2}\{(\varepsilon_F - \varepsilon^*)/(k\,T)\}$ is the dimensionless integral of the Fermi function. In Eqn. 2–30, ε^* represents the localized electron site level for the hopping model and the conduction band edge level for the band model. In the ideal case ($\gamma = 1$) in which the Fermi function can be approximated by the Boltzmann function, Eqn. 2–29 yields Eqn. 2–31:

$$n = N^* \exp \frac{\varepsilon_F - \varepsilon^*}{k\,T} \;. \tag{2-31}$$

Consequently, comparing Eqn. 2–30 with Eqn. 2–31, we obtain the activity coefficient, γ, as shown in Eqn. 2–32:

$$\gamma = \frac{\exp\dfrac{\varepsilon_F - \varepsilon^*}{kT}}{F_{1/2}\left(\dfrac{\varepsilon_F - \varepsilon^*}{kT}\right)} . \tag{2-32}$$

Thus, for the band model the activity coefficients, γ_e and γ_h, of electrons and holes in the conduction and valence bands are, respectively, given in Eqn. 2–33:

$$\gamma_e = \frac{\exp\dfrac{\varepsilon_F - \varepsilon_C}{kT}}{F_{1/2}\left(\dfrac{\varepsilon_F - \varepsilon_C}{kT}\right)} , \quad \gamma_h = \frac{\exp\dfrac{\varepsilon_V - \varepsilon_F}{kT}}{F_{1/2}\left(\dfrac{\varepsilon_V - \varepsilon_F}{kT}\right)} . \tag{2-33}$$

Simple calculation based on Eqn. 2–33 yields the relationship between the activity coefficient of electrons and the molar fraction, n/N^*, of the state density occupied by electrons [Rosenberg, 1960]: the result of calculation is illustrates in Fig. 2–19. It appears that the activity coefficient of electrons is greater than one ($\gamma > 1$) in the range of relatively high concentrations of electron occupation ($n > 0.1\,N^*$). For the band model in which the effective electron state density N_C is of the order of $N^* = N_C \doteq 10^{19}$ cm^{-3} at room temperature, the activity coefficient of electrons is greater than one ($\gamma > 1$) in the range of electron concentration greater than $n > 10^{18}$ cm^{-3}: this corresponds to the case in which the Fermi level is either in the conduction band or close to the conduction band edge level in the band gap ($\varepsilon_C - \varepsilon_F < 2\,kT$), as shown in Fig. 2–20. For the hopping model, on the other hand, the density of electron sites N is of the order of $N^* = N \doteq 10^{22}$ cm^{-3} at room temperature and, therefore, the range of electron concentration in which the activity coefficient of electrons holds unity ($\gamma \doteq 1$, $a \doteq n$) extends to the electron concentration of the order of $n \doteq 10^{21}$ cm^{-3}.

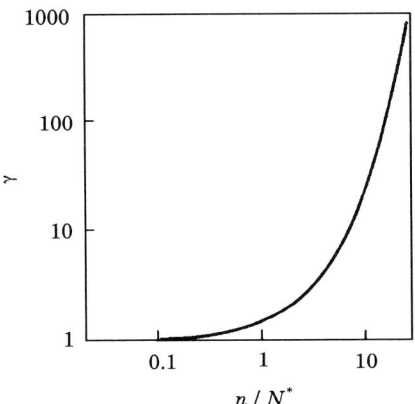

Fig. 2-19. Activity coefficient, γ, and molar fraction, n/N^*, of electrons in an electron ensemble in condensed phases: n = electron concentration; N^* = total energy state density available for electrons. [From Rosenberg, 1960.]

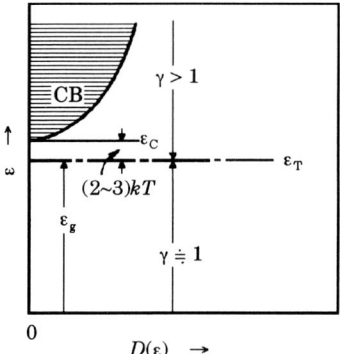

Fig. 2–20. Electron state density and the range of the Fermi energy where the electron occupation probability in the conduction band of an electron ensemble of low electron density (e.g., semiconductor) follows the Boltzmann function ($\gamma \doteq 1$) or the Fermi function ($\gamma > 1$): γ = electron activity coefficient; ε_T = transition level from $\gamma \doteq 1$ to $\gamma > 1$; $D(\varepsilon)$ = electron energy state density; CB = conduction band. [From Rosenberg, 1960.]

2.6 Metal Oxides

2.6.1 Formation of electron energy bands

Most metal oxides are ionic crystals and belong to either the class of semiconductors or insulators, in which the valence band mainly comprises the frontier orbitals of oxygen ions and the conduction band contains the frontier orbitals of metal ions. In forming an ionic metal oxide crystal from metal ions and oxygen ions, as shown in Fig. 2–21, the crystalline field shifts the frontier electron level

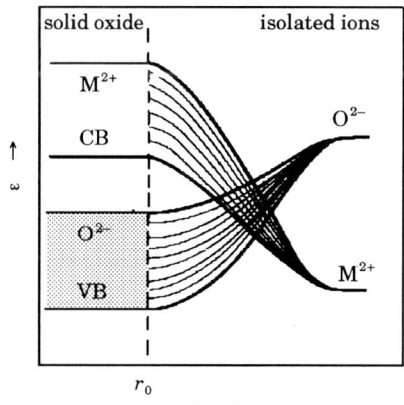

Fig. 2–21. Formation of electron energy bands in metal oxides from isolated metal ions and oxygen ions.

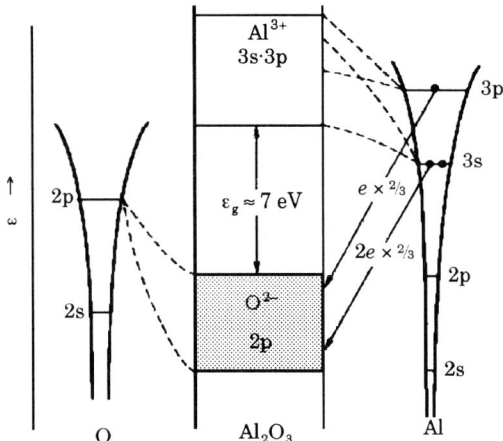

Fig. 2–22. Formation of electron energy bands in aluminum oxide. [From Vijh, 1970.]

of metal ions to higher energies to form an antibonding band (the conduction band) and shifts the frontier electron level of oxygen ions to lower energies to form a bonding band (the valence band).

Fig. 2–22 shows a schematic energy diagram for formation of the valence and conduction bands of aluminum oxide, Al_2O_3, from the frontier 3s and 3p orbitals of an aluminum atom, which form the conduction band by donating electrons, and the frontier 2p orbital of an oxygen atom, which forms the valence band filling its two vacant levels with electrons coming from the 3s and 3p orbitals of the aluminum atom. The band gap of Al_2O_3 is 7 eV and, hence, it is an insulator. In the case of transition metal oxides, the d-orbital band appearing in otherwise a wide band gap makes the band gap relatively narrow, as shown with Cu_2O (ε_g = 1.8 eV) in Fig. 2–23: the valence band comprises the fully occupied 3d orbital and the conduction band comprises the vacant 4s orbital of copper atoms.

In iron oxide of $\alpha\text{-}Fe_2O_3$ (hematite), the conduction band of the Fe^{3+} 3d antibonding orbital confronts two valence bands: one comprises a hybridized band (VB_1) of the Fe^{3+} 3d bonding orbital and the O^{2-} 2p antibonding orbital, and the other comprises a band (VB_2) of the O^{2-} 2p bonding orbital. As shown in Fig. 2–24, the band gap between the conduction band and the hybridized valence band VB_1 is 2.2 eV, which is the direct band gap; and the band gap between the conduction band and the oxygen valence band VB_2 is 3.1 to 3.5 eV, which is the indirect band gap of $\alpha\text{-}Fe_2O_3$. In the case of Fe_3O_4 the hybridized valence band VB_1 contains vacant levels making the oxide electronically conductive.

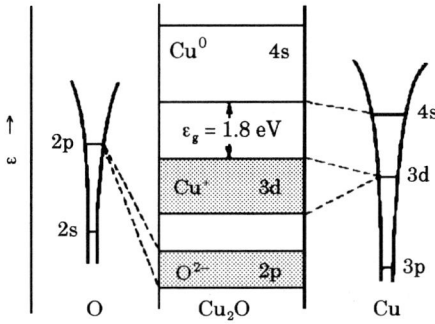

Fig. 2-23. Formation of electron energy bands in cuprous oxide. [From Shive, 1959.]

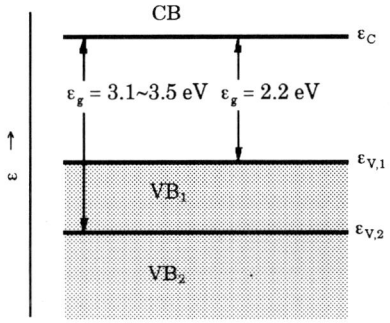

Fig. 2-24. Electron energy bands in α-Fe_2O_3: CB = conduction band of the vacant 3d orbitals of Fe^{3+}; VB_1 = valence band comprising the occupied 3d orbitals of Fe^{3+} and the antibonding 2p orbitals of O^{2-}; VB_2 = valence band of the occupied 2p orbitals of O^{2-}. [From Anderman-Kennedy, 1988.]

The band gap, ε_g, of ionic crystals appears to be a linear function of the bond energy of ionic crystals as shown in Eqn. 2-34 and in Fig. 2-25:

$$\varepsilon_g = 2\,(\Delta H_a - R)\,, \tag{2-34}$$

where ΔH_a is the atom–atom bond energy in ionic crystals, and R is an energy related to the crystal field.

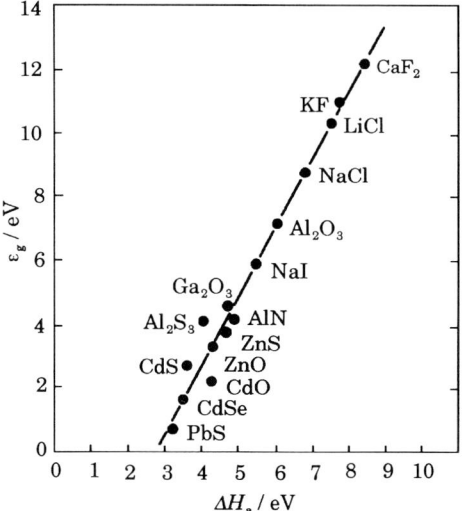

Fig. 2–25. Band gap ε_g and bond energy ΔH_a of binary ionic compounds. [From Vijh, 1970.]

2.6.2 Localized electron levels

Lattice defects and impurities introduce localized electron levels in the band gap of metal oxides. As shown in Fig. 2–26, interstitial metal ions introduce the donor levels ($M_i \rightarrow M_i^+ + e$, $M_i^+ \rightarrow M_i^{2+} + e$, etc.), metal ion vacancies introduce the acceptor levels ($V_M \rightarrow V_M^- + h$, $V_M^- \rightarrow V_M^{2-} + h$, etc.), and oxygen ion vacancies give the donor levels ($V_O \rightarrow V_O^+ + e$, $V_O^+ \rightarrow V_O^{2+} + e$, etc.). Further, donor levels are introduced by impurity cations with the positive ionic valence greater than the metal ion valence (In in ZnO) and by impurity anions with the negative ionic valence greater than the oxygen ion valence (N in ZnO); while on the contrary acceptor levels are introduced by impurity cations with the positive ionic valence smaller than the metal ion valence (Li in ZnO) and by impurity anions with the negative ionic valence smaller than the oxygen ion valence (Cl in ZnO).

Fig. 2–27 illustrates the electron levels of lattice defects in zinc oxide. The interstitial zinc ions and oxygen ion vacancies create the donor levels close to the edge level of the conduction band. These donor levels are lowered stepwise with increasing charge valence of the defects. Similarly, the acceptor levels of metal ion vacancies, which appear close to the edge level of the valence band, are raised stepwise with increasing charge valence of the zinc ion vacancies. Zinc oxide contains interstitial zinc ions as its main defects creating the donor levels;

thus, it is an n-type semiconductor. Nickel oxide contains nickel ion vacancies creating the acceptor levels so that it is a p-type semiconductor.

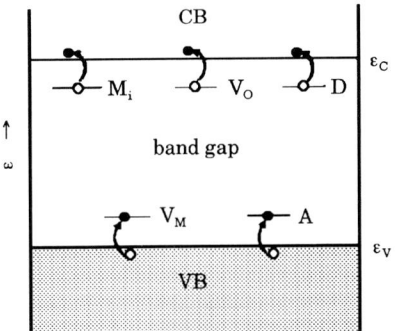

Fig. 2–26. Localized electron levels of lattice defects and impurities in metal oxides: M_i = interstitial metal ion; V_M = metal ion vacancy; V_O = oxygen ion vacancy; D = donor impurity (cations and anions with greater charge valence than metal ions and oxygen ions); A = acceptor impurity (cations and anions with smaller charge valence than metal ions and oxygen ions).

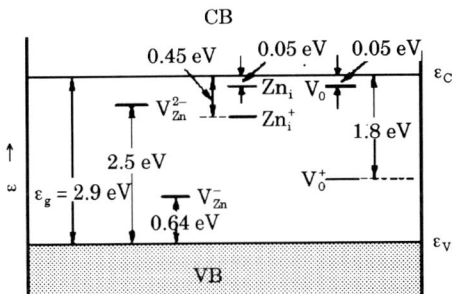

Fig. 2–27. Localized electron levels of lattice defects in zinc oxide. [From Kröger, 1974.]

2.7 The Surface of Semiconductors

2.7.1 The surface state

In addition to the electron energy bands and impurity levels in the semiconductor interior, which are three dimensional, two dimensional localized levels in the band gap exist on the semiconductor surface as shown in Fig. 2–28. Such electron levels associated with the surface are called *surface states* or *interfacial states*, ε_{ss}. The surface states are classified according to their origin into the following two categories; (a) the surface dangling state, and (b) the surface ion-induced state.

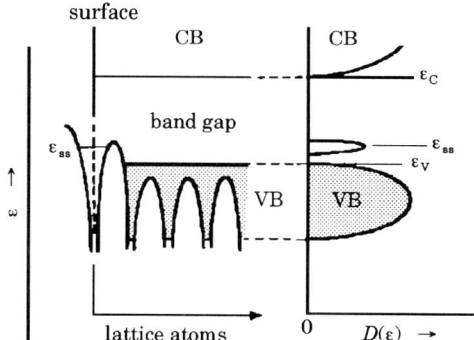

Fig. 2–28. Localized surface states on semiconductors: ε_{ss} = surface state level.

(a) *Surface dangling states*: The surface dangling state (called the Shockley surface state) is caused by the dangling bonds of surface atoms on covalently bonded semiconductors. Fig. 2–29 illustrates for semiconductor silicon crystals the formation of the bonding band (VB), the antibonding band (CB), and the surface dangling states from the hybridized sp³ orbitals of silicon atoms. The surface dangling states may act as donors or acceptors. To reduce the surface energy, a dangling bond tends to combine with an adjacent dangling bond to form a bonding level (donor level) or an antibonding level (acceptor level). Usually, their surface reconstruction takes place on covalently bonded solids to reduce the number of dangling bonds producing the surface donor levels (bonding levels) near the valence band edge and the surface acceptor levels near the conduction band edge as shown in Fig. 2–30 (a).

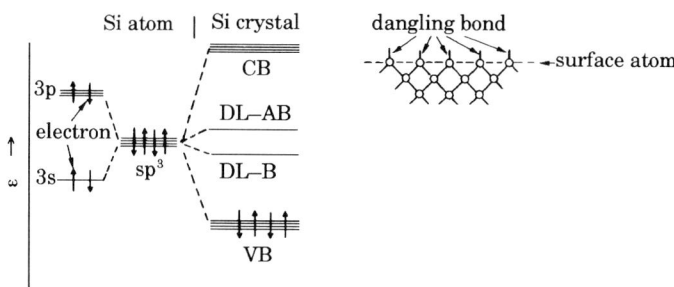

Fig. 2–29. Formation of electron energy bands and surface dangling states of silicon crystals: DL-B = dangling level in bonding; DL-AB = dangling level in antibonding.

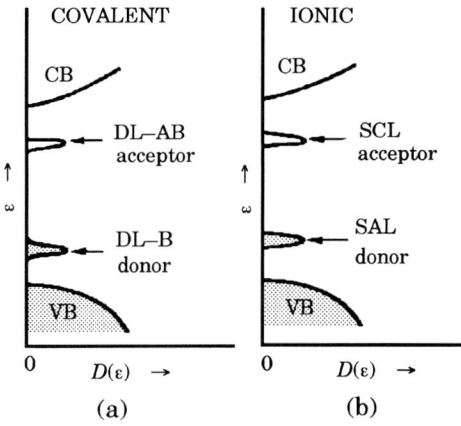

Fig. 2–30. Surface dangling states and surface ion-induced states: (a) the surface dangling donor (DL-B) and acceptor (DL-AB) levels on covalent bonding semiconductors, (b) the surface cation-induced acceptor (SCL) and surface anion-induced donor (SAL) levels on ionic bonding semiconductors.

(b) *Surface ion-induced states*: The surface ion-induced state (called the Tamm state) is caused by the surface lattice potential different from the internal lattice potential of ionic compound semiconductors such as metal oxides. As shown in Fig. 2–30 (b), the surface cation level, which is associated with the conduction band, provides an acceptor level close to the conduction band; and the surface anion level, which is associated with the valence band, provides a donor level close to the valence band.

On semiconductors that are partially ionic and partially covalent, such as transition metal oxides, the surface ion-induced and dangling states may coexist together.

The dangling and surface ion-induced states are intrinsic surface states that are characteristic of individual semiconductors. In addition, there are extrinsic surface states produced by adsorbed particles and surface films that depend on the environment in which the surface is exposed. In general, adsorbed particles in the covalently bonded state on the semiconductor surface introduce the dangling-like surface states; and those in the ionically bonded state introduce the adsorption ion-induced surface states. In electrochemistry, the adsorption-induced surface states are important.

The concentration of the surface states is in the range from 1×10^{10} to 1×10^{14} cm^{-2}, which is 1/10 to 1/100000 of the concentration of surface atoms ($\sim 1\times10^{15}$ cm^{-2}). Usually, the surface state concentration is greater on the rough surface than on the smooth surface.

2.7.2 The space charge layer

Since the Fermi level at the semiconductor surface containing the surface states differs from the Fermi level of the semiconductor interior, electrons (or holes) may transfer between the surface state level and the conduction band (or the valence band) until the electron transfer equilibrium is established between the surface and the interior of semiconductor. As a result, an electric charge, Q_{SC}, equivalent but opposite in the sign to the charge, Q_{ss}, captured at the surface state, is generated inside the semiconductor forming near the surface a *space charge layer*. The thickness of the space charge layer increases with decreasing concentration of the mobile charge carriers (concentration of electrons or holes), usually being 10 to 1000 nm.

The Fermi level, ε_F^s, at the surface can be derived in the same way as the interior Fermi level of extrinsic semiconductors shown in Eqns. 2–22 and 2–24 to give Eqn. 2–35 for the surface with a donor surface state at the energy level ε_{ss}:

$$\varepsilon_F^s = \frac{1}{2}(\varepsilon_C + \varepsilon_{ss}) - \frac{1}{2}kT\ln\frac{N_C}{\zeta N_{ss}}, \qquad (2\text{--}35)$$

and Eqn. 2–36 for the surface containing an acceptor surface state at the energy level ε_{ss}:

$$\varepsilon_F^s = \frac{1}{2}(\varepsilon_V + \varepsilon_{ss}) + \frac{1}{2}kT\ln\frac{N_V}{\zeta N_{ss}}, \qquad (2\text{--}36)$$

where N_{ss} is the concentration of surface states, and ζ is the conversion factor between the volume concentration and the surface concentration.

The surface Fermi level, ε_F^s, which depends on the surface state, is not the same as the interior Fermi level, ε_F, which is determined by the bulk impurity and its concentration. As the electron transfer equilibrium is established, the two Fermi levels are equilibrated each other ($\varepsilon_F^s = \varepsilon_F$); and the band level bends downward or upward near the surface forming a space charge layer as shown in Fig. 2–31.

In cases in which the surface state density is high ($N_C/\zeta N_{ss}, N_V/\zeta N_{ss} \to 1$), the electron distribution in the surface state conforms the Fermi function (the state of degeneracy); and the Fermi level is *pinned* at the surface state level. This is what is called the *Fermi level pinning* at the surface state.

The downward bending of the band level near the surface of n-type semiconductors gives rise to the accumulation of electrons at the surface, which is called the *accumulation layer* of electrons. On the other hand, the upward band bending leads to the depletion of electrons at the surface and is called the *depletion layer* of electrons. For p-type semiconductors, on the contrary, the downward band bending leads to the depletion of holes at the surface and is called the depletion layer of holes; and the upward band bending produces the accumulation of holes at the surface and is called the accumulation layer of holes.

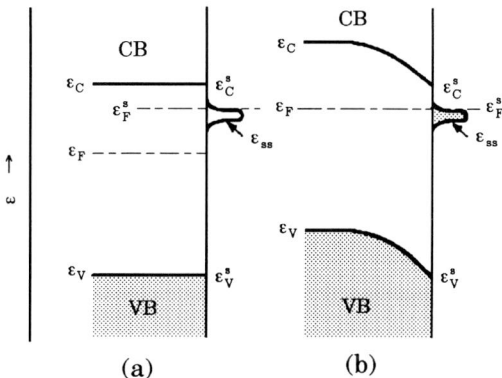

Fig. 2–31. Surface degeneracy caused by Fermi level pinning at a surface state of high state density: (a) in the flat band state ($\varepsilon_F^s \neq \varepsilon_F$), (b) in the electron equilibrium ($\varepsilon_F^s = \varepsilon_F$). ε_F^s = surface Fermi level; ε_C^s = surface conduction band edge level.

The space charge layer caused by the band bending generates a potential difference between the interior and the surface, $V_{SC} = (\varepsilon_C^s - \varepsilon_C)/e$, called the *potential* or *potential difference of the space charge layer*, where ε_C^s and ε_C are the edge levels of the conduction band at the surface and in the interior, respectively. In the case of the depletion layer, the potential difference of the space charge layer, V_{SC}, provides a *potential barrier* (called the *Schottky barrier*) for electrons or holes to transfer from the interior to the exterior of semiconductors. In the presence of a space charge layer, the surface concentrations of electrons, n_s, and holes, p_s, are different from the bulk concentrations of electrons, n, and holes, p, as given by Eqns. 2–37 and 2–38:

$$n_s = n \exp\left(-\frac{\varepsilon_C^s - \varepsilon_C}{kT}\right) = n \exp\left(-\frac{eV_{SC}}{kT}\right), \tag{2–37}$$

$$p_s = p \exp\left(\frac{\varepsilon_V^s - \varepsilon_V}{kT}\right) = p \exp\left(\frac{eV_{SC}}{kT}\right). \tag{2–38}$$

These surface concentrations of electrons and holes are important in electrochemistry.

2.7.3 Surface degeneracy (Quasi-metallization of surfaces)

The semiconductor surface where the Fermi level is pinned at the surface state of high density (Fig. 2–31) is in the state of *degeneracy* of electron levels, because of the high electron state density at the surface Fermi level. Similarly, the surface degeneracy is also established when the band bending becomes so great that the Fermi level is pinned either in the conduction band or in the valence band as shown in Fig. 2–32.

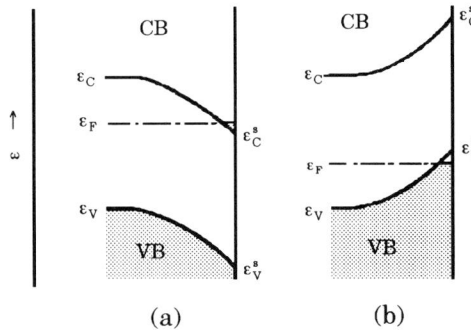

Fig. 2–32. Surface degeneracy caused by Fermi level pinning (a) in the conduction band and (b) in the valence band.

Since the electron state density near the Fermi level on the degenerated surface (Fermi level pinning) is so high as to be comparable with that of metals, the Fermi level pinning at the surface state, in the conduction band and in the valence band, is often called the *quasi-metallization* of semiconductor surfaces. As is described in Chap. 8, the quasi-metallized surface occasionally plays an important role in semiconductor electrode reactions.

2.8 Amorphous Semiconductors

In amorphous semiconductors, the atomic order in the long range ceases to exist; but the short range order remains to some extent, giving rise thereby to a band-like structure of electron energy states similar to that of crystalline semiconductors. Therefore, the electron energy band structure of amorphous semiconductors is basically the same as that of crystalline semiconductors. It appears, however, that the band edge of amorphous semiconductors becomes indistinct, diffusing into the band gap, because the distance and bond energy between the atoms vary to some extent depending on the location in amorphous semiconductors.

As shown in Fig. 2–33, localized electron levels arise (A and C in the figure) near the band edges at relatively high state densities tailing into the band gap; these are called the *diffuse band tail states*. Further, localized electron levels may occur due to dangling bonds and impurities (B in the figure) in the band gap, which are called the *gap states*.

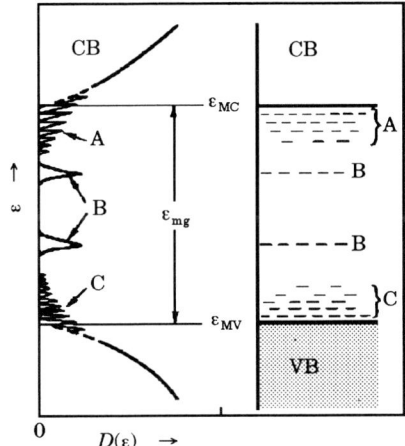

Fig. 2–33. Electron energy and state density in amorphous semiconductors: A and C = diffuse band tail states; B = gap states, ε_{MC} = mobility edge level for electrons; ε_{MV} = mobility edge level for holes; ε_{mg} = mobility gap. [From Mott-Davis, 1979.]

Electrons in the diffuse band tail states migrate in accord with the hopping mechanism rather than the band mechanism. The energy gap between the mobility edge ε_{MC} for electrons and the mobility edge ε_{MV} for holes is called the *mobility gap*, ε_{mg}, instead of the band gap, ε_g.

2.9 Electron Energy Bands of Liquid Water

Solid ice forms a crystal of the diamond structure, in which one water molecule is hydrogen-bonded with four adjacent water molecules. Most (85%) of the hydrogen bonds remain even after solid ice melts into liquid water. The structure of electron energy bands of liquid water (hydrogen oxide) is basically similar to that of metal oxides, although the band edges are indefinite due to its amorphous structure.

Fig. 2–34 shows the diagram of electron energy bands of liquid water. As in the case of metal oxides, the oxygen 2p orbital constitutes the valence band, and the hydrogen 1s orbital constitutes the conduction band of liquid water. It has been reported in the literature [Watanabe-Gerischer, 1981] that the band gap (mobility gap) of liquid water is about $\varepsilon_{mg} \geq 8$ eV; the lower mobility edge level of the conduction band is about $\varepsilon_C \geq -1.2$ eV; and the upper mobility edge level of the valence band is about $\varepsilon_V \geq -9.3$ eV referred to the vacuum electron level.

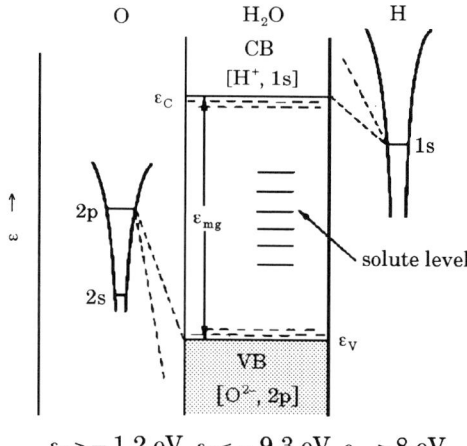

Fig. 2-34. Electron energy bands of liquid water formed from atomic orbitals of hydrogen and oxygen atoms.

The conduction band electron of liquid water, which could be produced by photoelectron emission from metal electrodes, is very unstable with its lifetime being 10^{-11} seconds; it is readily captured by water molecules to form a hydrated electron. The hydrated electron is also very unstable being rapidly absorbed in electron scavenger particles such as H_3O^+, NO_2^- and O_2. The level of the hydrated electron has been estimated at 0.3 to 0.5 eV below the conduction band edge [Battisi-Trasatti, 1977; Watanabe-Gerischer, 1981].

On the free surface of liquid water there is a surface electric dipole due to the orientation of surface water molecules with the negative charge (oxygen) outside and the positive charge (hydrogen) inside; this generates a positive *surface potential difference*, χ_{H_2O}. The surface potential difference of liquid water has been reported to be in the range from 0.025 eV to 0.7 eV in the literature [Gomer-Tryson, 1977; Notoya-Matsuda, 1982]. The following value is frequently used [Trasatti, 1980]:

$$\chi_{H_2O} = 0.13 \pm 0.02 \text{ V} . \tag{2-39}$$

The value of χ_{H_2O} is important for estimating theoretically the energy levels of hydrated ions and redox electrons in aqueous solutions.

In the case of aqueous solutions containing dissolved particles (solutes), a number of localized electron levels associated with solute particles arise in the mobility gap of aqueous solutions as shown in Fig. 2–34. These *localized* electron levels of solutes may be compared with the localized impurity levels in semiconductors. In electrochemistry, the electron levels of the solutes of general interest are those located within the energy range from -4 eV to -6 eV (around the electron levels of the hydrogen and oxygen electrode reactions) in the mobility gap.

2.10 Redox Electrons in Aqueous Solution

2.10.1 Electron levels of gaseous redox particles

A reaction in which an electron transfer takes place between two different chemical particles is called a *redox* reaction; this is a combination of an *electron donating* reaction and an *electron accepting* reaction as shown in Eqn. 2–40. The reaction in which a chemical particle donates electrons is called the *oxidation*, and the particle produced by the oxidation is the *oxidant*; while on the other hand the electron accepting reaction is called the *reduction*, and its product particle is called the *reductant*.

$$\text{RED} \rightleftarrows \text{OX} + e_{(\text{REDOX})} , \tag{2–40}$$

where RED is the reductant, OX is the oxidant, and $e_{(\text{REDOX})}$ is the *redox electron*.

The redox reaction is illustrated with gaseous iron ions: $Fe^{2+}_{(gas)} \rightleftarrows Fe^{3+}_{(gas)} + e_{(\text{STD})}$, where $e_{(\text{STD})}$ is the gaseous electron in the standard state. The occupied electron level $\varepsilon_{Fe^{2+}, D}$ of the reductant $Fe^{2+}_{(gas)}$ can release an electron (donor level); and the vacant electron level $\varepsilon_{Fe^{3+}, A}$ of the oxidant $Fe^{3+}_{(gas)}$ can accept an electron (acceptor level). The donor level $\varepsilon_{Fe^{2+}, D}$ of the reductant $Fe^{2+}_{(gas)}$ is represented by the negative ionization energy $-I$; and the acceptor level $\varepsilon_{Fe^{3+}, A}$ of the oxidant $Fe^{3+}_{(gas)}$ is represented by the negative electron affinity $-A$, as shown in Fig. 2–35. These donor and acceptor levels are equal to each other, because they are the same level of an identical frontier orbital of gaseous iron ions and are not affected by the presence and absence of electrons in the levels. Then, the electrochemical potential, $\bar{\mu}_{e(\text{gas, REDOX})}$, of a gaseous redox electron is given by Eqn. 2–41:

$$\bar{\mu}_{e(\text{gas, REDOX})} = \varepsilon_{Fe^{2+}, D} = \varepsilon_{Fe^{3+}, A} = -I = -A . \tag{2–41}$$

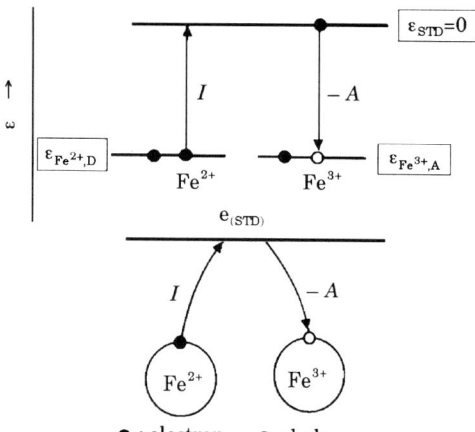

Fig. 2–35. Localized electron levels of gaseous redox particles, Fe^{3+}/Fe^{2+}: I = ionization energy of Fe^{2+}; A = electron affinity of Fe^{3+}; $e_{(STD)}$ = standard gaseous electrons; ε_{STD} = standard gaseous electron level (reference zero level).

2.10.2 Electron levels of hydrated redox particles

In contrast to gaseous redox particles, the donor and acceptor levels in hydrated redox particles are not equal to each other, although they are in the identical frontier orbital. In aqueous solutions a hydrated particle includes coordinated water molecules in its outer sphere to form the hydrate structure, which depends on different charge valences of the hydrated particle. Since the electron level of hydrated particles contains the energy required for organizing the hydrate structure, the donor level *differs* from the acceptor level by an amount equivalent to the reorganization energy of the hydrate structure due to the different charge valences of the reductant and oxidant particles, *even though* the two levels are in the identical orbital.

We consider a redox reaction of hydrated iron ions shown in Eqn. 2–42:

$$Fe^{2+}_{(aq)} \rightleftarrows Fe^{3+}_{(aq)} + e_{(STD)} \,. \tag{2-42}$$

Fig. 2–36 shows the occupied electron level (donor level) of reductant $Fe^{2+}_{(aq)}$ and the vacant electron level (acceptor level) of oxidant $Fe^{3+}_{(aq)}$ referred to the standard gaseous electron level at the outer potential of aqueous solution.

The electron level in hydrated redox particles consists of the energy ΔG_{int} (<0) required for the standard gaseous electron to combine with or to be released from the gaseous redox particles and the energy ΔG_{org} (>0) required for the redox particles to form their hydrate structures. Since the donor and acceptor levels of gaseous redox particles $Fe^{2+}_{(gas)}/Fe^{3+}_{(gas)}$ equal each other, the difference between

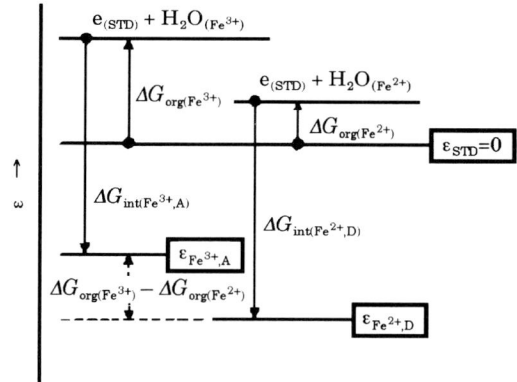

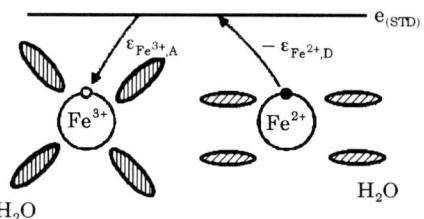

Fig. 2–36. Electron energy levels in hydrated oxidant Fe^{3+} and reductant Fe^{2+}: ΔG_{org} = energy to organize hydrate structures; ΔG_{int} = energy required for dehydrated redox ions to donate or accept gaseous electrons; $\varepsilon_{Fe^{2+}, D}$ = most probable electron donor level of Fe^{2+}; $\varepsilon_{Fe^{3+}, A}$ = most probable electron acceptor level of Fe^{3+}; $H_2O_{(Fe^{2+} \text{ or } Fe^{3+})}$ = hydrated structures; ε_{STD} = standard gaseous electron level (= 0).

the donor and the acceptor level of hydrated redox particles $Fe^{2+}_{(aq)}/Fe^{3+}_{(aq)}$ arises from the difference in the organization energy of the hydrate structures between the reductant $Fe^{2+}_{(aq)}$ and the oxidant $Fe^{3+}_{(aq)}$ as shown in Fig. 2–36. In general, the energy for organizing the hydrate structure increases with increasing charge valence of the particle; hence, this energy is greater for the oxidant $Fe^{3+}_{(aq)}$ than for the reductant $Fe^{2+}_{(aq)}$ ($\Delta G_{int(Fe^{3+}, A)} > \Delta G_{int(Fe^{2+}, D)}$), giving rise to the acceptor level being higher than the donor level as shown in Eqn. 2–43 and in Fig. 2–36:

$$\varepsilon_{Fe^{3+}, A} - \varepsilon_{Fe^{2+}, D} = \Delta G_{int(Fe^{3+}, A)} - \Delta G_{int(Fe^{2+}, D)} , \qquad (2\text{–}43)$$

where the occupied electron level $\varepsilon_{Fe^{2+}, D}$ of the reductant corresponds to the ionization energy of the hydrated ferrous ion; and the vacant electron level $\varepsilon_{Fe^{3+}, A}$ of the oxidant corresponds to the electron affinity of the hydrated ferric ion.

The reorganization of the hydrate structure occurs immediately after the electron transfer. Since the rate of electron transfer (about 10^{-15} seconds) is much faster than the rate of molecular vibration (about 10^{-11} seconds), the electron transfer occurs *adiabatically* while the hydrate structure is *frozen*. This is the reason why the electron level of the same state splits into the two different donor and acceptor levels. Such a splitting of an electron level is called the Franck-Condon *level splitting*.

The redox reaction of hydrated particles referred to the standard gaseous electron may be represented by the following four steps, as shown for the redox reactions of hydrated iron ions in Fig. 2–37:

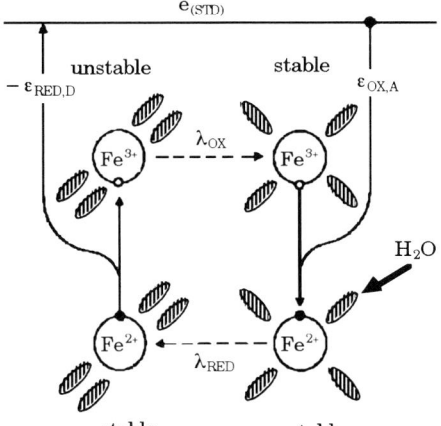

Fig. 2–37. Redox reaction cycle $Fe_{aq}^{2+} \to Fe_{aq}^{3+} + e_{(STD)} \to Fe_{aq}^{2+}$ in aqueous solution: solid arrow = adiabatic electron transfer; dotted arrow = hydrate structure reorganization; λ = reorganization energy; $\varepsilon_{RED,D}$ = most probable donor level; $\varepsilon_{OX,A}$ = most probable acceptor level.

(1) $RED_{(red)} \to OX_{(red)} + e_{(STD)}$ adiabatic electron release, $\Delta G = -\varepsilon_{RED,D}$
(2) $OX_{(red)} \to OX_{(ox)}$ reorganization of hydrate structure, $\Delta G = -\lambda_{OX}$
(3) $OX_{(ox)} + e_{(STD)} = RED_{(ox)}$ adiabatic electron absorption, $\Delta G = -\varepsilon_{OX,A}$
(4) $RED_{(ox)} \to RED_{(red)}$ reorganization of hydrate structure, $\Delta G = -\lambda_{RED}$

where $OX_{(red)}$ is the unstable oxidant particle with the same hydrate structure as the reductant; $RED_{(ox)}$ is the unstable reductant particle with the same hydrate structure as the oxidant; ΔG is the free enthalpy change of the step; $\varepsilon_{RED,D}$ is the donor level of the reductant; $\varepsilon_{OX,A}$ is the acceptor level of the oxidant; and λ is the reorganization energy of the hydrate structure. Fig. 2–38 schematically illustrates the electron levels in the foregoing redox reaction cycle.

From the energy balance of the redox reaction cycle shown in Fig. 2–38, we obtain Eqn. 2–44:

$$\varepsilon_{OX,A} - \varepsilon_{RED,D} = \lambda_{OX} + \lambda_{RED} , \qquad (2\text{--}44)$$

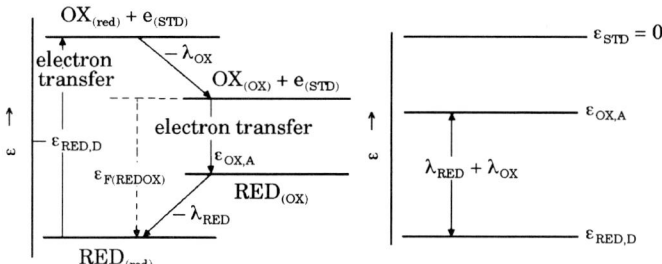

Fig. 2–38. Electron energy levels for the redox reaction of hydrated particles $RED_{aq} \rightarrow OX_{aq} + e_{(STD)} \rightarrow RED_{aq}$: $OX_{(red)}$ ($RED_{(ox)}$) = unstable oxidant (reductant) particles with the hydrate structure of reductant (oxidant).

which indicates that the acceptor level $\varepsilon_{OX,A}$ is higher than the donor level $\varepsilon_{RED,D}$ and that the gap between the two levels is equivalent to the sum of the two reorganization energies $\lambda_{OX} + \lambda_{RED}$ (usually, $\lambda_{OX} \doteq \lambda_{RED}$).

2.10.3 Fluctuation of electron energy levels

The localized electron level of hydrated particles in aqueous solutions, different from that of particles in solids, does not remain constant; but it *fluctuates* in the range of reorganization energy, λ, because of the thermal (rotational and vibrational) motion of coordinated water molecules in the hydration structure. The electron levels $\varepsilon_{OX,A}$ and $\varepsilon_{RED,D}$ are the most probable levels of oxidants and reductants, respectively.

The reorganization energy, λ, which is of the same order of magnitude as the hydration energy of ions, can be given by Eqn. 2–45 [Marcus, 1956]:

$$\lambda = \frac{e^2}{8\pi\varepsilon_o a}\left(\frac{1}{\kappa_{op}} - \frac{1}{\kappa_s}\right), \tag{2–45}$$

where a is the ionic radius, ε_o is the permittivity of vacuum, κ_s is the relative dielectric constant of water ($\kappa_s = 78.5$ at room temperature), and κ_{op} is the relative dielectric constant of water at the light wave frequency ($\kappa_{op} = 1.8$). The first term in the right hand side of Eqn. 2–45 is the electronic polarization, and the second term is the dipole polarization due to the reorganization of coordinated water molecules. The magnitude of the reorganization energy λ of simple hydrated ions has been estimated to be in the range from 1 to 2 eV; λ is 1.2 eV for hydrated redox iron ions Fe^{3+}/Fe^{2+} [Memming-Möllers, 1972].

The fluctuation in the vibrational and rotational energy $\Delta\lambda$, as shown in Eqn.

2–45, is proportional to the square of the particle charge, in contrast to the fluctuation in the electron energy level $\Delta\varepsilon$ which is directly proportional to the ionic charge. To the first approximation, therefore, we may assume that $(\Delta\varepsilon)^2 = \gamma\,\Delta\lambda$, where γ is a proportionality constant [Marcus, 1964; Dogonadze, 1970]. From the energy balance in the transfer cycle of redox electrons, it has been found that the proportionality constant is four times greater than the reorganization energy λ; ($\gamma = 4\,\lambda$) [Morrison, 1980]. Since the energy fluctuation $\Delta\lambda$ in the hydrated structure follows the Boltzmann distribution, the resulting fluctuation $\Delta\varepsilon$ in the electron energy level is represented by a *Gaussian normal distribution* as shown in Eqn. 2–46:

$$W(\varepsilon) = W_0 \exp\left(\frac{-(\varepsilon-\varepsilon_0)^2}{4\,\lambda\,k\,T}\right) = \frac{1}{\sqrt{4\,\lambda\,k\,T}}\exp\left(\frac{-(\varepsilon-\varepsilon_0)^2}{4\,\lambda\,k\,T}\right), \qquad (2\text{--}46)$$

where $W(\varepsilon)$ is the probability density of electron energy fluctuation, ε_0 is the *most probable* electron level of hydrated particles, and W_0 is the normalizing factor making $\int W(\varepsilon)\,d\varepsilon = 1$ which yields $W_0 = 1/\sqrt{4\,\lambda\,k\,T}$. Fig. 2–39 shows the Gaussian probability density of electron energy fluctuation in hydrated redox particles. The donor level of reductant particles fluctuates around its *most probable* level ε_{RED}, and the acceptor level of oxidant particles fluctuates around its *most probable* level ε_{OX}. The total probability density $W_{REDOX}(\varepsilon)$ of the donor and acceptor levels is the sum of the two probability densities of $W_{RED}(\varepsilon)$ and $W_{OX}(\varepsilon)$ as shown in Eqn. 2–47 and in Fig. 2–39:

$$W_{REDOX}(\varepsilon) = W_{RED}(\varepsilon) + W_{OX}(\varepsilon)\,. \qquad (2\text{--}47)$$

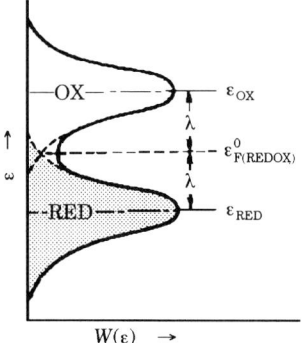

Fig. 2–39. Gaussian normal distribution of the probability density of redox electron levels due to the thermal fluctuation of hydrate structures: $\varepsilon^0_{F(REDOX)}$ = standard Fermi level of redox electrons.

As the localized electron level of hydrated redox particles distributes in a rather wide range, we may assume the presence of energy bands in which the redox electron level fluctuates; the reductant particles form a donor band, and the oxidant particles form an acceptor band. The donor and acceptor bands overlap in their probability density tailing as shown in Fig. 2–39.

2.10.4 The Fermi level of hydrated redox electrons

The electron state densities $D_{RED}(\varepsilon)$ and $D_{OX}(\varepsilon)$ in the donor and acceptor bands of hydrated redox particles are given by the product of the probability densities $W_{RED}(\varepsilon)$ and $W_{OX}(\varepsilon)$ and the concentrations N_{RED} and N_{OX}, respectvely, in Eqns. 2–48 and 2–49:

$$D_{RED}(\varepsilon) = W_{RED}(\varepsilon) N_{RED} = \frac{N_{RED}}{\sqrt{4\lambda k T}} \exp\left(\frac{-(\varepsilon - \varepsilon_{RED})^2}{4\lambda k T} \right), \qquad (2\text{--}48)$$

$$D_{OX}(\varepsilon) = W_{OX}(\varepsilon) N_{OX} = \frac{N_{OX}}{\sqrt{4\lambda k T}} \exp\left(\frac{-(\varepsilon - \varepsilon_{OX})^2}{4\lambda k T} \right). \qquad (2\text{--}49)$$

The total state density $D_{REDOX}(\varepsilon)$ is the sum of $D_{RED}(\varepsilon)$ and $D_{OX}(\varepsilon)$ as shown in Eqn. 2–50:

$$D_{REDOX}(\varepsilon) = D_{RED}(\varepsilon) + D_{OX}(\varepsilon) = W_{RED}(\varepsilon) N_{RED} + W_{OX}(\varepsilon) N_{OX} . \qquad (2\text{--}50)$$

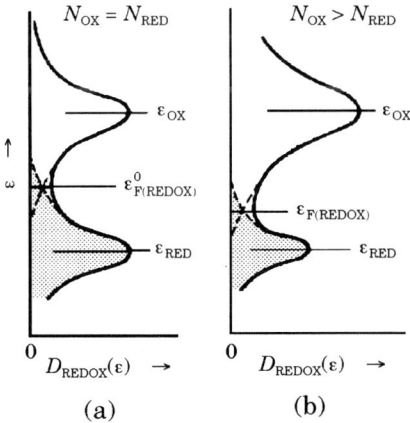

Fig. 2–40. Distribution of the electron state density of hydrated redox particles: (a) oxidant concentration N_{OX} equal to reductant concentration N_{RED}, (b) oxidant concentration N_{OX} higher than reductant concentration N_{RED}. $\varepsilon_{F(REDOX)}$ = Fermi level of redox electrons.

Fig. 2–40 shows the state density distribution curves of hydrated redox particles in two cases in which (a) $N_{RED} = N_{OX}$ and (b) $N_{RED} < N_{OX}$.

In the fluctuation band of electron energy of hydrated redox particles, the donor band of the reductant is an occupied band, and the acceptor band of the oxidant is a vacant band. The level $\varepsilon_{F(REDOX)}$ at which the donor state density equals the acceptor state density ($D_{RED}(\varepsilon) = D_{OX}(\varepsilon)$) is called the *Fermi level of the redox electron* by analogy with the Fermi level ε_F of metal electrons [Gerischer, 1961]. From Eqns. 2–48 and 2–49 with $D_{RED}(\varepsilon) = D_{OX}(\varepsilon)$, the Fermi level $\varepsilon_{F(REDOX)}$ (the *redox electron level*) can be derived as shown in Eqn. 2–51:

$$\varepsilon_{F(REDOX)} = \frac{1}{2}(\varepsilon_{OX} + \varepsilon_{RED}) + kT \ln \frac{N_{RED}}{N_{OX}} = \varepsilon^0_{F(REDOX)} + kT \ln \frac{N_{RED}}{N_{OX}}, \quad (2\text{–}51)$$

where $\varepsilon^0_{F(REDOX)}$ is the *standard* Fermi level of the redox electron at which the reductant and oxidant particles are at the same standard concentration ($N_{RED} = N_{OX}$). Then, we obtain Eqn. 2–52:

$$\varepsilon^0_{F(REDOX)} = \frac{1}{2}(\varepsilon_{OX} + \varepsilon_{RED}). \quad (2\text{–}52)$$

The standard Fermi level $\varepsilon^0_{F(REDOX)}$ is located midway between the most probable donor level ε_{RED} of the reductant particles and the most probable acceptor level ε_{OX} of the oxidant particles.

Fig. 2–41 compares the electron level diagrams of intrinsic semiconductors with hydrated redox particles at the standard concentration. The two diagrams resemble each other in that the Fermi level is located midway between the occupied level and the vacant level. It is, however, obvious that the occupied and vacant bands for semiconductors are of delocalized electron states, whereas they are the fluctuation bands of localized electron states for hydrated redox particles.

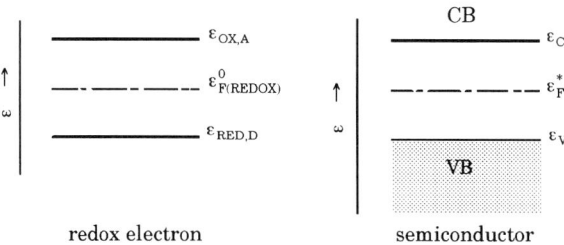

Fig. 2–41. Electron energy levels of hydrated redox particles and intrinsic semiconductors.

The most probable donor level, ε_{RED}, the most probable acceptor level, ε_{OX}, and the standard Fermi level, $\varepsilon^0_{F(REDOX)}$, of redox electrons are characteristic of individual redox particles; but the Fermi level, $\varepsilon_{F(REDOX)}$, of redox electrons depends on the concentration ratio of the reductants to the oxidants which is similar to the Fermi level of extrinsic semiconductors depending on the concentration ratio of the donor to the acceptor.

The Fermi level $\varepsilon_{F(REDOX)} (= \bar{\mu}_{e(REDOX)})$ of the redox electrons may also be obtained thermodynamically from the reaction equilibrium ($RED_{aq} = OX_{aq} + e_{(REDOX)}$), i.e. $\varepsilon_{F(REDOX)} = \bar{\mu}_{e(REDOX)} = \bar{\mu}_{RED} - \bar{\mu}_{OX}$, as shown in Eqn. 2–53:

$$\bar{\mu}_{e(REDOX)} = (\bar{\mu}^0_{RED} - \bar{\mu}^0_{OX}) + kT \ln \frac{N_{RED}}{N_{OX}} = \bar{\mu}^0_{e(REDOX)} + kT \ln \frac{N_{RED}}{N_{OX}} , \qquad (2-53)$$

where $\bar{\mu}^0$ is the standard electrochemical potential of particles. Eqns. 2–51 and 2–53 correspond to the well-known Nernst equation for the equilibrium potential of redox reactions.

The most probable levels ε_{RED} and ε_{OX} of the reductants and oxidants may be derived from the standard Fermi level, $\varepsilon^0_{F(REDOX)}$, and the reorganization energy, λ. As an example, Fig. 2–42 shows the standard Fermi level $\varepsilon^0_{F(Fe^{3+}/Fe^{2+})} (= -5.27 \text{ eV})$ with the most probable levels $\varepsilon_{Fe^{2+},D}$ and $\varepsilon_{Fe^{3+},A}$ of hydrated Fe^{2+} and Fe^{3+} ions which are obtained from the observed standard equilibrium redox potential and $\lambda (= 1.2 \text{ eV})$.

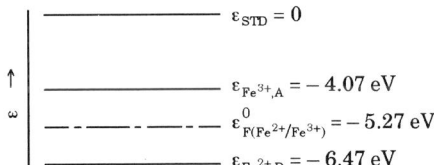

Fig. 2–42. The most probable donor level, $\varepsilon_{Fe^{2+},D}$, the most probable acceptor level, $\varepsilon_{Fe^{3+},A}$, and the standard Fermi level, $\varepsilon^0_{F(Fe^{3+}/Fe^{2+})}$, of hydrated iron ion redox reaction

2.11 The Electron Level of Normal Hydrogen Electrode

In electrochemistry, the electron level of the *normal hydrogen electrode* is important, because it is used as the reference zero level of the electrode potential of redox reactions in aqueous solutions. The reaction of normal hydrogen electrode in the standard state (temperature 25°C, hydrogen pressure 1 atm, and unit activity of hydrated protons) is written in Eqn. 2–54:

$$\frac{1}{2} H_{2\,(gas)} = \frac{1}{2} H_{2\,(aq)} = H^+_{(aq)} + e_{(NHE)} \ , \tag{2-54}$$

where $e_{(NHE)}$ is the equilibrium electron in the reaction of normal hydrogen electrode.

Taking the standard gaseous electron $e_{(STD)}$ as the reference electron at the zero level, we write the reaction in Eqn. 2–55:

$$\frac{1}{2} H_{2\,(gas)} \to H^+_{(aq)} + e_{(STD)} \ , \qquad \text{reaction energy} \ \ \Delta G_{NHE} \ , \tag{2-55}$$

where the free enthalpy ΔG_{NHE} of the reaction represents the level of the equilibrium electron $e_{(NHE)}$ in the normal hydrogen electrode, referred to the energy levels of $H_{2\,(gas)}$ and $e_{(STD)}$, both set to zero in the standard gaseous state. In order to compute the energy of the equilibrium electron $e_{(NHE)}$ in the normal hydrogen electrode, we consider the following reaction steps, as shown in Fig. 2–43.

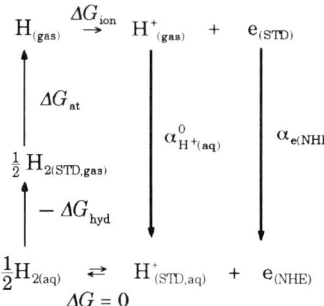

Fig. 2–43. Energy balance in the reaction of normal hydrogen electrode: $H_{2\,(STD,\,gas)}$ = hydrogen molecule in the gaseous standard state (at 1 atm); $H^+_{(STD,\,aq)}$ = hydrated proton of unit activity; $\alpha^0_{H^+(aq)}$ = real potential of the hydrated proton of unit activity; $\alpha_{e(NHE)}$ = real potential of the equilibrium electron of NHE (= Fermi level $\varepsilon_{F(NHE)}$ of NHE).

(1) $\frac{1}{2} H_{2\,(aq)} \to \frac{1}{2} H_{2\,(gas)}$ hydration energy of $\frac{1}{2} H_{2\,(gas)}$ $-\Delta G_{hyd} = 0.2$ eV

(2) $\frac{1}{2} H_{2\,(gas)} \to H_{(gas)}$ dissociation energy of $\frac{1}{2} H_{2\,(gas)}$ $\Delta G_{at} = 2.1$ eV

(3) $H_{(gas)} \to H^+_{(gas)} + e_{(STD)}$ ionization energy of $H_{(gas)}$ $\Delta G_{ion} = 13.6$ eV

(4) $H^+_{(gas)} \to H^+_{(aq)}$ standard hydration energy of $H^+_{(gas)}$ $\alpha^0_{H^+(aq)} = -11.4$ eV

(5) $e_{(STD)} \to e_{(NHE)}$ electron energy of NHE $\Delta G_{NHE} = \alpha_{e(NHE)}$

$\frac{1}{2} H_{2\,(aq)} = H^+_{(aq)} + e_{(NHE)}$ $\Delta G = 0$

In the foregoing reaction steps $\alpha_{e(NHE)}$ is the *real potential of an equilibrium redox electron* of the reaction of normal hydrogen electrode (NHE), which is the energy required for transferring the standard gaseous electron $e_{(STD)}$ at the outer potential of the aqueous solution to the equilibrium redox electron $e_{(NHE)}$ of the reaction of normal hydrogen electrode ($\Delta G_{NHE} = \alpha_{e(NHE)}$); and $\alpha^0_{H^+(aq)}$ is the standard real potential of the hydrated proton, which is the energy required for transferring the standard gaseous proton to the state of the hydrated proton of unit activity. The numerical energy values given in the foregoing are those reported in the literature [Goodisman, 1987].

The energy balance in the foregoing reaction cycle gives the real potential $\alpha_{e(NHE)}$ of the equilibrium redox electron in the reaction of normal hydrogen electrode as shown in Fig. 2–44; $\alpha_{e(NHE)}$ represents the Fermi level $\varepsilon_{F(NHE)}$ of the normal hydrogen electrode. From the forgoing we obtain the numerical value of the equilibrium redox electron level of the reaction of normal hydrogen electrode as shown in Fig. 2–44 and Eqn. 2–56:

$$\alpha_{e(NHE)} = \varepsilon_{F(NHE)} = (11.4 - 13.6 - 2.1 - 0.2)\,eV = -4.5\,eV. \tag{2-56}$$

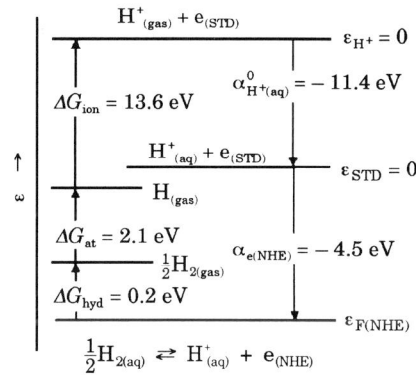

Fig. 2–44. Energy balance and electron energy levels in the NHE reaction: ε_{H^+} = standard gaseous proton level; ε_{STD} = standard gaseous electron level.

As some numerical energy values such as $\alpha^0_{H^+(aq)}$ given in the foregoing involve a certain degree of inaccuracy, there have been several values reported for the Fermi level of the equilibrium redox electron of NHE. For instance, the value of $\alpha_{e(NHE)} = \varepsilon_{F(NHE)} = -4.44\,eV$ has been reported in the International Union of Pure and Applied Chemistry (IUPAC) [Trasatti, 1986].

From $\alpha_{e(NHE)} = -4.5\,eV$ we obtain the chemical potential, $\mu_{e(NHE)}$, of the equilibrium redox electron of NHE as shown in Eqn. 2–56, assuming $\chi = 0.13\,V$:

$$\mu_{e(NHE)} = \alpha_{e(NHE)} + e\chi = -4.35\,eV. \tag{2-57}$$

References

[Anderson-Kennedy, 1988]: M. Anderson and J. H. Kennedy, *Semiconductor Electrode, Studies in Physical and Theoretical Chemistry* 55, (Edited by H. O. Finklea), p. 147, Elsever, Tokyo, (1980).
[Battisi-Trasatti, 1977]: A.Battisi and S. Trasatti, *J. Electroanal. Chem.*, **79**, 251(1977).
[Blakemore, 1985]: J. S. Blakemore, *Solid State Physics*, Cambridge University Press, London, (1985).
[Dogonadze, 1970]: R. R. Dogonadze, *Reactions of Molecules at Electrodes*, (Edited by N. S. Hush), p. 463, John Wiley & Sons, (1970).
[Fujita, 1996]: E. Fujita , *Bull. Iron Steel Inst. Japan*, **1**, 337(1996).
[Gerischer, 1961]: H. Gerischer, *Advance in Electrochemistry and Electrochemical Engineering*, **Vol. 1**, (Edited by P. Delahay), p. 139, John Wiley & Sons, (1961).
[Gomer-Tryson, 1977]: R. Gomer and G. Tryson, *J. Chem. Phys.*, **66**, 4413(1977).
[Goodisman, 1987]: J. Goodisman, *Electrochemistry, Theoretical Fundamentals*, John Wily & Sons, New York, (1987).
[Kröger, 1974]: F. A. Kröger, *The Chemistry of Imperfect Crystals*, 2nd Ed., **Vol. 2**, p. 31, p. 744, North-Holland Publishing, Amsterdam, (1974).
[Lange-Kohn, 1970]: N. D. Lange and W. Kohn, *Physical Rev.*, **B 1**, 4555(1970); **B 3**, 1215(1971).
[Marcus, 1956]: R. A. Marcus, *J. Chem. Phys.*, **24**, 966(1956).
[Marcus, 1964]: R. A. Marcus, *Ann. Rev. Phys. Chem.*, **15**, 115(1964).
[Memming-Möllers, 1972]: R. M. Memming and F. Möllers, *Ber. Bunsenges. Phys. Chem.*, **76**, 475(1972).
[Morrison, 1980]: S. R. Morrison, *Electrochemistry at Semiconductor and Oxidized Metal Electrodes*, Plenum Press, New York, (1980).
[Mott-Davis, 1979]: N. F. Mott and E. A. Davis, *Electronic Processes in Non-Crystalline Materials*, 2 Ed., Clarendon Press, Oxford, (1979).
[Notoya-Matsuda, 1982]: R. Notoya and A. Matsuda, *J. Res. Inst. Catalysis, Hokkaido University*, **30**, 107(1982).
[Rosenberg, 1960]: A. J. Rosenberg, *J. Chem. Phys.*, **33**, 665(1960).
[Sato, 1993]: N. Sato, *Electrode Chemistry* **Vol. 1**, p. 47-55, Japan Technical Information Service, Tokyo, (1993).
[Shive, 1959]: J. N. Shive, *The Properties, Physics and Design of Semiconductor Devices*, D. Van Nostrand Co., Inc., Princeton, (1959).
[Trasatti, 1980]: S. Trasatti, *Comprehensive Treatise of Electrochemistry*, **Vol. 1**, (Edited by J. O'M. Bockris, B. E. Conway and E. Yeager), p. 45, (1980).
[Trasatti, 1986]: S. Trasatti, *Pure and Applied Chemistry*, **58**, No.7, 956(1986).
[Vijh, 1970]: A. K. Vijh, *J. Electrochem. Soc.*, **117**, 173C(1970).

[Watanabe-Gerischer, 1981]: T. Watanabe and H. Gerischer, *J. Electroanal. Chem.*, **122**, 73(1981).

CHAPTER 3

THE ENERGY LEVEL OF IONS

3.1 Ionic Dissociation of Gaseous Molecules

The dissociation of an isolated gaseous molecule XY_{vac} into gaseous ions X^+_{vac} and Y^-_{vac} is written in Eqn. 3–1:

$$XY_{vac} \rightarrow X^+_{vac} + Y^-_{vac} , \qquad \text{ionic dissociation energy } D_{XY} . \qquad (3-1)$$

Assuming that the isolated gaseous ion X^+_{vac} is at the reference zero energy level as shown in Fig. 3–1, we define the *occupied* energy level of the X^+ ion (the unitary X^+ ion donor level, $\mu^*_{X^+(XY,D)}$) in a molecule XY_{vac} by the negative ionic dissociation energy $-D_{XY}$, which in turn is equivalent to the *vacant* X^+ ion level (the unitary X^+ ion acceptor level, $\mu^*_{X^+(Y^-,A)}$) of the gaseous ion Y^-_{vac} to form a molecule XY_{vac} as shown in Eqn. 3–2:

$$\mu^*_{X^+(XY,D)} = \mu^*_{X^+(Y^-,A)} = -D_{XY} , \qquad (3-2)$$

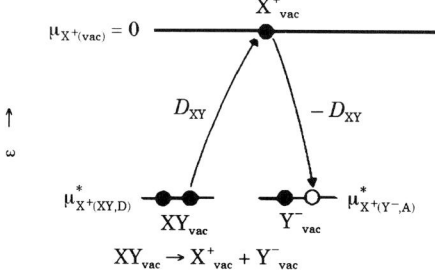

Fig. 3–1. Ionic dissociation of a gaseous molecule XY and the X^+ ion level in molecule XY: ε = energy; $\mu^*_{X^+}$ = unitary X^+ ion level (unitary chemical potential); D_{XY} = ionic dissociation energy of XY.

where the chemical potential, μ, instead of the electrochemical potential, $\bar{\mu}$, is used to represent the ion energy level of gaseous molecules not subjected to an electric potential gradient.

The dissociation reaction given in Eqn. 3–1 is composed of the following steps:

(1) $XY_{vac} \rightarrow X_{vac} + Y_{vac}$	atomic dissociation energy	D_{at}
(2) $X_{vac} \rightarrow X^+_{vac} + e_{vac}$	ionization energy	I_X
(3) $Y_{vac} + e_{vac} \rightarrow Y^-_{vac}$	negative electron affinity	$-A_Y$
(4) $X^+_{vac} + Y^-_{vac} \rightarrow XY_{vac}$	negative ionic dissociation energy	$-D_{XY}$

The energy balance in the foregoing reaction cycle yields Eqn. 3–3:

$$-D_{XY} = D_{at} + I_X - A_Y = \mu^*_{X^+(XY,D)} = \mu^*_{X^+(Y^-,A)} \ . \tag{3–3}$$

The energy equation 3–3 may be used to define the ion energy level of ionic compound molecules. Further, this energy equation may also be used to define the ion energy level in covalently bonded compound molecules although the physical meaning of the ion level in covalent compounds is a matter of debate.

As for the proton level of a gaseous molecule of hydrogen chloride, HCl_{vac}, the following reaction cycle (Fig. 3–2) may be used to estimate the occupied proton level of HCl_{vac} at $\mu^*_{H^+(HCl,D)} = -14.4$ eV that also represents the vacant proton level $\mu^*_{H^+(Cl^-,A)}$ of a gaseous chloride ion, Cl^-_{vac}, referred to the isolated gaseous proton level.

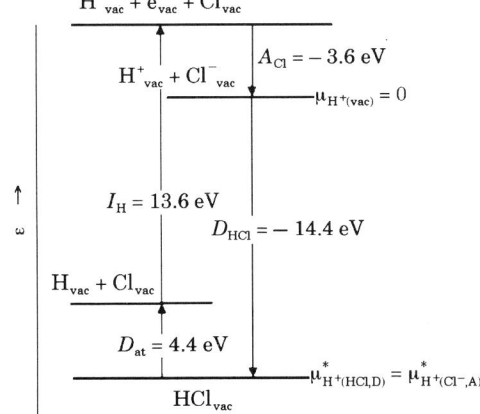

Fig. 3–2. Energy of ionic dissociation of gaseous HCl molecules and proton levels: μ_{H^+} = proton level; $\mu^*_{H^+(HCl,D)}$ = unitary occupied proton level (donor level) in gaseous HCl molecules; $\mu^*_{H^+(Cl^-,A)}$ = unitary vacant proton level (acceptor level) of gaseous Cl^- ions.

(1) $HCl_{vac} \to H_{vac} + Cl_{vac}$	atomic dissociation energy	$D_{at} = 4.4$ eV
(2) $H_{vac} \to H^+_{vac} + e_{vac}$	ionization energy	$I_H = 13.6$ eV
(3) $Cl_{vac} + e_{vac} \to Cl^-_{vac}$	negative electron affinity	$-A_{Cl} = -3.6$ eV
(4) $H^+_{vac} + Cl^-_{vac} \to HCl_{vac}$	negative ionic dissociation energy	$-D_{HCl} = -14.4$ eV

3.2 Metal Ion Levels in Solid Metals

3.2.1 The unitary energy level of surface metal ions

The sublimation of the surface metal ion M^+_s of solid metals to form the gaseous metal ion $M^+_{(STD)}$ in the standard state requires energy equivalent to the standard free enthalpy, Y_M, of metal ion sublimation as shown in Eqn. 3–4 and in Fig. 3–3 (a):

$$M^+_s \to M^+_{(STD)}, \qquad \text{surface metal ion sublimation energy } Y_M, \qquad (3\text{--}4)$$

where the metal ion valence is assumed to be monovalent, subscript s denotes the metal surface site, and subscript (STD) denotes the standard gaseous state.

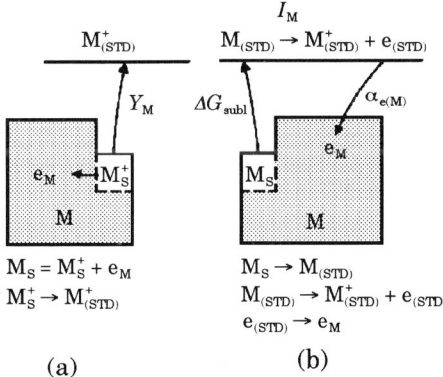

Fig. 3–3. Formation of a gaseous metal ion $M^+_{(STD)}$ in the standard state from the surface metal atom of solid metals: (a) surface atom ionization followed by surface ion sublimation, (b) surface atom sublimation followed by gaseous atom ionization.

The energy level of surface metal ions is defined as the *real potential* $\alpha^*_{M^+_s}$ of the ion which is the energy released to form the surface metal ion from the standard gaseous metal ion located at the outer potential of the metal phase; this is the negative ion sublimation energy, $-Y_M$ ($=\alpha^*_{M^+_s}$). The real potential, $\alpha_{M^+_s}$, of the surface metal ion is independent of the electrostatic charge on the metal phase (Sec. 1.6); hence, it is characteristic of individual solid metals (the unitary level), in contrast to the electrochemical potential $\bar{\mu}^*_{M^+_s}$ $(= \alpha^*_{M^+_s} + e\,\psi)$ of the ion which depends on the electrostatic charge and thus on the outer potential ψ of the metal phase. Here, the star mark superscript, *, indicates the unitary level (the unitary quantities in general).

The metal ion sublimation of Eqn. 3–4 may also be equivalent to the process that consists of the sublimation of the surface metal atom M_s followed by the ionization of the gaseous metal atom $M_{(STD)}$ and the injection of the gaseous electron $e_{(STD)}$ into the metal phase to produce the standard gaseous metal ion $M^+_{(STD)}$ leaving an electron e_M in the solid metal as shown in Eqn. 3–5 and in Fig. 3–3 (b):

$$M_s \rightarrow M^+_{(STD)} + e_M, \qquad \text{metal atom sublimation–ionization energy} \quad Y_M . \quad (3\text{–}5)$$

This surface metal atom sublimation–ionization process may be divided into the following steps:

(1) $M_s \rightarrow M_{(STD)}$	metal sublimation energy	ΔG_{subl}
(2) $M_{(STD)} \rightarrow M^+_{(STD)} + e_{(STD)}$	ionization energy	I_M
(3) $e_{(STD)} \rightarrow e_M$	the real potential of electron in solid metals	$\alpha_{e(M)}$
(4) $M_s \rightarrow M^+_{(STD)} + e_M$		Y_M

where the metal ionization, $M_s = M^+_s + e_M$, at the metal surface is assumed to be in equilibrium. The energy balance in the foregoing processes gives the *unitary level* (the unitary real potential) $\alpha^*_{M^+_s}$ of the surface metal ion as follows:

$$\alpha^*_{M^+_s} = -Y_M = -(\Delta G_{subl} + I_M + \alpha_{e(M)}) . \tag{3–6}$$

The unitary real potential, $\alpha^*_{M^+_s}$, of the surface metal ion consists of the chemical potential, $\mu^*_{M^+_s}$, and the electrostatic surface term $e\,\chi$ as shown in Eqn. 3–7:

$$\alpha^*_{M^+_s} = \mu^*_{M^+_s} + e\,\chi , \tag{3–7}$$

where χ is the surface potential difference due to the surface dipole of the metal phase. The two terms of $\mu^*_{M^+_s}$ and $e\,\chi$ are both characteristic of individual solid metals and are independent of the electrostatic charge on the metal phase. Consequently, the unitary real potential $\alpha^*_{M^+_s}$ of the surface metal ion is the inherently characteristic of individual solid metals.

Instead of the real potential, we may use the unitary electrochemical potential,

$\bar{\mu}^*_{M_s^+}$, to represent the unitary energy level of the surface metal ion as shown in Eqn. 3–8:

$$\bar{\mu}^*_{M_s^+} = \alpha^*_{M_s^+} + e\,\psi = \mu^*_{M_s^+} + e\,\phi \; , \tag{3-8}$$

where ϕ is the inner potential and ψ is the outer potential of the metal phase; $\bar{\mu}^*_{M_s^+}$ depends on ψ and on the electrostatic charge of the metal phase. Obviously, it is advantageous to use the real potential rather than the electrochemical potential for representing the ion level in condensed phases, because the ion level in terms of the real potential is independent of the electrostatic charge of the condensed phase.

Fig. 3–4 shows the energy level of a surface silver ion on the solid silver metal from which the unitary surface silver ion level is estimated to be $\alpha^*_{Ag^+} = -Y_{Ag^+} = -5.84$ eV referred to the standard gaseous silver ion at the outer potential of the metal phase.

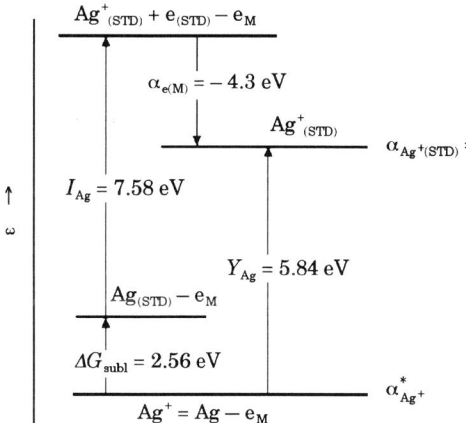

Fig. 3–4. Energy for the formation of gaseous silver ions in the standard state from the surface silver atoms of solid silver metal: $\alpha^*_{Ag^+}$ = unitary level (unitary real potential) of surface silver ions.

3.2.2 Metal ion levels at the surface and in the interior

The metal ion level at the surface would equal the metal ion level in the metal interior, if the ionic equilibrium is established between the surface and the interior of the metal phase. However, the unitary metal ion level, α^*_s, at the metal surface differs in general from the unitary metal ion level, α^*_L, in the interior of the solid metal. The metal ion in the interior is located at a lattice site or at an interstitial site; while the surface metal ion is either in the terrace, in the step, at the kink site, or at the adsorption site. For simplicity, we consider only the occupied and vacant lattice sites in the interior and only the kink site on the surface as shown in Fig. 3–5.

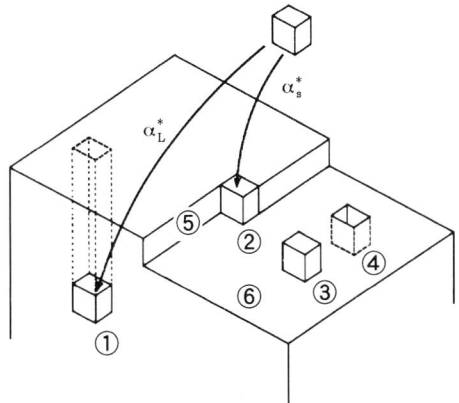

Fig. 3–5. Ions on the surface and in the interior of solids: ①= occupied or vacant lattice site; ②= surface kink site; ③ = surface adsorption site; ④ = surface lattice vacancy; ⑤ = step plane; ⑥ = terrace; α_L^* = unitary level of occupied or vacant lattice site ions; α_s^* = unitary level of surface kink site ions.

The energy level of metal ions (the *real potential*) $\alpha_{M_s^+}$ at the kink site on the surface is given as a function of the concentration of the kink sites (molar fraction) $x_k = n_k / N_s$ in Eqn. 3–9:

$$\alpha_{M_s^+} = \alpha_{M_s^+}^* + k\, T \ln\left(\frac{n_k}{N_s} \right) = \alpha_{M_s^+}^* + k\, T \ln x_k \ , \tag{3–9}$$

where $\alpha_{M_s^+}^*$ is the unitary metal ion level, n_k is the number of kink sites, and N_s is the number of surface metal atoms in unit surface area; usually, $x_k \doteq 10^{-2}$ to 10^{-4} in the molar fraction. On the other hand, the energy level of metal ions in the interior of the metal phase (the energy released in forming a metal ion at the lattice site , $M^+_{(STD)} + V_{M^+} \rightarrow M^+_L$) is given as a function of the molar fraction $x_V = n_V / N_L$ of vacant lattice site V_{M^+} as shown in Eqn. 3–10:

$$\alpha_{M_L^+} = \alpha_{M_L^+}^* - k\, T \ln\left(\frac{n_V}{N_L} \right) = \alpha_{M_L^+}^* - k\, T \ln x_V \ , \tag{3–10}$$

where $\alpha_{M_L^+}^*$ is the unitary level of metal ions at the metal lattice site, n_V is the number of metal ion vacancies (vacant lattice sites), and N_L is the number of total lattice sites in the unit volume of solid metal. The difference in the unitary level of metal ions between the surface and the interior is equivalent to the energy required for the formation of a pair of the vacant lattice site and the surface metal ion from the lattice site metal ion in the interior ($M_L^+ \rightarrow M_s^+ + V_{M^+}$); their energy is about 1 eV, giving $x_V \doteq 10^{-16}$ to 10^{-17} in the molar fraction.

In the case in which the ionic equilibrium is established between the surface and the interior, the two metal ion levels, $\alpha_{M_s^+}$ and $\alpha_{M_L^+}$, become equal to each other; and the energy level of metal ions $\alpha_{M^+(M)}$ at the surface and in the interior of the metal phase is given by Eqn. 3–11:

$$\alpha_{M^+(M)} = \alpha_{M_L^+}^* - kT \ln x_V = \alpha_{M_s^+}^* + kT \ln x_k = \frac{1}{2}\left(\alpha_{M_L^+}^* + \alpha_{M_s^+}^*\right) + \frac{1}{2} kT \ln\left(\frac{x_k}{x_V}\right). \quad (3\text{–}11)$$

Since the concentration of kink sites ($x_k \doteq 10^{-2}$ to 10^{-4} in the molar fraction) on the metal surface is much greater than the concentration of vacant lattice site ($x_V \doteq 10^{-16}$ to 10^{-17} in the molar fraction) in the interior of the metal phase, the metal ion level $\alpha_{M^+(M)}$ of the metal phase is close to the unitary level of surface metal ions $\alpha_{M_s^+}^*$ as shown in Eqn. 3–12 and in Fig. 3–6:

$$\alpha_{M^+(M)} \doteq \alpha_{M_s^+}^* . \quad (3\text{–}12)$$

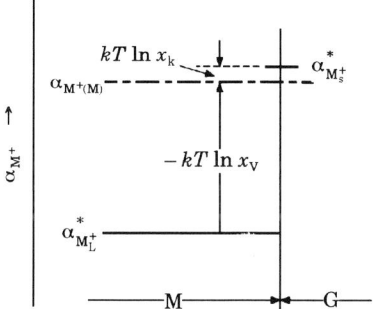

Fig. 3–6. Metal ion levels at the surface and in the interior of a solid metal: α_{M^+} = metal ion energy; $\alpha_{M^+(M)}$ = metal ion level of solid metal; $\alpha_{M_L^+}^*$ = unitary metal ion level at interior lattice sites; $\alpha_{M_s^+}^*$ = unitary metal ion level at surface kink sites; x_k = molar fraction of surface kink site; x_V = molar fraction of interior lattice vacancy; G = gaseous phase.

3.3 Ion Levels of Covalent Semiconductors

3.3.1 The unitary level of surface ions

In covalent semiconductors of single substance S such as silicon, the covalent bonding electron is in the valence band; and the valence band hole participates in the surface atom ionization as shown in Eqn. 3–13 and in Fig. 3–7:

$$S_s + h = S_s^+$$
$$S_s^+ \to S_{(STD)}^+ \qquad \text{surface ion sublimation energy} \quad Y_S$$

$$S_s + h \to S_{(STD)}^+ \, , \tag{3-13}$$

where S_s is the surface atom, S_s^+ is the surface ion, $S_{(STD)}^+$ is the gaseous ion, h is the hole in the valence band of the covalent semiconductor S, and Y_S is the energy to form the standard gaseous ion $S_{(STD)}^+$ from the surface ion S_s^+.

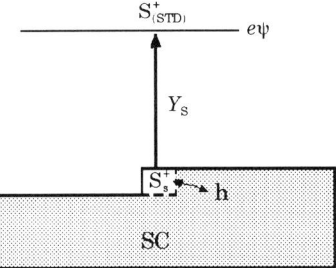

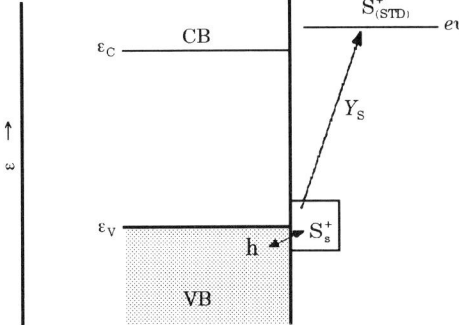

Fig. 3–7. Formation of the standard gaseous ion, $S_{(STD)}^+$, from the surface atom of a semiconductor of single substance S: S_s^+ = surface ion; h = hole; SC = semiconductor; CB = conduction band; VB = valence band; Y_S = sublimation energy of the surface ion; ψ = outer potential.

In the case in which the ionization of surface atoms ($S_s + h = S_s^+$) is in equilibrium, the unitary level of the surface ions S_s^+ referred to the standard gaseous ion $S_{(STD)}^+$ at the outer potential of the semiconductor is represented by the unitary *real potential*, $\alpha_{S_s^+}^* \, (= -Y_S)$. This unitary real potential is equivalent to the sum of the standard free enthalpy ΔG_{subl} of sublimation of the semiconductor S, the ionization energy I_S of the gaseous atom S, and the electron energy ε_V at the upper edge level of the valence band as shown in Eqn. 3–14:

$$\alpha^*_{S^+_s} = -Y_S = -(\Delta G_{subl} + I_S + \varepsilon_V) = \alpha^*_{S_s} - \varepsilon_V , \qquad (3\text{–}14)$$

where $\alpha^*_{S_s}$ is the unitary real potential of the surface atom S_s referred to the gaseous ion $S^+_{(STD)}$ and electron $e_{(STD)}$ in the standard state.

Fig. 3–8 shows the unitary level of the surface ion of a semiconductor of single substance. Simple calculation gives the unitary ion level of the surface ion Ge^+ of semiconductor germanium to be $\alpha^*_{Ge^+_s} = -6.8$ eV.

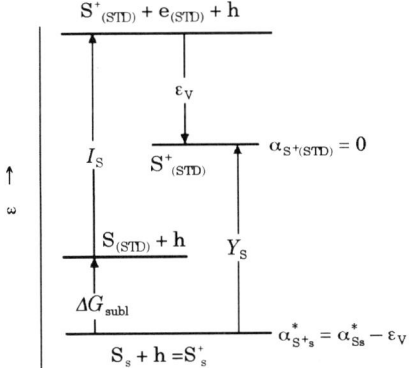

Fig. 3–8. Energy for formation of the standard gaseous ions, $S^+_{(STD)}$, from the surface atoms of a semiconductor of single substance S: ΔG_{subl} = standard free enthalpy of the surface atom sublimation; I_S = ionization energy of gaseous atoms; $\alpha^*_{S^+_s}$ = unitary level of the surface ion; $\alpha^*_{S_s} = -(\Delta G_{subl} + I_S)$ = unitary level of the surface atom referred to the standard gaseous ions and electrons.

3.3.2 Ion levels at the surface and in the interior

We consider only the kink site for the surface ion and the occupied and vacant lattice sites for the interior ion. The ion level $\alpha_{S^+_s}$ at the kink site on the surface is represented by the energy released in the process, $S^+_{(STD)} \to S^+_s = S_s + h_s$; and the ion level $\alpha_{S^+_L}$ at the interior lattice site is represented by the energy released in the process, $S^+_{(STD)} + V_L \to S^+_L = S_L + h_L$, where V_L is the lattice vacancy site, h_s and h_L are the holes at the surface and in the bulk interior, respectively. The ion level $\alpha_{S^+_s}$ at the surface is given as a function of the energy level of holes at the surface ε^s_V and the concentration of the kink sites x_k in the same way as Eqn. 3–9 as shown in Eqn. 3–16:

$$\alpha_{S^+_s} = \alpha^*_{S^+_s} + kT \ln x_k = \alpha^*_{S_s} - \varepsilon^s_V + kT \ln x_k , \qquad (3\text{–}15)$$

Similarly, the ion level $\alpha_{S^+_L}$ in the interior of the semiconductor is given as a function of the energy level of holes in the interior ε^l_V and the concentration of the vacant lattice sites x_V in the same way as Eqn. 3–10 as shown in Eqn. 3–16:

$$\alpha_{S_L^+} = \alpha_{S_L^+}^* - kT \ln x_V = \alpha_{S_L}^* - \varepsilon_V^b - kT \ln x_V \; . \tag{3–16}$$

In Eqns. 3–15 and 3–16, $\alpha_{S_s^+}^*$ and $\alpha_{S_L^+}^*$ are the unitary levels of the ion at the surface kink site and at the interior lattice site; $\alpha_{S_s}^*$ and $\alpha_{S_L}^*$ are the unitary levels of the atom at the surface kink sit and at the interior lattice site; ε_V^s and ε_V^L are the levels of the valence band edge at the surface and in the interior, respectively. The ion levels of $\alpha_{S_s^+}^*$ and $\alpha_{S_L^+}^*$ are dependent on the hole level; but the atom levels of $\alpha_{S_s}^*$ and $\alpha_{S_L}^*$ are constant and characteristic of individual semiconductors.

In the case in which the ionic equilibrium is established between the surface and the interior of the semiconductor, the two ion levels of $\alpha_{S_s^+}$ and $\alpha_{S_L^+}$ are equilibrated each other, and the ion level $\alpha_{S^+(SC)}$ of the semiconductor is given by Eqn. 3–17:

$$\alpha_{S^+(SC)} = \frac{\alpha_{S_L^+}^* + \alpha_{S_s^+}^*}{2} + kT \ln\left(\frac{x_k}{x_V}\right) = \frac{\alpha_{S_L}^* + \alpha_{S_s}^*}{2} - \frac{\varepsilon_V^L + \varepsilon_V^s}{2} + kT \ln\left(\frac{x_k}{x_V}\right) \; . \tag{3–17}$$

Since the concentration of the surface kink sites is much greater than the concentration of the lattice vacancies $(x_k / x_V) \gg 1$, it follows from Eqn. 3–17 that the ion level $\alpha_{S^+(SC)}$ of the semiconductor is *close to* the unitary level $\alpha_{S_s^+}^*$ of the surface ion as shown in Eqn. 3–18 and in Fig. 3–9:

$$\alpha_{S^+(SC)} \doteq \alpha_{S_s^+}^* \; . \tag{3–18}$$

It also appears in Eqn. 3–17 that the ion level $\alpha_{S^+(SC)}$ of the semiconductor is to some extent dependent on the potential of the space charge layer.

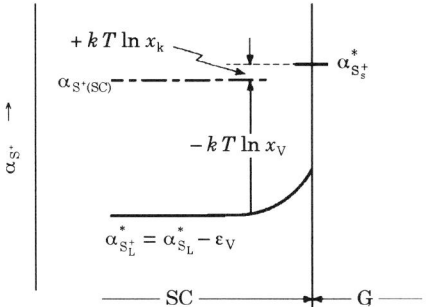

Fig. 3–9. Ion levels at the surface and in the interior of a semiconductor of single substance: α_{S^+} = energy level of ion S$^+$; $\alpha_{S^+(SC)}$ = ion level of the semiconductor; $\alpha_{S_L^+}^*$ = unitary ion level in the semiconductor interior, $\alpha_{S_L}^*$ = unitary atom level in the semiconductor interior.

3.4 Ion Levels of Compound Semiconductors

3.4.1 The unitary level of surface ions

The formation of the standard gaseous ions from a semiconductor of compound AB may be written in Eqns. 3–19 and 3–20 based on Fig. 3–10:

$$A^+_{s(AB)} \rightarrow A^+_{(STD)} , \qquad \text{surface ion sublimation energy } L_A \qquad (3\text{–}19)$$

$$B^-_{s(AB)} \rightarrow B^-_{(STD)} , \qquad \text{surface ion sublimation energy } L_B \qquad (3\text{–}20)$$

where L_A and L_B are the sublimation energy of the surface ions $A^+_{s(AB)}$ and $B^-_{s(AB)}$ produced by the ionization of the surface atoms $A_{s(AB)}$ and $B_{s(AB)}$. To simplify the treatment, we assume the electron and hole levels of $A_{s(AB)}$ and $B_{s(AB)}$ to be in the conduction band and in the valence band, respectively. Then, the unitary levels $\alpha^*_{A^+_s(AB)}$ and $\alpha^*_{B^-_s(AB)}$ of the surface ions $A^+_{s(AB)}$ and $B^-_{s(AB)}$ are, respectively, given by Eqns. 3–21 and 3–22:

$$\alpha^*_{A^+_s(AB)} = - L_A = - (\Delta G_A + I_A + \varepsilon_C) = \mu^*_{A^+_s(AB)} + e \chi , \qquad (3\text{–}21)$$

$$\alpha^*_{B^-_s(AB)} = - L_B = - (\Delta G_B - A_B - \varepsilon_V) = \mu^*_{B^-_s(AB)} - e \chi , \qquad (3\text{–}22)$$

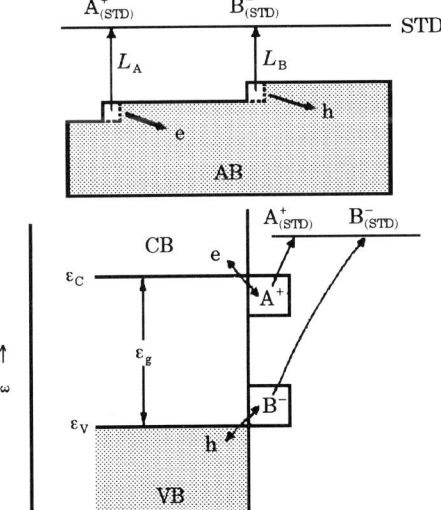

Fig. 3–10. Formation of the standard gaseous ions $A^+_{(STD)}$ and $B^-_{(STD)}$ from the surface atoms of compound semiconductor AB.

where ΔG, I, A, ε_C (or ε_V), μ^*, and χ are the surface atom sublimation energy, the ionization energy of the gaseous atom, the electron affinity of the gaseous atom, the electron level at the conduction (or valence) band edge, the unitary chemical potential, and the surface potential difference of the semiconductor, respectively.

Then, we consider the following reaction cycle:

Reaction step	Energy
(1) $A_{s(AB)} \rightarrow A^+_{(STD)} + e_{CB}$	L_{Ae}
(2) $B_{s(AB)} \rightarrow B^-_{(STD)} + h_{VB}$	L_{Bh}
(3) $A^+_{(STD)} + e_{(STD)} \rightarrow A_{(STD)}$	$-I_A$
(4) $B^-_{(STD)} \rightarrow B_{(STD)} + e_{(STD)}$	A_B
(5) $A_{(STD)} + B_{(STD)} \rightarrow AB_{s(AB)}$	$-\Delta G_{AB}$
(6) $0 \rightarrow e_{CB} + h_{VB}$,	ε_g

where $-\Delta G_{AB}$ is the free enthalpy of the formation of solid compound AB, and ε_g is the energy for the formation of an electron–hole pair (the band gap energy). From the energy balance in the foregoing reaction cycle we obtain that ε_g equals $(L_{Ae} + L_{Bh} - I_A + A_B - \Delta G_{AB})$. Obviously, this band gap energy, ε_g, corresponds to the difference of energy between the decomposing ionization of solid compound AB via the decomposing sublimation followed by the ionization of gaseous atoms which requires the energy of $(\Delta G_{AB} + I_A - A_B)$ and that via the formation of an *electron–hole pair* followed by the sublimation of surface ions which requires the energy of $(L_{Ae} + L_{Bh})$. Therefore, this energy difference, ε_g, equals the excess energy in the decomposing ionization via the electron–hole pair formation over the decomposing sublimation-ionization. Since this excess energy is released in

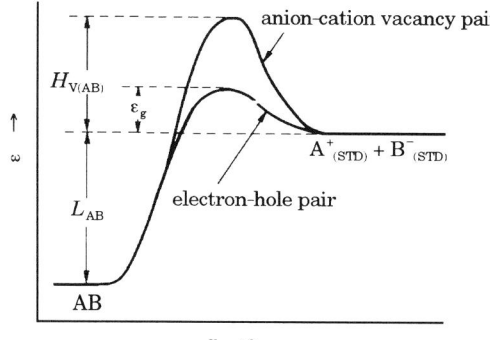

Fig. 3–11. Energy for decomposing ionization of compound AB to form gaseous ions $A^+_{(STD)}$ and $B^-_{(STD)}$ via the electron–hole pair formation and via the cation–anion vacancy pair formation: r = reaction coordinate of decomposing ionization; ε_g = semiconductor band gap; $H_{V(AB)}$ = cation–anion vacancy pair formation energy (V_{A^+} V_{B^-}); L_{AB} = decomposing ionization energy of compound AB.

the recombination of an electron–hole pair, it may be regarded as the *activation energy* in the decomposing ionization via the electron–hole pair formation as shown in Fig. 3–11.

The decomposing ionization of solid compound AB may also take place via the formation of a cation–anion vacancy pair as follows:

Reaction step	Energy
(1) $A_{s(AB)} \rightarrow A^+_{(STD)} + V^-_A$	L_{AV}
(2) $B_{s(AB)} \rightarrow B^-_{(STD)} + V^+_B$	L_{BV}
(3) $A^+_{(STD)} + e_{(STD)} \rightarrow A_{(STD)}$	$-I_A$
(4) $B^-_{(STD)} \rightarrow B_{(STD)} + e_{(STD)}$	A_B
(5) $A_{(STD)} + B_{(STD)} \rightarrow AB_{s(AB)}$	$-\Delta G_{AB}$
(6) $0 \rightarrow V^-_A + V^+_B$	$H_{V(AB)}$

where V^-_A and V^+_B are the A^+ ion vacancy with an electron and the B^- ion vacancy with a hole, respectively. The energy balance gives the formation energy of vacancy pairs $H_{V(AB)}$ equal to $(L_{AV} + L_{BV} - I_A + A_B - \Delta G_{AB})$. This formation energy of cation–anion vacancy pairs $H_{V(AB)}$ corresponds to the difference of energy between the decomposing ionization of solid compound AB via the decomposing sublimation of AB followed by the gaseous atom ionization, which requires the energy of $(\Delta G_{AB} + I_A - A_B)$, and that via the cation–anion vacancy pair formation followed by the surface ion sublimation, which requires the energy of $(L_{AV} + L_{BV})$. Thus, $H_{V(AB)}$ equals the excess energy in the decomposing ionization via the cation–anion pair formation over the decomposing sublimation-ionization. Since this excess energy is released in the recombination of the cation–anion vacancy pairs, it may be regarded as the *activation energy* in the decomposing ionization via the cation–anion vacancy pair formation, as shown in Fig. 3–11.

The decomposing ionization would take place preferentially via the electron–hole pair formation, if the formation energy of the electron–hole pair, ε_g, is smaller than the formation energy of the cation–anion vacancy pair, $H_{V(AB)}$, and vice versa. In general, compound semiconductors in which the band gap is small ($\varepsilon_g < H_{V(AB)}$) would prefer the formation of electron–hole pairs; whereas, compound insulators such as sodium chloride, in which the band gap is great ($\varepsilon_g > H_{V(AB)}$), would prefer the formation of cation–anion vacancy pairs [Fumi-Tosi, 1964].

In the case in which the formation of cation–anion vacancy pairs is preferential, the ion levels of A^+ and B^- ions in solid compound AB are obtained in the same way as Eqns. 3–21 and 3–22 by Eqns. 3–23 and 3–24, respectively:

$$\alpha^*_{A^+_s(AB)} = -L_{AV} = \mu^*_{A^+_s(AB)} + e\chi , \qquad (3\text{–}23)$$

$$\alpha^*_{B^-_s(AB)} = -L_{BV} = \mu^*_{B^-_s(AB)} - e\chi \quad . \tag{3-24}$$

In the ionic crystal of NaCl, the unitary ion level is $\alpha^*_{Na^+(NaCl)} = -4.80$ eV for the sodium ion and $\alpha^*_{Cl^-(NaCl)} = -5.14$ eV for the chloride ion [Fumi-Tosi, 1964; Sato, 1993].

3.4.2 Ion levels at the surface and in the interior

The lattice defects in ionic crystals are interstitial ions and ion vacancies. In crystalline sodium chloride NaCl a cation vacancy V_{Na^+} is formed by producing a surface cation Na^+_s, ($Na^+_L \rightarrow Na^+_s + V_{Na^+}$); this is called the *Schottky defect*. On the other hand, in crystalline silver chloride AgCl a pair of cation vacancy V_{Ag^+} and interstitial cation Ag^+_I is formed, ($Ag^+_L \rightarrow Ag^+_I + V_{Ag^+}$); this is called the *Frenkel defect*; where the subscripts L and I denote the lattice site and the interstitial site, respectively. The formation energy of both kinds of defects is about 2 eV (2.1 eV for NaCl).

We consider the ionic compound AB containing lattice defects of the Frenkel type (*a pair of ion vacancy and interstitial ion*) as shown in Fig. 3–12. In the case in which the defect formation is in equilibrium ($A^+_L = A^+_I + V_{A^+}$), the A^+ ion level $\alpha_{A^+(AB)}$ in the compound AB is given by Eqn. 3–25:

$$\alpha_{A^+(AB)} = \alpha^*_{A^+_L} - kT \ln\left(\frac{n_V}{N_L}\right) = \alpha^*_{A^+_I} + kT \ln\left(\frac{n_I}{N_I}\right), \tag{3-25}$$

where $\alpha^*_{A^+_L}$ and $\alpha^*_{A^+_I}$ are the unitary ion levels at the lattice sites and at the interstitial sites, N_L and N_I are the numbers of the lattice sites and the interstitial sites in unit volume ($N_L \doteq N_I$), n_V and n_I are the numbers of the ion vacancies and the interstitial ions in unit volume, respectively. In the formation of Frenkel defects, n_V equals n_I; hence, Eqn. 3–25 yields Eqn. 3–26 [Blakely, 1973]:

$$\alpha_{A^+(AB)} = \frac{\alpha^*_{A^+_L} + \alpha^*_{A^+_I}}{2} - kT \ln\left(\frac{N_I}{N_L}\right). \tag{3-26}$$

Therefore, it follows that the A^+ ion level $\alpha_{A^+(AB)}$ in the interior of crystalline compound AB is located midway between the unitary ion level $\alpha^*_{A^+_L}$ at the lattice sites and the unitary ion level $\alpha^*_{A^+_I}$ at the interstitial sites as shown in Fig. 3–12.

In the case in which the ionic equilibrium is established between the surface and the interior, the surface ion level, $\alpha_{A^+_s}$, equals the interior ion level, $\alpha_{A^+(AB)}$. Consequently, the unitary ion levels at the lattice and interstitial sites bend either upward or downward forming a space charge layer in a region adjacent to the surface as shown in Fig. 3–13. When the surface ion level, $\alpha_{A^+_s}$, is lower than the interior ion level, $\alpha_{A^+(AB)}$, the A^+ ions move from the interior to the surface

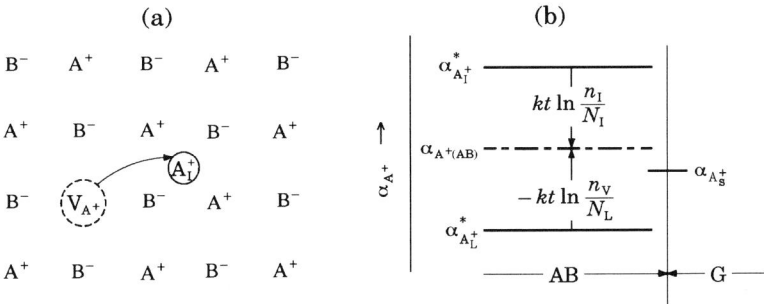

Fig. 3–12. Lattice defects and ion levels of an ionic compound AB: (a) formation of a pair of ion vacancy and interstitial ion, (b) A⁺ ion levels in ionic crystals. V_{A^+} = A⁺ ion vacancy; A_I^+ = interstitial A⁺ ion; α_{A^+} = A⁺ ion level; $\alpha^*_{A_L^+}$ = unitary A⁺ ion level at the lattice site; $\alpha^*_{A_I^+}$ = unitary A⁺ ion level at the interstitial site; $\alpha_{A_s^+}$ = A⁺ ion level at the surface site; N_L = total number of lattice sites; $\alpha_{A^+(AB)}$ = A⁺ ion level in ionic crystal AB; N_I = total number of interstitial sites; n_V = number of vacant lattice sites; n_I = number of interstitial ions.

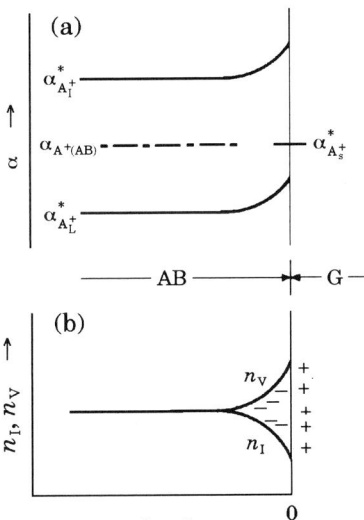

Fig. 3–13. (a) A⁺ ion levels at the surface and in the interior of an ionic compound AB, and (b) concentration profile of lattice defects in a surface space charge layer: since the energy scales of occupied and vacant ion levels are opposite to each other, ion vacancies accumulate and interstitial ions deplete in the space charge layer giving excess A⁺ ions on the surface.

forming an excess of A⁺ ions on the surface (a positive surface charge) and a space charge layer of ion vacancies V_{A^+} (a negative space charge to compensate the positive surface charge) as shown in Fig. 3–13. Such a negative space charge layer of ion vacancies is found to form in silver chloride crystal AgCl with the thickness of several tens nanometers and the potential difference of several tenth volts across the space charge layer [Blakely, 1973].

Conversely, when the surface ion level, $\alpha_{A_s^+}$, is higher than the interior ion level, $\alpha_{A^+(AB)}$, the surface charge is negative; and an accumulation layer of interstitial ions (a positive space charge layer) is formed. Obviously, when the surface ion level, $\alpha_{A_s^+}$, is close to the interior ion level, $\alpha_{A^+(AB)}$, almost no surface charge (no excess cations and anions on the surface) is formed; hence, no space charge layer.

The same theoretical examination as applies to the Frenkel defect may also be apply to the Schottky defect in ionic compounds, in which an accumulation layer of vacancies is formed with the surface ion excess.

3.5 Ion Levels in Aqueous Solution

3.5.1 Levels of hydrated ions

Hydrated ions are formed by introducing gaseous ions into aqueous solutions, i.e. the hydration of gaseous ions. Since liquid water is a polar solvent of large permittivity, the hydration reduces the energy level of ions. We consider the hydration of gaseous $A_{(STD)}^{z+}$ ions in the standard state as shown in Eqn. 3–27:

$$A_{(STD)}^{z+} \rightarrow A_{(aq)}^{z+}, \qquad \text{hydration energy of an ion} \quad J \qquad (3-27)$$

where z is the ion valence, and J is the hydration energy of the ion. The energy level of the hydrated ion that is referred to the standard gaseous state of the ion at the outer potential of the aqueous solution phase is represented by the *real potential*, $\alpha_{A^{z+}(aq)}$, of the hydrated ion which corresponds to the negative hydration energy of the ion, $-J$.

The ion energy level $\alpha_{A^{z+}(aq)}$ consists of the unitary term $\alpha^*_{A^{z+}(aq)}$ (the unitary real potential) and the communal term $k\,T\ln x_A$, due mainly to the entropy of mixing; this communal term depends on the ion concentration where x_A denotes the molar fraction of hydrated $A_{(aq)}^{z+}$ ions. The concentration-independent unitary energy level $\alpha^*_{A^{z+}(aq)}$, therefore, is given by Eqn. 3–28:

$$\alpha^*_{A^{z+}(aq)} = \alpha_{A^{z+}(aq)} - k\,T\ln x_A = -J^*, \qquad (3-28)$$

where J^* is the unitary energy of the hydration of the ion.

In electrochemistry, the concentration of ions is usually expressed either by the *molality*, m, (the number of moles of solute ions in solvent water of 1 kg:

mol/kg) or the *molarity*, M, (the number of moles of solute ions in an aqueous solution of a volume of 1 dm^3). Since the solvent water of 1 kg contains 55.51 moles of H$_2$O, the mol fraction x_A is expressed as a function of the molality, m_A, as follows: $x_A = m_A/(m_A + 55.51) \doteq m_A/55.51$ with $m_A \ll 55.51$. Therefore, Eqn. 3–28 is expressed as a function of the molality, m_A, of hydrated $A^{z+}_{(aq)}$ ions as follows: $\alpha_{A^{z+}(aq)} = \alpha^*_{A^{z+}(aq)} + k\,T \ln m_A - k\,T \ln 55.51$. Then, we define the standard ion level (the *standard real potential* of the ion) $\alpha^0_{A^{z+}(aq)}$ at the unit molality of the ion ($m_A = 1$ mol/kg) as shown in Eqn. 3–29:

$$\alpha^0_{A^{z+}(aq)} = \alpha^*_{A^{z+}(aq)} - k\,T \ln 55.51 = \alpha_{A^{z+}(aq)} - k\,T \ln m_A = -J^0\,, \qquad (3\text{--}29)$$

where J^0 is the standard hydration energy of the ion; the value of $k\,T \ln 55.51$ amounts to 0.103 eV at room temperature. In electrochemistry, the ion level $\alpha_{A^{z+}(aq)}$ in aqueous solution is expressed usually as a function of the molality m_A with the standard real potential $\alpha^0_{A^{z+}(aq)}$ as shown in Eqn. 3–30:

$$\alpha_{A^{z+}(aq)} = \alpha^0_{A^{z+}(aq)} + k\,T \ln m_A\,. \qquad (3\text{--}30)$$

The standard energy level, $\alpha^0_{A^{z+}(aq)}$, of hydrated $A^{z+}_{(aq)}$ ions may be divided into two terms: the chemical potential term, $\mu^0_{A^{z+}(aq)}$, and the electrostatic term, $z\,e\,\chi$, due to the surface potential difference χ as shown in Eqn. 3–31 and in Fig. 3–14:

$$\alpha^0_{A^{z+}(aq)} = \mu^0_{A^{z+}(aq)} + z\,e\,\chi = \bar{\mu}^0_{A^{z+}(aq)} - z\,e\,\psi\,, \qquad (3\text{--}31)$$

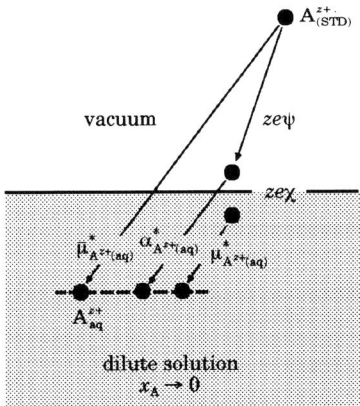

Fig. 3–14. Energy change in the hydration of A^{z+} ions: ψ = outer potential of aqueous solution; χ = surface potential difference; $\bar{\mu}^*_{A^{z+}(aq)}$ ($\mu^*_{A^{z+}(aq)}$) = unitary electrochemical (chemical) potential of hydrated A^{z+} ions; $\alpha^*_{A^{z+}(aq)}$ = unitary real potential of hydrated A^{z+} ions ($\alpha^*_{A^{z+}(aq)} = \bar{\mu}^*_{A^{z+}(aq)} - z\,e\,\psi = \mu^*_{A^{z+}(aq)} + z\,e\,\chi$).

Table 3–1. The standard real potential, α^0, of hydrated ions referred to the standard gaseous ions at room temperature: z = ionic valence. [From Trasatti, 1980.]

Ion	$-\alpha^0/z$	Ion	$-\alpha^0/z$	Ion	$-\alpha^0/z$
H^+	11.28	Ca^{2+}	8.94	F^-	-4.32
Li^+	5.28	Cu^{2+}	10.76	Cl^-	-3.08
Na^+	4.24	Zn^{2+}	10.49	Br^-	-2.83
Cu^+	5.89	Al^{3+}	15.93	I^-	-2.50
Ag^+	4.95	Fe^{3+}	14.95	S^{2-}	-6.60

where $\mu^0_{A^{z+}(aq)}$ and $\bar{\mu}^0_{A^{z+}(aq)}$ are the standard chemical potential and the standard electrochemical potential of the hydrated $A^{z+}_{(aq)}$ ion, respectively. Table 3–1 shows the standard real potential of hydrated ions estimated in the literature.

3.5.2 Proton levels in aqueous solution

The formation of the hydrated proton in acidic aqueous solution from the standard gaseous proton is written as follows:

$$H_2O_{aq} + H^+_{(STD)} \rightarrow H_3O^+_{aq}, \qquad \text{hydration energy of acidic protons } J^*_{H^+(H_2O)}, \quad (3\text{–}32)$$

where $J^*_{H^+(H_2O)}$ is the unitary energy of acidic proton hydration. In the acidic proton hydration, as shown in Fig. 3–15, a proton in the standard gaseous state combines with a water molecule to occupy the unitary *vacant proton level* (the unitary proton acceptor level, $\alpha^*_{H^+(H_2O,A)}$) of the water molecule to form the unitary *occupied proton level* (the unitary proton donor level, $\alpha^*_{H^+(H_3O^+,D)}$) in the hydrated *acidic proton*, i.e. hydroxonium ion H_3O^+ or hydronium ion $H(H_2O)_n^+$. In contrast to the electron accepting or donating step of hydrated particles, which is an adiabatic step followed by a thermal change in the hydrated structure of the redox particles, the proton accepting or donating step is a thermal process in which the proton transfer and the reorganization of hydrated structure proceed at the same time and *without any* Franck-Condon level splitting. Therefore, the proton acceptor level of H_2O is equal to the proton donor level of H_3O^+ as shown in Eqn. 3–33:

$$\alpha^*_{H^+(H_2O,A)} = \alpha^*_{H^+(H_3O^+,D)} = -J^*_{H^+(H_2O)}, \qquad (3\text{–}33)$$

where subscripts A and D denote the acceptor and the donor levels, respectively.

In addition to the acidic proton level, there is the *basic proton* level in basic aqueous solution which is represented by the unitary vacant proton level (the

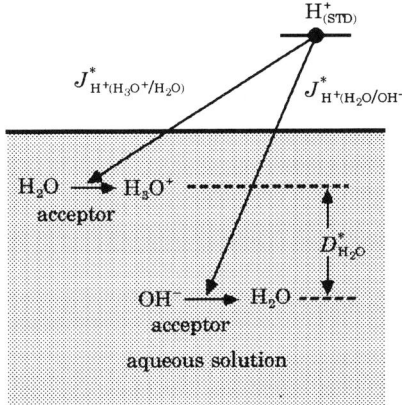

Fig. 3–15. Acidic and basic proton levels in aqueous solutions: $J^*_{H^+(H_3O^+/H_2O)}$ = unitary energy of hydration of the standard gaseous proton to occupy the unitary vacant acidic proton level; $J^*_{H^+(H_2O/OH^-)}$ = unitary energy of hydration of the standard gaseous proton to occupy the unitary vacant basic proton level; $D^*_{H_2O}$ = ionic dissociation energy of H_2O.

unitary proton acceptor level, $\alpha^*_{H^+(OH^-, A)}$) of the hydroxide ion OH^- and the unitary occupied proton level (the unitary proton donor level, $\alpha^*_{H^+(H_2O, D)}$) of the water molecule H_2O as shown in Eqn. 3–34 and in Fig. 3–15:

$$OH^-_{aq} + H^+_{(STD)} \rightarrow H_2O_{aq} , \qquad \text{hydration energy of basic protons } J^*_{H^+(OH^-)} . \quad (3\text{–}34)$$

In the same way as the acidic proton level, the unitary proton acceptor level $\alpha^*_{H^+(OH^-, A)}$ of the OH^- ion is equal to the unitary proton donor level $\alpha^*_{H^+(H_2O, D)}$ of the H_2O molecule:

$$\alpha^*_{H^+(OH^-, A)} = \alpha^*_{H^+(H_2O, D)} = -J^*_{H^+(OH^-)} , \qquad (3\text{–}35)$$

where $J^*_{H^+(OH^-)}$ is the unitary energy of basic proton hydration.

The combination of the acidic proton hydration 3–32 and the basic proton hydration 3–34 gives rise to the ionic dissociation of the water molecule as shown in Eqn. 3–36:

$$H_2O_{aq} + H_2O_{aq} \rightarrow H_3O^+_{aq} + OH^-_{aq} , \qquad D^*_{H_2O} = J^*_{H^+(H_2O)} - J^*_{H^+(OH^-)} , \quad (3\text{–}36)$$

where $D^*_{H_2O}$ is the ionic dissociation energy of the water molecule. Therefore, the ionic dissociation energy $D^*_{H_2O}$ is equal to the difference of energy between the unitary proton acceptor level $\alpha^*_{H^+(H_2O, A)}$ and the unitary proton donor level $\alpha^*_{H^+(H_2O, D)}$ of the water molecule; hence, there is a difference of the unitary proton level between the acidic proton and the basic proton which is estimated to be

$D^*_{H_2O} = 1.03$ eV at room temperature [Gurney, 1953].

As mentioned in Chap. 1, the ion level represented by the *real potential* α of an ion consists of the chemical potential μ and the electrostatic surface energy $z_i e \chi$ ($\alpha = \mu + z_i e \chi$). Since the surface potential difference, χ, of aqueous solutions is constant ($\chi \doteq 0.13$ V), the relative energy level of hydrated ions may be expressed in terms of the chemical potential μ.

Fig. 3–16 shows the ionic dissociation of the water molecule and both the vacant (right hand side) and the occupied (left hand side) proton levels represented in terms of the chemical potential of vacant and occupied protons. Following the solid arrow, a proton moves from the donor level of H_2O molecules to the vacant level of H_2O molecules absorbing an amount of energy of $D^*_{H_2O}$. The dotted arrow indicates a proton transfer from the occupied level of H_3O^+ ions to the vacant level of OH^- ions thereby liberating an amount of energy of $D^*_{H_2O}$.

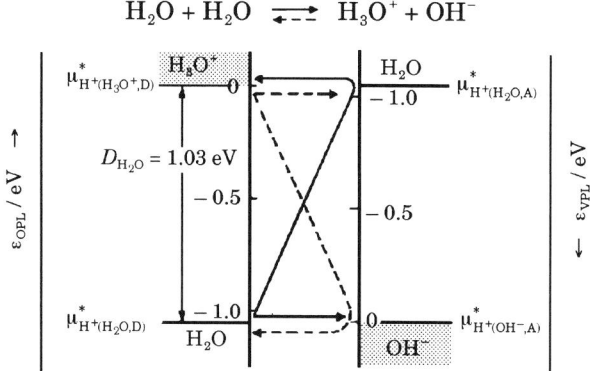

Fig. 3–16. The unitary levels of the acidic proton H_3O^+ / H_2O and basic proton H_2O / OH^- in pure water and the proton transfer between the two levels: ε_{OPL} = occupied proton level; ε_{VPL} = vacant proton level; $\mu^*_{H^+(H_3O^+, D)}$ = unitary occupied (donor) level of the acidic proton, $\mu^*_{H^+(H_2O, A)}$ = unitary vacant (acceptor) level of the acidic proton.

The energy level of the hydrated proton depends on the concentration of hydrated protons. For an acidic proton in Eqn. 3–32 and a basic proton in Eqn. 3–34, the proton levels μ_{H^+} are, respectively, given in Eqns. 3–37 and 3–38:

$$\mu_{H^+(H_3O^+/H_2O)} = \mu^*_{H^+(H_3O^+/H_2O)} + k\,T \ln x_{H_3O^+}, \qquad (3\text{–}37)$$

$$\mu_{H^+(H_2O/OH^-)} = \mu^*_{H^+(H_2O/OH^-)} - kT \ln x_{OH^-}, \qquad (3\text{–}38)$$

where $\mu^*_{H^+}$ denotes the unitary chemical potential of the hydrated proton. In terms of the molality m, instead of the molar fraction x, Eqns. 3–37 and 3–38 are expressed, respectively, by Eqns. 3–39 and 3–40:

$$\mu_{H^+(H_3O^+/H_2O)} = \mu^*_{H^+(H_3O^+/H_2O)} - kT \ln 55.51 + kT \ln m_{H_3O^+}$$

$$= \mu^0_{H^+(H_3O^+/H_2O)} + kT \ln m_{H_3O^+}, \qquad (3\text{–}39)$$

$$\mu_{H^+(H_2O/OH^-)} = \mu^*_{H^+(H_2O/OH^-)} + kT \ln 55.51 - kT \ln m_{OH^-}$$

$$= \mu^0_{H^+(H_2O/OH^-)} - kT \ln m_{OH^-}, \qquad (3\text{–}40)$$

where $\mu^0_{H^+}$ denotes the standard chemical potential of the hydrated proton. These relationships are shown in Fig. 3–17 which is an energy diagram both for the occupied proton level of H_3O^+ ions as a function of the concentration of H_3O^+ ions and for the vacant proton level of OH^- ions as a function of the concentration of OH^- ions.

In general, the acidic and basic proton hydration processes may occur simultaneously giving rise to the same proton level for both the acidic and the basic protons. In pure liquid water where $m_{H_3O^+} = m_{OH^-}$ in the electroneutrality, the proton level is obtained from Eqns. 3–39 and 3–40 as shown in Eqn. 3–41:

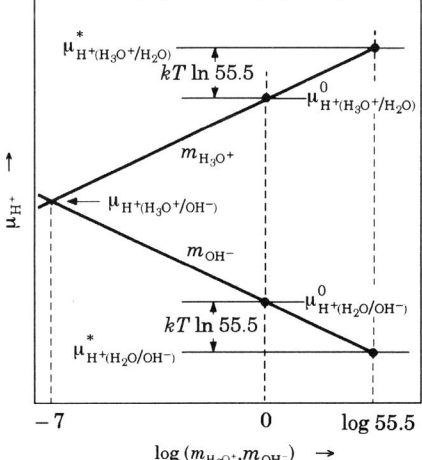

Fig. 3–17. Proton level vs. concentration curves ($\mu_{H^+} - m$) of an acidic proton (H_3O^+/H_2O) and a basic proton (H_2O/OH^-) in pure liquid water: m = concentration in molality; $\mu^*_{H^+(H_3O^+/H_2O)}$ = unitary acidic proton level; $\mu^0_{H^+(H_3O^+/H_2O)}$ = standard acidic proton level; $\mu_{H^+(H_3O^+/OH^-)}$ = proton level in pure water; $m_{H_3O^+}$ = occupied proton level–concentration curve of acidic protons; m_{OH^-} = vacant proton level–concentration curve of basic protons.

$$\mu_{H^+(H_3O^+/OH^-)} = \frac{1}{2} (\mu^*_{H^+(H_3O^+/H_2O)} + \mu^*_{H^+(H_2O/OH^-)}) . \tag{3-41}$$

It follows, therefore, that the proton level in pure water is located midway between the unitary levels of the acidic proton and the basic proton, giving rise to the hydrated proton concentration at pH 7.

Further, we consider a proton-donating and -accepting solute of acetic acid HAc/Ac$^-$, in which the proton association and dissociation processes occur as shown in Eqn. 3–42:

$$Ac^-_{aq} + H^+_{(STD)} \rightarrow HAc_{aq} , \qquad \text{proton solvation energy in acetic acid } J^*_{HAc} . \tag{3-42}$$

The proton level $\mu_{H^+(HAc/Ac^-)}$ in the association-dissociation of the acetic acid is given by Eqn. 3–43:

$$\mu_{H^+(HAc/Ac^-)} = \mu^*_{H^+(HAc/Ac^-)} + k T \ln \frac{x_{HAc}}{x_{Ac^-}} = \mu^*_{H^+(HAc/Ac^-)} + k T \ln m_{HAc} - k T \ln m_{Ac^-}, \tag{3-43}$$

where the molar fractions of HAc and Ac$^-_{aq}$ are expressed, respectively, to a first approximation by $x_{HAc} = m_{HAc}/(m_{H_2O} + m_{HAc} + m_{H^+} + m_{Ac^-})$ and $x_{Ac^-} = m_{Ac^-}/(m_{H_2O} + m_{HAc} + m_{H^+} + m_{Ac^-})$. Note that the dissociation constant of acetic acid is very small: $m_{Ac^-} \doteq m_{H^+} \rightarrow 0$ and $m_{HAc} \doteq m_{HAc} + m_{Ac^-} + m_{H^+}$.

In the presence of acetic acid (HAc/Ac$^-$) in an aqueous solution containing the acidic and basic protons (H$_3$O$^+$/H$_2$O and H$_2$O/OH$^-$), the following two proton transfer processes of Eqns. 3–44 and 3–45 take place in addition to the ionic dissociation of the water molecule of Eqn. 3–36 as shown in Fig. 3–18:

$$HAc + H_2O \rightarrow Ac^- + H_3O^+ , \qquad \text{energy of acidic proton transfer } D^*_{H_3O^+/HAc} , \tag{3-44}$$

$$Ac^- + H_2O \rightarrow HAc + OH^- , \qquad \text{energy of basic proton transfer } D^*_{HAc/OH^-} , \tag{3-45}$$

where D^* is the unitary energy of proton transfer; these energies have been estimated at $D^*_{H_3O^+/HAc} = 0.83$ eV and $D^*_{HAc/OH^-} = 0.65$ eV [Gurney, 1953].

For the *acidic proton transfer* of Eqn. 3–44, the proton hydration processes of Eqns. 3–32 and 3–42 are represented by the proton level versus concentration curves of Eqns. 3–39 and 3–43, respectively, as shown in Fig. 3–19. In this proton level diagram the proton level in an acetic acid solution is given by the intersecting point ($m_{H_3O^+} = m_{Ac^-}$) of the occupied proton level versus concentration curve of the H$_3$O$^+$ ion and the vacant proton level versus concentration curve of the Ac$^-$ ion as expressed in Eqn. 3–46:

$$\mu_{H^+(H_3O^+/HAc)} = \frac{1}{2} (\mu^*_{H^+(HAc/Ac^-)} + \mu^*_{H^+(H_3O^+/H_2O)}) + \frac{1}{2} k T \ln \frac{m_{HAc}}{55.51} . \tag{3-46}$$

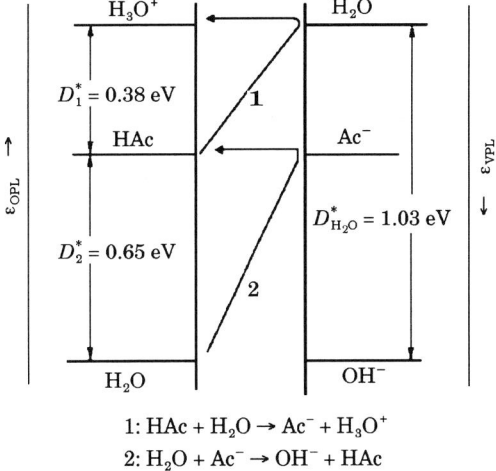

1: HAc + H$_2$O → Ac$^-$ + H$_3$O$^+$
2: H$_2$O + Ac$^-$ → OH$^-$ + HAc

Fig. 3-18. The unitary proton level of acetic acid and the proton transfer: D_1^* = acidic dissociation energy of acetic acid; D_2^* = basic dissociation energy of acetic acid.

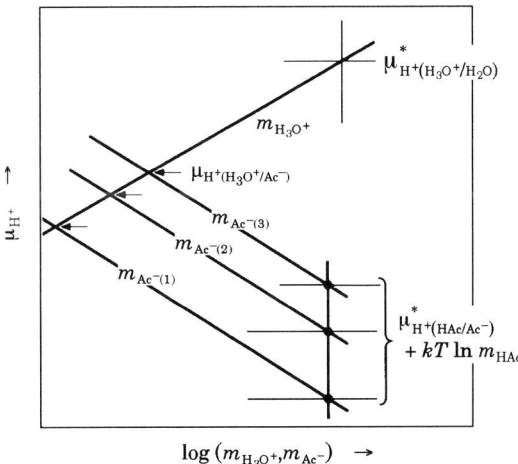

Fig. 3-19. Proton level vs. concentration curves of the acidic proton (H$_3$O$^+$/H$_2$O) and acetic proton (HAc/Ac$^-$): $\mu_{H^+(HAc/Ac^-)}^*$ = unitary acetic proton level; $\mu_{H^+(H_3O^+/Ac^-)}$ = proton levels in acetic acid solution; m_{Ac^-} = vacant proton level–concentration curve of acetic acid (as m_{HAc} increases; curve $m_{Ac^-(1)}$ moves to $m_{Ac^-(3)}$).

Eqn. 3–46 shows that the proton level $\mu_{H^+(H_3O^+/HAc)}$ in acetic acid solution is a function of acetic acid concentration.

Since acetic acid is a weak acid and its unitary proton level (HAc/Ac$^-$) is lower than the unitary acidic proton level (H$_3$O$^+$/H$_2$O), the proton moves from the unitary occupied acidic proton level to the unitary vacant proton level of acetic acid, thereby reducing the concentration of H$_3$O$^+$ ions toward the acetic acid. Contrastively, in strong acids such as hydrochloric acid of which the unitary proton level (HCl/Cl$^-$) is higher than the unitary acidic proton level, the proton moves from the hydrochloric acid to the vacant level of the acidic proton (H$_3$O$^+$/H$_2$O), thereby increasing the concentration of H$_3$O$^+$ ions.

It is interesting to point out the similarity between the proton level diagram of aqueous solutions and the electron level diagram of semiconductors as shown in Fig. 3–20. The ionic dissociation energy (1.03 eV) of the water molecule H$_2$O to form an ion pair of H$_3$O$^+$-OH$^-$ is the energy gap between the unitary acidic proton level and the unitary basic proton level, which corresponds to the band gap of semiconductors. The concentration product, $m_{H_3O^+} \times m_{OH^-}$, of the acidic occupied proton and the basic vacant proton is constant in aqueous solution;

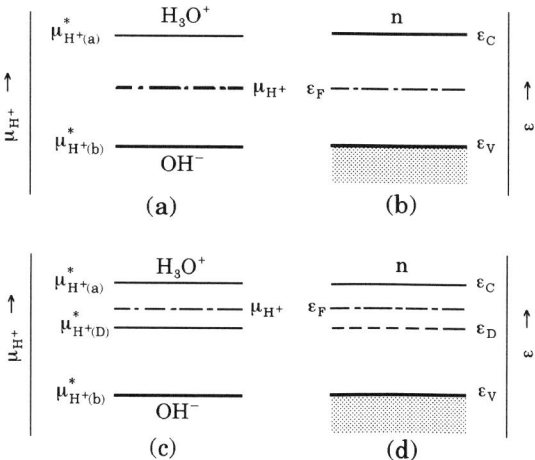

Fig. 3–20. Comparison of the proton level diagram of aqueous solutions with the electron level diagram of semiconductors: (a) proton levels in pure water, (b) electron levels in intrinsic semiconductors, (c) proton levels in week acid solutions, (d) electron levels in n type semiconductors. μ_{H^+} = proton level in aqueous solutions; $\mu^*_{H^+(a)}$ = unitary acidic proton level of H$_3$O$^+$/H$_2$O; $\mu^*_{H^+(b)}$ = unitary basic proton level of H$_2$O/OH$^-$; $\mu^*_{H^+(D)}$ = unitary proton level of acidic (donor) impurity in aqueous solutions; ε_F = Fermi level in semiconductors; ε_D = donor level in semiconductors.

while on the other hand the concentration product, $n \times p$, of electrons and holes is also constant in a semiconductor. In the case of pure liquid water containing no impurity solutes, the proton level is midway between the unitary acidic proton level and the unitary basic proton level giving rise to $m_{H_3O^+} = m_{OH^-}$ which corresponds the Fermi level of intrinsic semiconductors located midway in the band gap giving rise to $n = p$.

In aqueous solutions containing impurity solutes, the proton level approaches either the unitary acidic proton level ($m_{H_3O^+} > m_{OH^-}$) with increasing concentration of acidic impurity solutes or the unitary basic proton level ($m_{H_3O^+} < m_{OH^-}$) with increasing concentration of basic impurity solute. Similarly, the Fermi level of extrinsic semiconductors approaches either the conduction band ($n > p$) with increasing concentration of donors or the valence band ($n < p$) with increasing concentration of acceptors.

3.6 Thermodynamic Reference Level for Ions

The ion level in condensed phases has been represented by the real potential, α, of ions referred to the standard gaseous state of the ion at the electrostatic outer potential of the condensed phases. The reference level, therefore, is not common to all ions but differs with different ions. In chemical thermodynamics, the conventional energy scale is based on the assumption that all atoms in the stable form in the standard state are at the zero energy level which is the thermodynamic reference level of energy for chemical substances. In the following, we discuss the relationship between the ion level scale represented by the *real potential* of ions and the *conventional energy scale* of particles in chemical thermodynamics.

We consider, as an example, the gaseous hydrogen ion (proton) in the standard state, which is at the reference zero of proton level. The ionization of the gaseous hydrogen molecule in the standard state to produce the standard gaseous proton is shown in Eqn. 3–47:

$$\frac{1}{2} H_{2\,(STD)} \rightarrow H^+_{(STD)} + e_{(STD)}, \qquad \text{molecule ionization energy } \Delta G_{H^+/H_2}, \qquad (3\text{--}47)$$

where $\Delta G_{H^+/H_2}$ is the standard free enthalpy of the dissociation of gaseous hydrogen molecules followed by the ionization of gaseous hydrogen atoms. Then, the energy balance gives Eqn. 3–48:

$$\alpha_{H^+(STD)} + \alpha_{e(STD)} = \frac{1}{2}\alpha_{H_2(STD)} + \Delta G_{H^+/H_2}, \qquad (3\text{--}48)$$

where $\alpha_{H_2(STD)}$ is zero in the energy scale of chemical thermodynamics. The energy level $\alpha_{e(STD)}$ of the standard gaseous electron is set at zero in the electrochemical energy scale (Sec. 1.4 and Chap. 2); hence, the energy level of the standard gaseous hydrogen ion is given in the chemical thermodynamic energy scale as shown in Eqn. 3–49:

$$\alpha_{H^+(STD)} = \Delta G_{H^+/H_2} . \tag{3–49}$$

In deriving the energy level of anions, we may use the electron affinity for anions instead of the ionization energy for cations; for the energy level of ions in solids we may use the sublimation energy instead of the molecular dissociation energy. The ion level at the standard gaseous state in the energy scale of chemical thermodynamics can thus be estimated to be $\alpha_{H^+(STD)} = 15.72$ eV for the standard gaseous proton, $\alpha_{Cl^-(STD)} = -2.51$ eV for the standard gaseous chloride ion, $\alpha_{Ag^+(STD)} = 10.14$ eV for the standard monovalent silver ions, and $\alpha_{Fe^{2+}(STD)} = 20.04$ eV for the standard gaseous divalent iron ion.

The chemical thermodynamic energy scale of ions described in this section is not the same as the conventional energy scale of hydrated ions in aquatic electrochemistry (Refer to Sec. 6.4.) which is referred to the ion level of hydrated protons of unit activity.

References

[Blakely, 1973]: J. M. Blakely, *Introduction to the Properties of Crystal Surfaces*, p. 79-106, Pergamon Press, Oxford, (1973).
[Fumi-Tosi, 1964]: F. G. Fumi and M. P. Tosi, *J. Phys. Chem. Solids*, **25**, 31, 45(1964).
[Gurney, 1953]: R. W. Gurney, *Ionic Processes in Solution*, McGraw-Hill Book Co., Inc., New York, (1953).
[Sato, 1993]: N. Sato, *Electrode Chemistry* **Vol. 1**, p. 115-122, Japan Technical Information Service, Tokyo, (1993).
[Trasatti, 1980]: S. Trasatti, *Comprehensive Treatise of Electrochemistry*, **Vol. 1**, (Edited by J. O'M. Bockris, B. E. Conway and E. Yeager), p. 45, Plenum Press, New York, London, (1980).

CHAPTER 4

ELECTRODE POTENTIAL

4.1 Electrode

4.1.1 Electrode

In general, an electronic conductor which is used to introduce electric field or electric current is generally called the *electrode*. Specifically, in electrochemistry, the electronic conductors immersed in electrolytes of ionic conductors are conventionally called the *electrodes*. Since the function of an electrode to provide electric current does not work in isolation but requires the presence of electrolyte in contact with the electrode, the term of *"electrode"* is defined as a combination of an electronic conductor and an ionic electrolyte. Usually, an electrode is used in the form of its partial immersion in an electrolyte as shown in Fig. 4–1 (a). It is, however, more common to define the electrode in the form of complete immersion in electrolyte as shown in Fig. 4–1 (b).

The electronic conductor of an electrode may be either a metal or a semiconductor, and the electrolyte may be either an aqueous solution, fused salt, solid electrolyte, or gaseous electrolyte (gaseous plasma). In this textbook, we deal mainly with metal and semiconductor electrodes in aqueous electrolytes.

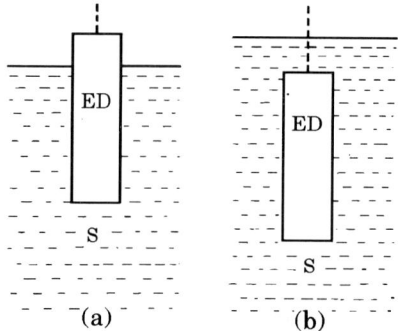

Fig. 4–1. Electrodes (a) partially immersed and (b) completely immersed in an electrolyte: ED = electrode; S = aqueous electrolyte solution; bold broken line = terminal lead.

4.1.2 Anode and cathode

In general, two electrodes are used in an electrolytic cell as shown in Fig. 4–2: Positive charge flows into the electrolyte from the "*anode*", and negative charge flows into electrolyte from the "*cathode*". *Anode* means an upper (inflow) entrance and *cathode* means a lower (outflow) exit in Greek. The ion drawn toward the anode in electrolyte is called the "*anion*" (negative ion), and the ion attracted to the cathode in electrolyte is called the "*cation*" (positive ion). The anode and cathode are defined in terms of the inflow and outflow, respectively, of electric current of *positive charge*, irrespective of the electrical polarity (positive or negative pole) of the electrode: the anode is the positive pole in material-producing electrolytic cells, but it is the negative pole in electricity-producing voltaic cells.

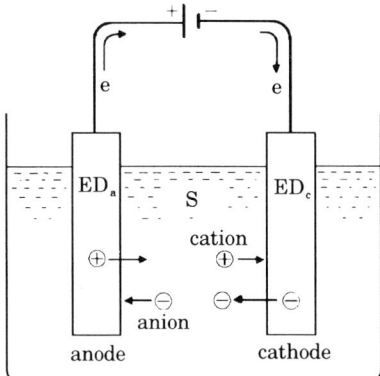

Fig. 4–2. Electrodes in an electrolytic cell: ED_a = anode (inflow entrance of positive charge current); ED_c = cathode (outflow exit of positive charge current); e = electron.

4.1.3 Electronic electrode and ionic electrode

Electrodes may be classified into the following two categories as shown in Fig. 4–3: one is the *electronic electrode* at which the transfer of electrons takes place, and the other is the *ionic electrode* at which the transfer of ions takes place. The electronic electrode corresponds, for instance, to the case in which the transfer of redox electrons in reduction–oxidation reactions, such as $Fe^{2+}_{aq} = Fe^{3+}_{aq} + e$, occurs; and the ionic electrode corresponds to the case in which the transfer of ions, such as $Fe^{2+}_{metal} = Fe^{2+}_{aq}$, occurs across the electrode interface. Usually, the former is found with insoluble electrodes such as platinum electrodes in aqueous solution containing redox particles; and the latter is found with soluble metal electrodes such as iron and nickel. In practice, both the electron transfer and the ion transfer can take place simultaneously across the electrode interface.

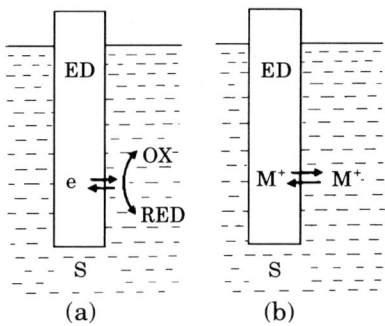

Fig. 4-3. (a) electronic electrodes and (b) ionic electrodes: OX = oxidant particle; RED = reductant particle; M⁺ = ion transferring across the electrode interface

4.1.4 Polarizable and nonpolarizable electrodes

In electrochemistry, the electrode at which no transfer of electrons and ions occurs is called the *polarizable electrode*; and the electrode at which the transfer of electrons and/or ions takes place is called the *nonpolarizable electrode* as shown in Fig. 4-4. The term of *polarization* in electrochemistry, different from the dipole polarization in physics, indicates the deviation in the electrode potential from a specific potential; this specific potential is usually the potential at which no electric current flows across the electrode interface. "*To polarize*" means to shift the electrode potential from a specific potential in the anodic (anodic polarization) or in the cathodic (cathodic polarization) direction.

With nonpolarizable electrodes the polarization (the shift of the electrode potential) does not occur, because the charge transfer reaction involves a large electric current without producing appreciable change in electrode potential. The nonpolarizable electrode, therefore, cannot be polarized to a significant extent.

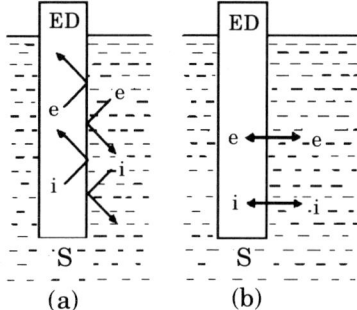

Fig. 4-4. (a) polarizable electrode and (b) nonpolarizable electrode: e = electron; i = ion.

The nonpolarizable electrode may also be defined as the electrode at which the electron or ion transfer reaction is essentially in equilibrium: i. e. the electron or ion level in the electrode is pinned at the electron level of hydrated redox particles or at the hydrated ion level in an aqueous electrolyte. In order for the electrode reaction to be in equilibrium at the interface of nonpolarizable electrode, an appreciable concentration of redox particles or potential determining ions must exist in the electrolyte.

With polarizable electrodes, the charge transfer (electron or ion transfer) across the electrode interface does not occur, and the electrode can be readily polarized to change the electrode potential without causing any appreciable increase in electrode current. No charge transfer equilibrium is realized at polarizable electrodes; but the electrostatic equilibrium is established between the electrode and the electrolyte thereby producing an electric charge of equal amount but opposite sign on both sides of the "interfacial condenser" (the interfacial electric double layer).

In electrochemistry, the electrode current is conventionally classified into the *faradaic current* and *nonfaradaic current*. The former is the electric current associated with charge transfer reactions at nonpolarizable electrodes; and the latter is the current that is required to establish the electrostatic equilibrium at the interfacial double layer on both polarizable and nonpolarizable electrodes. The nonfaradaic current, sometimes called a transient current, flows also in the course of establishing the adsorption of ions on electrodes.

4.2 The Interface of Two Condensed Phases

4.2.1 Potential difference between two contacting phases

In general, an electrostatic potential difference arises between two phases when they are brought into contact. We deal with the electrostatic potential in two condensed phases being in contact. The electrostatic *inner potential*, ϕ, of a condensed phase (solid or liquid) is given by the sum of the *outer potential*, $\psi_{(\sigma)}$, associated with the *electric charge*, σ, of the condensed phase and the *surface potential difference*, $\chi_{(dip)}$, associated with the *surface dipole* as shown in Eqn. 4–1 and Fig. 4–5:

$$\phi = \psi_{(\sigma)} + \chi_{(dip)} . \tag{4-1}$$

For noncharged phases ($\sigma = 0$), the outer potential $\psi_{(\sigma=0)}$ is zero and, hence, the inner potential equals the surface potential difference ($\phi = \chi_{(dip)}$). As described in Chap. 2, the surface potential difference arises from the surface dipole due to the

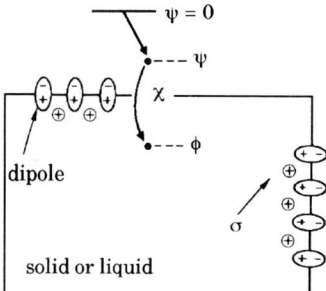

Fig. 4–5. The inner potential, ϕ, the outer potential, ψ, and the surface potential difference, χ, of a condensed phase: σ = surface charge; dipole = surface dipole.

spread-out of electrons from the surface in the case of metals and due to the dipole orientation of water molecules at the surface in the case of aqueous solutions.

When two condensed phases A and B are brought into contact, as shown in Fig. 4–6, the surface charge, σ, the inner potential, ϕ, and the outer potential, $\psi_{(\sigma)}$, of the initially separate two phases are changed to values characteristic of the joined solid A/B; on the other hand, the surface potential difference, $\chi_{(dip)}$, at the free surface of each phase remains unchanged except, of course, for the joining surface. The outer potential, ψ, of the free surface of the two phases is given by Eqn. 4–2, whether or not the two phases are in contact:

$$\psi_A = \phi_A - \chi_{A(dip)}, \qquad \psi_{B(s)} = \phi_B - \chi_{B(dip)} . \tag{4-2}$$

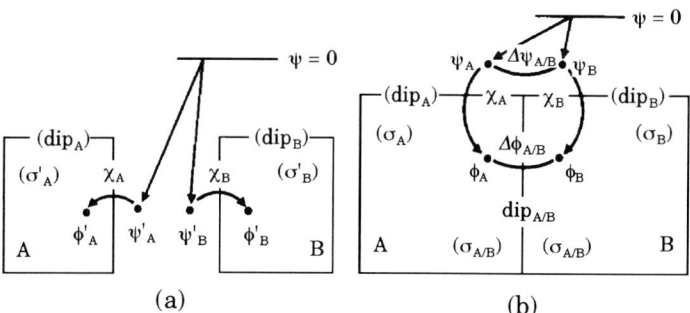

Fig. 4–6. The inner potential, ϕ, and the outer potential, ψ, of two condensed phases A and B before and after their contact: $\Delta\phi$ ($\Delta\psi$) = inner (outer) potential difference between two contacting phases; σ = surface or interface charge; dip = surface or interface dipole.

Bringing two phases into contact, as in Fig. 4–6, produces an electric charge σ on the free surface of the two phases which is different from its initial value σ'; hence, the outer potential of the two phases changes from its initial value ψ' to its final value ψ, giving rise to a change of the outer potential difference between the two phases from its initial value $\Delta\psi'_{A/B}$ before the contact to its final value $\Delta\psi_{A/B}$ after the contact as shown in Fig. 4–6.

An electrostatic potential difference, called the *inner potential difference*, arises across the interface of two contacting phases. This inner potential difference consists of a potential difference $g_{A/B(\sigma)}$ due to the *interfacial charge* (charge on both sides of the interface), $\sigma_{A/B}$, and a potential difference $g_{A/B(\mathrm{dip})}$ due to the *interfacial dipole*, $\mathrm{dip}_{A/B}$, as shown in Eqn. 4–3:

$$\Delta\phi_{A/B} = \phi_A - \phi_B = g_{A/B(\mathrm{dip})} + g_{A/B(\sigma)} \ . \tag{4-3}$$

Obviously, the *interfacial charge* $\sigma_{A/B}$ differs from the initial (before contact) surface charges σ_A and σ_B ($\sigma_{A/B} \neq \sigma_A + \sigma_B$); and the *interfacial dipole* $\mathrm{dip}_{A/B}$ is not the same as the algebraic sum of the initial surface dipoles ($g_{A/B(\mathrm{dip})} \neq \chi_{A(\mathrm{dip})} - \chi_{B(\mathrm{dip})}$). Thus, it follows that both the inner and the outer potential differences, $\Delta\phi_{A/B}$ and $\Delta\psi_{A/B}$, between the two contacting phases are not the same as those $\Delta\phi'_{A/B}$ and $\Delta\psi'_{A/B}$ before the contact. As a result, Eqn. 4–2 yields Eqn. 4–4:

$$\Delta\psi_{A/B} = \psi_A - \psi_B = \phi_A - \phi_B - (\chi_{A(\mathrm{dip})} - \chi_{B(\mathrm{dip})}) = \Delta\phi_{A/B} - \Delta\chi_{A/B} \ , \tag{4-4}$$

which indicates that the outer potential difference $\Delta\psi_{A/B}$ differs from the inner potential difference $\Delta\phi_{A/B}$ by a magnitude equal to the discrepancy, $\Delta\chi_{A/B}$, in the surface potential difference between the free surfaces of the two phases.

The *inner potential difference* of contacting two phases is called in electrochemistry the *Galvani potential difference*, and the *outer potential difference* is called the *Volta potential difference*. The outer potential difference corresponds to what is called the *contact potential* between two phases.

The outer potential difference between two contacting phases can be measured because it is a potential difference between two points in the same vacuum or gas phase outside the free surfaces of the two contacting phases. On the other hand, the inner potential difference can not be measured because the potential measuring probe introduces its interfacial potential difference that differs with the two phases and thus can not be canceled; this gives rise to an unknown potential difference in the potential measurement.

4.2.2 The interface of zero charge

The interface at which the interfacial charge, $\sigma_{A/B}$, is zero is called the interface of zero charge or the *zero charge interface*. The inner potential difference across the zero charge interface is determined by the interfacial dipole *only*; thus, it is characteristic of the contacting interface of the two phases as indicated in Eqn. 4–5:

$$\Delta\phi_{A/B(\sigma=0)} = g_{A/B(\text{dip})} . \tag{4–5}$$

The outer potential difference, $\Delta\psi_{A/B(\sigma=0)}$, between the free surfaces of the two phases contacting at the interface of zero charge with the inner potential difference $\Delta\phi_{A/B(\sigma=0)}$ between the two phases is given by Eqn. 4–4 as follows: $\Delta\psi_{A/B(\sigma=0)} = \Delta\phi_{A/B(\sigma=0)} - (\chi_{A(\text{dip})} - \chi_{B(\text{dip})})$.

Fig. 4–7 compares (a) the charged interface and (b) the zero charge interface of a metal electrode M in an aqueous solution S. For the charged interface, the inner potential difference, $\Delta\phi_{M/S}$, is the algebraic sum of the dipole-induced potential difference, $g_{M/S(\text{dip})}$, and the charge-induced potential difference, $g_{M/S(\sigma)}$, as shown in Eqn. 4–3. The dipole-induced interfacial potential difference, $g_{M/S(\sigma)}$, consists of a potential difference $g_{M/S(M,\text{dip})}$ due to the dipole on the metal side and a potential difference $g_{M/S(S,\text{dip})}$ due to the dipole on the solution side: $g_{M/S(\text{dip})} = g_{M/S(M,\text{dip})} - g_{M/S(S,\text{dip})}$. As described in the foregoing, the dipole potential differences, $g_{M/S(M,\text{dip})}$ and $g_{M/S(S,\text{dip})}$, at the interface differ, respectively, from the surface potential differences, $\chi_{M(\text{dip})}$ and $\chi_{S(\text{dip})}$, at the free surfaces of the two contacting phases. Namely, $g_{M/S(M,\text{dip})} = \chi_{M(\text{dip})} + \delta\chi_M$ and $g_{M/S(S,\text{dip})} = \chi_{S(\text{dip})} + \delta\chi_S$, where $\delta\chi$ is the change in the dipole potential difference resulting from the contact of the two phases.

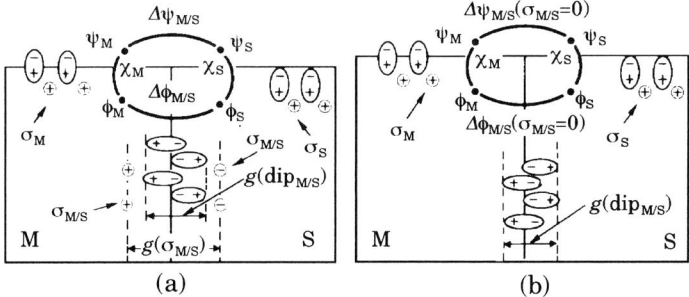

Fig. 4–7. The inner potential difference, $\Delta\phi_{M/S}$, and the outer potential difference, $\Delta\psi_{M/S}$, between a solid metal M and an aqueous solution S: (a) the charged interface where $\Delta\phi_{M/S} = g_{M/S(\text{dip})} + g_{M/S(\sigma)}$, (b) the zero charge interface where $\Delta\phi_{M/S} = g_{M/S(\text{dip})}$.

For the interface of zero charge the inner potential difference is given by $\Delta\phi_{M/S(\sigma=0)} = g_{M/S(dip)} = g_{M/S(M, dip)} - g_{M/S(S, dip)} = (\chi_{M(dip)} - \chi_{S(dip)}) + (\delta\chi_M - \delta\chi_S)$. Hence, the outer potential difference is given by Eqn. 4–6:

$$\Delta\psi_{M/S(\sigma=0)} = \Delta\phi_{M/S(s=0)} - (\chi_{M(dip)} - \chi_{S(dip)}) = \delta\chi_M - \delta\chi_S , \tag{4-6}$$

which indicates that the outer potential difference $\Delta\psi_{M/S(\sigma=0)}$ between the two phases contacting at the interface of zero charge equals the change $(\delta\chi_M - \delta\chi_S)$ in the surface dipole potential difference that occurs at the contact interface.

4.2.3 Interfaces in charge transfer equilibrium

In the case of the nonpolarizable interface A/B the inner and the outer potential differences, $\Delta\phi_{A/B}$ and $\Delta\psi_{A/B}$, are determined by the equilibrium of charge transfer that occurs across the interface. Fig. 4–8 shows the electron energy levels in two solid metals A and B before and after they are brought into contact with each other. As a result of contact, electrons in the metal B of the higher electron level (the lower work function) move into the metal A of the lower electron level (the higher work function), and the Fermi levels of the two metals eventually become equal to each other in the state of electron transfer equilibrium. Therefore, the

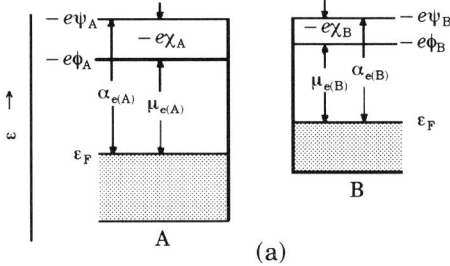

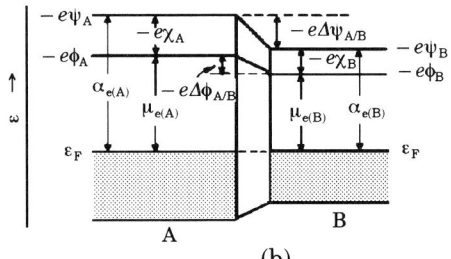

Fig. 4–8. Electron energy levels of two different metals A and B in (a) the isolated state and in (b) the contact state: ε = electron energy; α_e = real potential of electrons in metals; ε_F = Fermi level of electrons in metals; $\Delta\phi_{A/B}$ = inner potential difference; $\Delta\psi_{A/B}$ = outer potential difference.

electrochemical potential $\bar{\mu}_e$ of electrons in the metal A becomes equal to that in the metal B as shown in Eqn. 4–7:

$$\bar{\mu}_{e(A)} = \bar{\mu}_{e(B)} . \qquad (4\text{–}7)$$

Since the electrochemical potential of electrons in metals is a function of the inner potential of the metal ($\bar{\mu}_{e(A)} = \mu_{e(A)} - e\,\phi_A$ and $\bar{\mu}_{e(B)} = \mu_{e(B)} - e\,\phi_B$), the inner potential difference, $\Delta\phi_{A/B}$, across the interface where the electron transfer is in equilibrium is represented by the difference of the chemical potential μ_e of electrons between the two metal phases A and B as shown in Eqn. 4–8:

$$\Delta\phi_{A/B} = \phi_A - \phi_B = \frac{\mu_{e(A)} - \mu_{e(B)}}{e} . \qquad (4\text{–}8)$$

Further, since the electrochemical potential of electrons in metals is also a function of the outer potential of the metal ($\bar{\mu}_{e(A)} = \alpha_{e(A)} - e\,\psi_A$ and $\bar{\mu}_{e(B)} = \alpha_{e(B)} - e\,\psi_B$), the outer potential difference, $\Delta\psi_{A/B}$, between the two metal phases in electron transfer equilibrium is represented by the difference of the real potential of electrons, α_e, between the two metal phases and, hence, by the difference of the work function, Φ, as shown in Eqn. 4–9:

$$\Delta\psi_{A/B} = \psi_A - \psi_B = \frac{\alpha_{e(A)} - \alpha_{e(B)}}{e} = \frac{\Phi_B - \Phi_A}{e} . \qquad (4\text{–}9)$$

Next, we consider the interface M/S of a nonpolarizable electrode where the electron or ion transfer is in equilibrium between the solid metal M and the aqueous solution S. Here, the interfacial potential difference is determined by the charge transfer equilibrium. As shown in Fig. 4–9, the electron transfer equilibrium equates the Fermi level, $\varepsilon_{F(M)}$ ($= \bar{\mu}_{e(M)}$), of electrons in the metal with the Fermi level, $\varepsilon_{F(REDOX)}$ ($= \bar{\mu}_{e(S)}$), of redox electrons in hydrated redox particles; this gives rise to the inner and the outer potential differences $\Delta\phi_{M/S}$ and $\Delta\psi_{M/S}$, respectively, as shown in Eqn. 4–10:

$$\Delta\phi_{M/S} = \phi_M - \phi_S = \frac{\mu_{e(M)} - \mu_{e(S)}}{e}, \quad \Delta\psi_{M/S} = \psi_M - \psi_S = \frac{\alpha_{e(M)} - \alpha_{e(S)}}{e} . \qquad (4\text{–}10)$$

In the case of ion transfer equilibrium at the electrode interface M/S, the ion level, $\bar{\mu}_{i(M)}$, in the electrode equals the ion level, $\bar{\mu}_{i(S)}$, of hydrated ions in the solution as shown in Fig. 4–9. Hence, the inner potential difference, $\Delta\phi_{M/S}$, is given in terms of the chemical potential, μ_i, of the ion; and the outer potential difference, $\Delta\psi_{M/S}$, is given in terms of the real potential, α_i, of the ion as shown in Eqn. 4–11:

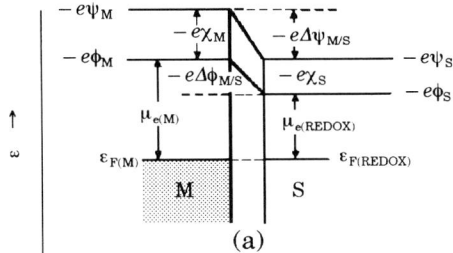

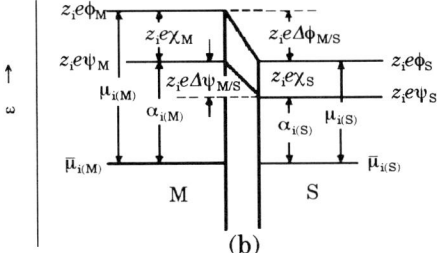

Fig. 4–9. Electron energy levels at the interface between metal M and aqueous solution S in (a) the state of electron transfer equilibrium and in (b) the state of ion transfer equilibrium: ε = electron or ion energy; μ_e = chemical potential of electrons; μ_i = chemical potential of ions; z_i = ionic valence.

$$\Delta\phi_{M/S} = \phi_M - \phi_S = -\frac{\mu_{i(M)} - \mu_{i(S)}}{z_i e}, \quad \Delta\psi_{M/S} = \psi_M - \psi_S = -\frac{\alpha_{i(M)} - \alpha_{i(S)}}{z_i e}. \quad (4\text{-}11)$$

where z_i is the ionic valence.

As is shown in Eqn. 4–3, the inner potential difference across the interface M/S consists of a charge-induced potential difference and a dipole-induced potential difference ($\Delta\phi_{M/S} = g_{M/S(dip)} + g_{M/S(\sigma)}$); this potential difference occurs whether or not the charge transfer is in equilibrium at the interface. Since the potential difference introduced by the surface dipole usually remains constant, the change in the interfacial potential difference across the interface is a function of the electric charge of the interfacial double layer.

4.3 Electrode Potential

4.3.1 Electrode potential defined by electron energy levels

The *electrode potential* is one of the most important concepts in electrochemistry. It represents the energy level of electrons or ions in electrodes rather than simply the electrostatic potential of electrodes [Trasatti, 1986, 1990; Sato, 1993, 1995].

In physics the electron level in an isolated solid metal is conventionally represented by the negative work function, $-\Phi_{M/V}$, which corresponds to the *real potential*, $\alpha_{e(M/V)}$, of electrons in the isolated metal. Similarly, in electrochemistry, the electron level in an electrode is represented by the negative work function, $-\Phi_{M/S/V}$, which is the *real potential*, $\alpha_{e(M/S/V)}$, of electrons in the electrode. Here, subscript e(M/V) indicates electrons in an isolated metal in vacuum, and subscript e(M/S/V) indicates electrons in a metal electrode in aqueous electrolyte solution. As shown in Fig. 4–10, the real potential of electrons in an electrode is expressed by the energy (free enthalpy) released in the electron transfer from the standard state of gaseous electrons at the outer potential, ψ_S, through the electrolyte solution into the electrode as shown in Eqn. 4–12:

$$\alpha_{e(M/S/V)} = \alpha_{e(M/S)} + \alpha_{e(S/V)} = -\Phi_{M/S/V} \quad , \tag{4-12}$$

where $\alpha_{e(S/V)}$ is the energy of electron transfer from the standard gaseous state into the electrolyte solution and $\alpha_{e(M/S)}$ is the energy of electron transfer from the electrolyte solution into the electrode.

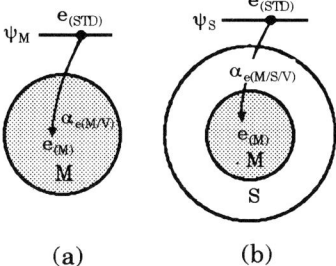

Fig. 4–10. Electron energy levels in (a) an isolated solid metal and in (b) a metal electrode immersed in an electrolyte solution: M = metal; S = electrolyte solution; $e_{(STD)}$ = gaseous electrons in the standard state; $e_{(M)}$ = electrons in metal; ψ_M = outer potential of an isolated solid metal; ψ_S = outer potential of the electrolyte solution.

Figs. 4–11 and 4–12 show schematic energy diagrams for the electron transfer from the standard gaseous state through the electrolyte solution into the metal electrode. As mentioned in Chap. 2, the electron level (the real potential of the electron) $\alpha_{e(S/V)}$ in an electrolyte solution is given by the sum of the electrostatic energy $-e\chi_{S/V}$ due to the surface dipole and the chemical potential $\mu_{e(S)}$ of the electron in the solution: $\alpha_{e(S/V)} = \mu_{e(S)} - e\chi_{S/V}$. Similarly, as shown in Figs. 4–11 and 4–12, the energy $\alpha_{e(M/S)}$ for the electron transfer across the electrode/solution interface is given by the sum of the electrostatic energy of $-e\,\Delta\phi_{M/S}$ required for an electron to transfer across the interface and the energy difference in the chemical potential of electrons $\mu_{e(M)} - \mu_{e(S)}$ between the electrolyte solution and

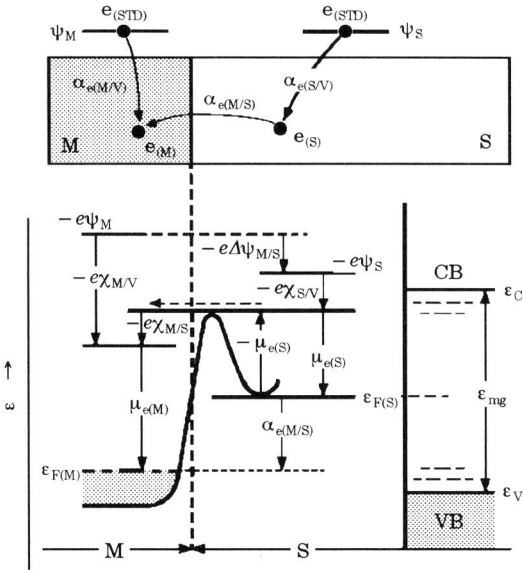

Fig. 4–11. Energy diagram for the electron transfer from the standard gaseous electron across the solution/vacuum interface, through the electrolyte solution, and across the metal/solution interface into a metal electrode: $\alpha_{e(S/V)}$ = real potential of electrons $e_{(S)}$ in the electrolyte solution (redox electron $e_{(REDOX, S)}$); $\alpha_{e(M/S)}$ = energy for an electron transfer from the electrolyte solution into the metal electrode; $\mu_{e(S)}$ = chemical potential of electrons $e_{(S)}$ in the electrolyte solution; $\varepsilon_{F(S)}$ = Fermi level of electrons in the electrolyte solution; $\varepsilon_{F(M)}$ = Fermi level of electrons in the metal; ε_{mg} = mobility gap of liquid water.

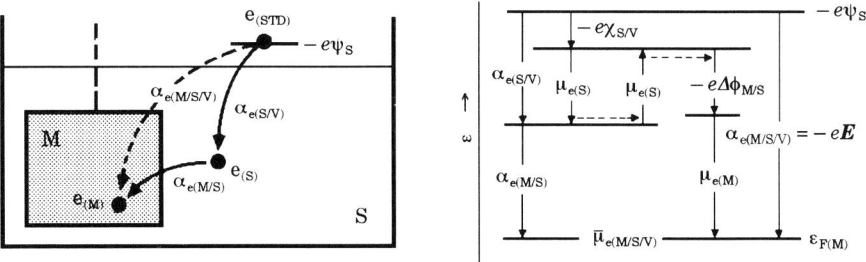

Fig. 4–12. Electron energy levels in the electron transfer from the standard gaseous electron through an electrolyte solution into an electrode: $\alpha_{e(M/S/V)}$ = real potential of electrons in the electrode ; $\boldsymbol{E}$ = electrode potential (absolute electrode potential).

the electrode; $\alpha_{e(M/S)} = \mu_{e(M)} - \mu_{e(S)} - e\,\Delta\phi_{M/S}$. Consequently, we obtain the *electron level* (the *real potential* of electrons) $\alpha_{e(M/S/V)}$ in the electrode as shown in Eqn. 4–13:

$$\alpha_{e(M/S/V)} = \alpha_{e(S/V)} + \alpha_{e(M/S)} = \mu_{e(M)} - e\,\Delta\phi_{M/S} - e\,\chi_{S/V} \quad , \tag{4–13}$$

where e is the elemental charge of electron.

It follows from Eqn. 4–13 that the electron level $\alpha_{e(M/S/V)}$ in the electrode is a function of the chemical potential $\mu_{e(M)}$ of electrons in the electrode, the interfacial potential difference (the inner potential difference) $\Delta\phi_{M/S}$ between the electrode and the electrolyte solution, and the surface potential difference $\chi_{S/V}$ of the electrolyte solution. It appears that the electron level $\alpha_{e(M/S/V)}$ in the electrode *depends* on the interfacial potential difference across the electrode interface and the chemical potential of electrons in the electrode *but does not depend upon* the chemical potential of electrons in the electrolyte solution. Eqn. 4–13 is valid when no electrostatic potential gradient exists in the electrolyte solution. In the presence of a potential gradient, an additional electrostatic energy is included in Eqn. 4–13.

The *electrode potential* E is defined in terms of the real potential $\alpha_{e(M/S/V)}$ of electrons in the electrode (i.e. the work function, $\Phi_{M/S/V}$) as shown in Eqn. 4–14:

$$E = -\frac{\alpha_{e(M/S/V)}}{e} = -\frac{\mu_{e(M)}}{e} + \Delta\phi_{M/S} + \chi_{S/V} = \frac{\Phi_{M/S/V}}{e} \quad . \tag{4–14}$$

Since $\mu_{e(M)}$ and $\chi_{S/V}$ are characteristic of specific combinations of electrodes and electrolyte solutions, they are constant. Therefore, for an electrode system, the electrode potential is eventually a function of the interfacial potential difference $\Delta\phi_{M/S}$ *only*. The electrode potential, E, defined in Eqn. 4–14 corresponds to what is called the *absolute electrode potential*. The reference zero level of the absolute electrode potential is set at the outer potential of the electrolyte solution in which the electrode is immersed.

For partially immersed electrodes, as shown in Fig. 4–13, an outer potential difference $\Delta\psi_{M/S}$ arises between the free surface of the electrode and the free surface of the electrolyte solution; $\Delta\psi_{M/S}$ equals the difference in the real potential of electrons ($\alpha_{e(M/S/V)} - \alpha_{e(M/V)}$) / e between the electrode immersed in the electrolyte solution and the electrode isolated from the electrolyte solution as in Eqn. 4–15:

$$\Delta\psi_{M/S} = \psi_{M/V} - \psi_{S/V} = \frac{\alpha_{e(M/S/V)} - \alpha_{e(M/V)}}{e} = -\frac{\Phi_{M/S/V} - \Phi_{M/V}}{e} \tag{4–15}$$

This potential difference is the *contact potential difference* between the electrode and the electrolyte solution.

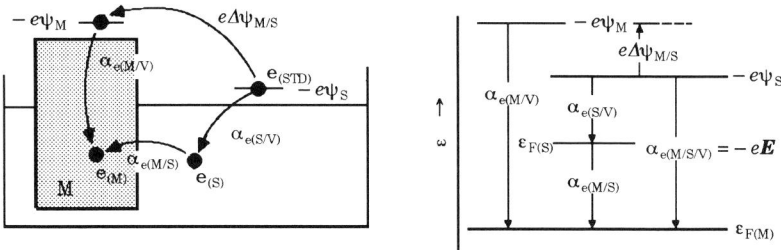

Fig. 4–13. Energy levels of electrons in an electrode partially immersed in an electrolyte solution: $\Delta\psi_{M/S}$ = outer potential difference (contact potential difference) between the electrode and the electrolyte solution.

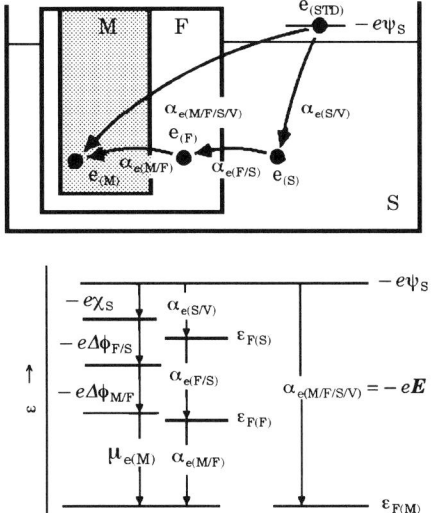

Fig. 4–14. Energy levels of electrons in a film-covered electrode: F = film on an electrode; $\alpha_{e(M/F/S/V)}$ = real potential of electrons in the electrode; ε_F = Fermi level of electrons.

Fig. 4–14 shows the electron energy levels in a film-covered metal electrode. The electrode potential E in this case is also given by the real potential $\alpha_{e(M/F/S/V)}$ of electrons in the electrode as shown in Eqn. 4–16:

$$E = -\frac{\alpha_{e(M/F/S/V)}}{e} = -\frac{\mu_{e(M)}}{e} + \Delta\phi_{M/F} + \Delta\phi_{F/S} + \chi_{S/V} \; , \tag{4–16}$$

where $\Delta\phi_{M/F}$ and $\Delta\phi_{F/S}$ are the interfacial potential differences between the electrode

4.3.2 Electrode potential and ion energy levels in electrodes

Next, we consider the metal ion M^{z+} in the metal electrode. In the same way as we considered the electron level, the ion energy level is represented by the *real potential* $\alpha_{M^{z+}(M/S/V)}$ of a metal ion in the metal electrode as shown in Eqn. 4–17 and Fig. 4–15:

$$\alpha_{M^{z+}(M/S/V)} = \alpha_{M^{z+}(S/V)} + \alpha_{M^{z+}(M/S)} = \mu_{M^{z+}(M)} + z\,e\Delta\phi_{M/S} + z\,e\,\chi_{S/V}\,. \tag{4–17}$$

From Eqns. 4–13 and 4–17, therefore, we obtain the relationship between the electrode potential, E, and the real potential, $\alpha_{M^{z+}(M/S/V)}$, of a metal ion in the electrode as shown in Eqn. 4–18:

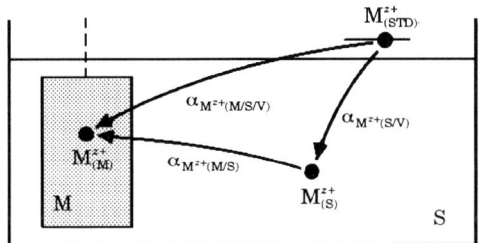

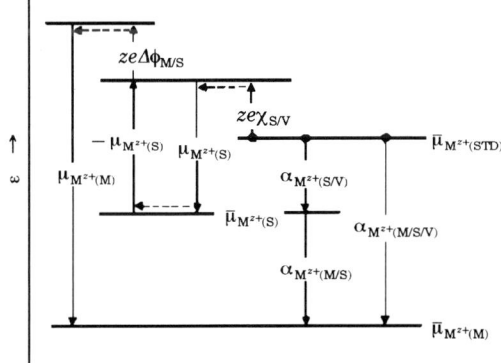

Fig. 4–15. Energy levels of a metal ion in a metal electrode: $M^{z+}_{(STD)}$ = gaseous metal ion in the standard state; $M^{z+}_{(M)}$ = metal ion in the electrode; $\alpha_{M^{z+}(M/S/V)}$ = real potential of metal ions in the metal electrode; $\alpha_{M^{z+}(S/V)}$ = real potential of hydrated metal ions in the aqueous solution; $\bar{\mu}_{M^{z+}}$ = electrochemical potential of metal ions.

$$\alpha_{M^{z+}(M/S/V)} = z\,e\,\boldsymbol{E} + \mu_{M^{z+}(M)} + z\,\mu_{e(M)} = z\,e\,\boldsymbol{E} + \mu_{M(M)} \quad , \tag{4–18}$$

where $\mu_{M(M)}$ is the energy for the formation of a solid metal from the standard gaseous metal ion and the standard gaseous electron; this energy $\mu_{M(M)}$ is given by the negative sum of the solid metal sublimation energy, ΔG_{subl}, and the metal atom ionization energy, $I_{M^{z+}}$: $\mu_{M(M)} = -\Delta G_{subl} - I_{M^{z+}}$. Fig. 4–16 illustrates the energy of metal ion transfer at a metal electrode.

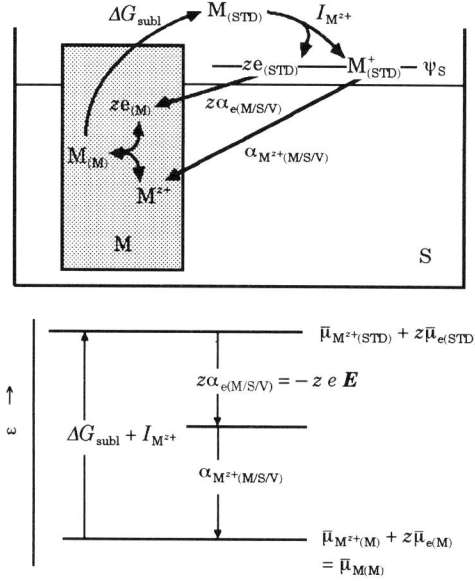

Fig. 4–16. Energy levels of a metal ion and an electron in an ionic electrode of metal ion transfer: ΔG_{subl} = sublimation energy of a solid metal; $I_{M^{z+}}$ = ionization energy of gaseous metal atoms; ψ_S = outer potential of the electrolyte solution; $\boldsymbol{E}$ = electrode potential (absolute electrode potential).

The electrode potential can be defined not only by the *energy level of electrons* (the real potential of electrons) but also by the *energy level of ions* (the real potential of ions) in the electrode. The former may be called the *"electronic electrode potential"* and the latter may be called the *"ionic electrode potential"* [Sato, 1995]. For instance, the electrode potential of a metal electrode can be defined in terms of the metal ion level (the real potential of metal ions), $\alpha_{M^{z+}(M/S/V)}$, in the electrode as follows: $E_{ion} = \alpha_{M^{z+}(M/S/V)} / z\,e$, where E_{ion} is the ionic electrode potential. Eqn. 4–18 gives the relationship between the electronic electrode potential, E, and the ionic electrode potential, E_{ion}; obviously both being in the same scale of units but with different zero levels.

For electrodes which have no electron energy levels in the energy range of general interest, such as ionic crystalline solid electrodes and membrane electrodes, *only the concept of ionic electrode potential* can be of practical significance.

4.4 Electrode Potential in Charge Transfer Equilibrium

4.4.1 Electrode potential in electron transfer equilibrium

The electrode potential defined in Sec. 4.3 applies to both nonpolarizable electrodes at which charge transfer reactions may take place and polarizable electrodes at which no charge transfer takes place. For nonpolarizable electrodes at which the charge transfer is in equilibrium, the interfacial potential difference $\Delta\phi_{M/S}$ is determined by the equilibrium of the charge transfer reaction.

For an electronic electrode at which the transfer of redox electrons is in equilibrium ($OX_{aq} + e_{(M)} = RED_{aq}$), as shown in Fig. 4–17, the Fermi level $\varepsilon_{F(REDOX, S)}$ of *redox electrons* $e_{(REDOX, S)}$ in hydrated redox particles equals the Fermi level $\varepsilon_{F(M)}$ of electrons $e_{(M)}$ in the electrode; the energy for the electron transfer across the electrode interface is, then, zero ($\alpha_{e(M/S)} = 0$). Consequently, the electron level $\alpha_{e(M/S/V)}$ in the electrode equals the electron level $\alpha_{e(S/V)}$ in the aqueous solution, i.e. the *redox electron level* $\alpha_{e(REDOX, S)}$ of the hydrated redox particles.

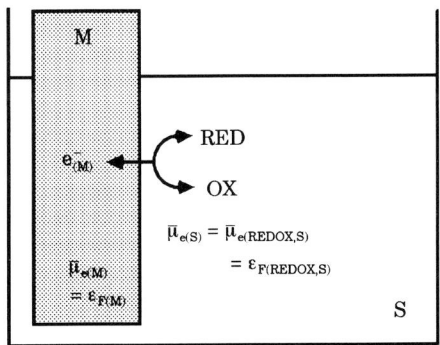

Fig. 4–17. Electronic electrode in the equilibrium of electron transfer: OX = hydrated oxidant particles; RED = hydrated reductant particles; $\varepsilon_{F(REDOX, S)}$ = Fermi level of redox electrons in hydrated redox particles in solution S; $\bar{\mu}_e$ = electrochemical potential of electrons.

The electrode potential, E_{eq}, which is represented by the real potential $\alpha_{e(M/S/V)}$ of electrons in the electrode, therefore, corresponds to the energy level $\alpha_{e(REDOX, S)}$ of redox electrons in the hydrated redox particles in aqueous solution as shown in Fig. 4–18 and defined in Eqn. 4–19:

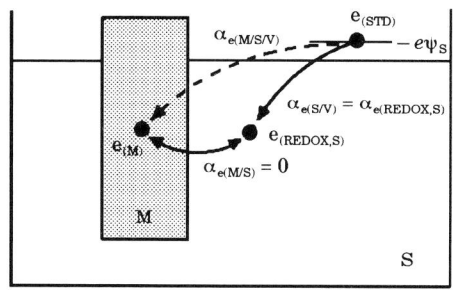

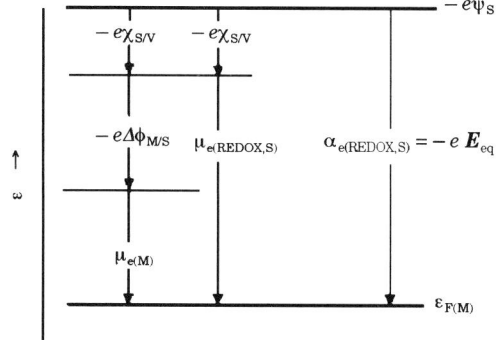

Fig. 4–18. Electron levels of an electronic electrode in the equilibrium of electron transfer: $e_{(REDOX, S)}$ = redox electron at equilibrium; $e_{(M)}$ = electrons in metal electrodes; E_{eq} = electrode potential in equilibrium of electron transfer.

$$E_{eq} = -\frac{\alpha_{e(M/S/V)}}{e} = -\frac{\alpha_{e(REDOX, S)}}{e} = -\frac{\mu_{e(REDOX, S)}}{e} + \chi_{S/V} \ . \tag{4–19}$$

It, thus, follows that the electrode potential in the electron transfer equilibrium represents the *redox electron level* of the redox particles in aqueous electrolyte solution. Further, it follows from Eqn. 4–19 that the electrode potential in the transfer equilibrium of redox electrons is characteristic of individual redox reactions but *independent* of the nature of the electrode materials.

The electrode potential in the equilibrium of redox electron transfer may also be defined by the free enthalpy change in the reaction of the hydrated redox particles with the standard gaseous electron $e_{(STD)}$ as shown in Eqn. 4–20:

$$OX_{aq} + e_{(STD)} \rightarrow RED_{aq}, \qquad \Delta G = \alpha_{e(REDOX, S)} = -eE_{eq} . \qquad (4\text{-}20)$$

The electrode potential, E_{eq}, in the electron transfer equilibrium *does not depend* on the nature of the electrode. However, the potential difference across the interface $\Delta\phi_{M/S, eq} = \mu_{e(M)} - \mu_{e(REDOX, S)}$, that is determined by the electron transfer equilibrium ($\bar{\mu}_{e(M)} = \bar{\mu}_{e(REDOX, S)}$), *does depend* upon the nature of electrodes involved, because the chemical potential $\mu_{e(M)}$ of electrons in the electrode differs among various electrode materials.

The chemical potential $\mu_{e(REDOX, S)}$ of *redox electrons* ($OX_{aq} + e_{(REDOX, S)} = RED_{aq}$) in the state of redox equilibrium is given by $\mu_{e(REDOX, S)} = \mu_{RED(S)} - \mu_{OX(S)}$. Hence, we obtain from Eqn. 4–19 the expression in Eqn. 4–21:

$$E_{eq} = \frac{\mu_{OX(S)} - \mu_{RED(S)}}{e} + \chi_{S/V} = \frac{\mu^0_{OX(S)} - \mu^0_{RED(S)}}{e} + \chi_{S/V} + \frac{kT}{e} \ln \frac{m_{OX}}{m_{RED}}$$

$$= E_0 + \frac{kT}{e} \ln \frac{m_{OX}}{m_{RED}} , \qquad (4\text{-}21)$$

where E_0 is the standard redox electrode potential, $\mu^0_{OX(S)}$ and $\mu^0_{RED(S)}$ are the standard chemical potentials of the oxidant and the reductant particles at unit concentration, respectively. Eqn. 4–21 shows that the redox electrode potential in the electron transfer equilibrium is a function of the ratio in concentration, m_{OX} / m_{RED}, of the oxidant to the reductant particles in agreement with what is called the Nernst equation for the equilibrium electrode potential of redox reactions in classical electrochemical thermodynamics.

4.4.2 Electrode potential in ion transfer equilibrium

For a metal electrode at which the metal ion transfer reaction $M^{z+}_{(S)} = M^{z+}_{(M)}$ is in equilibrium, as shown in Fig. 4–19, the metal *ion level* $\alpha_{M^{z+}(M/S/V)}$ in the electrode equals the hydrated metal ion level $\alpha_{M^{z+}(S/V)}$ in the aqueous solution; and the energy for the metal ion transfer across the electrode/solution interface equals zero ($\alpha_{M^{z+}(M/S)} = 0$). As shown in Fig. 4–20, therefore, we obtain Eqn. 4–22:

$$\alpha_{M^{z+}(M/S/V)} = \alpha_{M^{z+}(S/V)} = \mu_{M^{z+}(S)} + z\, e\, \chi_{S/V} . \qquad (4\text{-}22)$$

Further, from the ion transfer equilibrium ($\bar{\mu}_{M^{z+}(M)} = \bar{\mu}_{M^{z+}(S)}$) we obtain Eqn. 4–23:

$$\Delta\phi_{M/S, eq} = \frac{\mu_{M^{z+}(S)} - \mu_{M^{z+}(M)}}{z\, e} . \qquad (4\text{-}23)$$

Hence, the electrode potential, E_{eq}, in the equilibrium of metal ion transfer is derived from Eqn. 4–4 and 4–23 to obtain Eqn. 4–24:

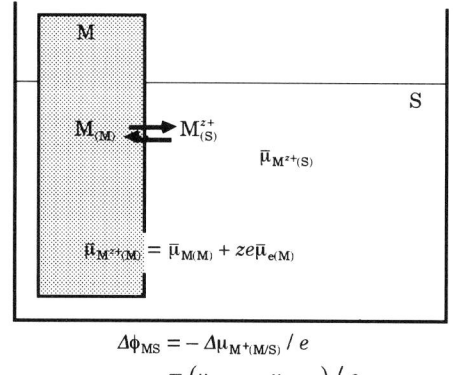

$$\Delta\phi_{MS} = -\Delta\mu_{M^+(M/S)}/e$$
$$= (\mu_{M^+(M)} - \mu_{M^+(S)})/e$$

Fig. 4–19. Ionic electrodes in equilibrium of metal ion transfer: M^{z+} = metal ion to transfer; $\bar{\mu}_{M^{z+}}$ = electrochemical potential of metal ions.

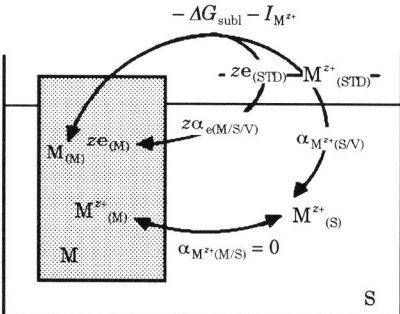

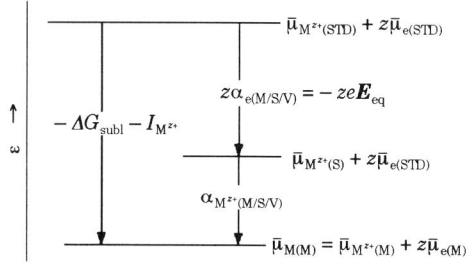

Fig. 4–20. Reaction cycle and energy levels of metal ions and electrons in the equilibrium of metal ion transfer: E_{eq} = equilibrium electrode potential.

$$E_{eq} = -\frac{\alpha_{e(M/S/V)}}{e} = -\frac{\mu_{M(M)}}{ze} + \frac{\mu_{M^{z+}(S)}}{ze} + \chi_{S/V} = -\frac{\mu_{M(M)}}{ze} + \frac{\alpha_{M^{z+}(S/V)}}{ze}, \qquad (4\text{–}24)$$

where $\alpha_{M^{z+}(S/V)}$ is the energy level of hydrated metal ions in terms of the real potential of the ions in aqueous solution ($\alpha_{M^{z+}(S/V)} = \mu_{M^{z+}(S)} + z\,e\,\chi_{S/V}$).

It, thus, follows that the electrode potential in the equilibrium of metal ion transfer is given by the free enthalpy for the formation of a solid metal from both the hydrated metal ion and the standard gaseous electron as shown in Eqn. 4–25:

$$M^{z+}_{(S)} + z\,e_{(STD)} \rightarrow M_{(solid)}, \qquad \Delta G = \alpha_{e(M^{z+}/M)} = -z\,e\,\boldsymbol{E}_{eq}, \qquad (4-25)$$

where $\alpha_{e(M^{z+}/M)}$ is the electron level *equivalent* to the metal ion level in the ion transfer equilibrium. The electron level of $\alpha_{e(M^{z+}/M)}$ may be called the *equivalent* Fermi level to or *hypothetical* Fermi level of the metal ion transfer equilibrium; i.e. the Fermi level of hypothetical electrons equivalent to the metal ion level in the ion transfer equilibrium.

The real potential $\alpha_{M^{z+}(S/V)}$ of hydrated metal ions given in Eqn. 4–24 depends upon the concentration of metal ions $m_{M^{z+}(S)}$: $\alpha_{M^{z+}(S/V)} = \alpha^0{}_{M^{z+}(S/V)} + k\,T \ln m_{M^{z+}(S)}$; the electrode potential is, thus, a function of the concentration of hydrated metal ions as shown in Eqn. 4–26:

$$\boldsymbol{E}_{eq} = \frac{-\mu_{M(M)} + \alpha^0{}_{M^{z+}(S/V)} + k\,T \ln m_{M^{z+}(S)}}{z\,e} = \boldsymbol{E}_0 + \frac{k\,T}{z\,e}\ln m_{M^{z+}(S)}, \qquad (4-26)$$

where $\alpha^0{}_{M^{z+}(S/V)}$ is the standard real potential of hydrated metal ions, and $\boldsymbol{E}_0$ is the standard equilibrium electrode potential of the metal ion transfer. Eqn. 4–26 corresponds to the Nernst equation of the equilibrium electrode potential for the metal ion transfer.

4.4.3 Potential of film-covered ionic electrodes in equilibrium

We consider a silver electrode covered with a silver chloride film in chloride solution. As shown in Fig. 4–21, the electron level of the silver–silver chloride electrode in the ion transfer equilibrium is expressed by the real potential $\alpha_{e(Ag/AgCl/S/V)}$ of electrons in the silver part of the electrode as shown in Eqn. 4–27:

$$\alpha_{e(Ag/AgCl/S/V)} = \mu_{e(Ag)} - e\,\Delta\phi_{Ag/AgCl} - e\,\Delta\phi_{AgCl/S} - e\,\chi_{S/V}. \qquad (4-27)$$

The potential difference, $\Delta\phi_{Ag/AgCl}$, between the silver metal and the film of silver chloride is determined by the equilibrium of silver ion transfer; the potential difference, $\Delta\phi_{AgCl/S}$, between the silver chloride film and the chloride solution is determined by the equilibrium of chloride ion transfer as shown in Eqn. 4–28:

$$e\,\Delta\phi_{Ag/AgCl} = \mu_{Ag^+(AgCl)} - \mu_{Ag^+(Ag)}, \qquad e\,\Delta\phi_{AgCl/S} = \mu_{Cl^-(AgCl)} - \mu_{Cl^-(S)}. \qquad (4-28)$$

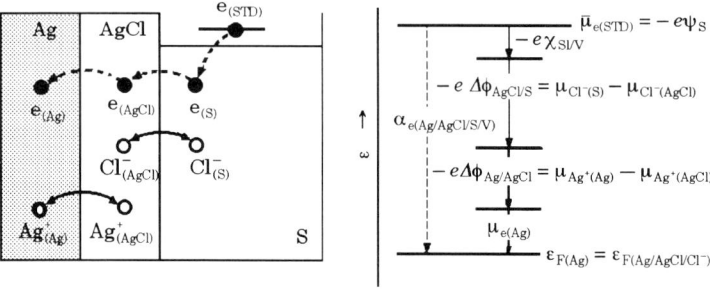

Fig. 4–21. Electron energy levels of the ionic electrode of silver-silver chloride in the ion transfer equilibrium: $\varepsilon_{F(Ag)}$ = Fermi level of electrons in the silver part of electrode; $\varepsilon_{F(Ag/AgCl/Cl^-)}$ = equivalent Fermi level to the transfer equilibrium of silver ions and chloride ions in the silver–silver chloride electrode.

From Eqns. 4–27 and 4–28, the equilibrium electrode potential, $\boldsymbol{E}_{eq}$, is obtained for the transfer of silver ions and chloride ions at the silver-silver chloride electrode as shown in Eqn. 4–29:

$$\boldsymbol{E}_{eq} = \frac{-\mu_{Ag(Ag)} + \mu_{AgCl(AgCl)} - \mu_{Cl^-(S)}}{e} = \boldsymbol{E}_0 - \frac{kT}{e} \ln m_{Cl^-(S)} \;, \tag{4-29}$$

which corresponds to the free enthalpy for the reaction of solid silver chloride with gaseous electrons to form silver metal and hydrated chloride ions as shown in Eqn. 4–30:

$$AgCl + e_{(STD)} \rightarrow Ag + Cl^-_{aq} \;, \qquad \Delta G = \alpha_{e(Cl^-/AgCl/Ag)} = -e\,\boldsymbol{E}_{eq} \;. \tag{4-30}$$

4.4.4 Potential of gas electrodes in equilibrium

Gas electrodes, such as the hydrogen electrode ($2\,H^+_{aq} + 2\,e_{(M)} = H_{2\,(gas)}$) and the oxygen electrode ($O_{2\,(gas)} + 4\,H^+_{aq} + 2\,e_{(M)} = 2\,H_2O_{aq}$), may be regarded as electronic electrodes at which the transfer reaction of redox electrons involving gaseous particles occurs. The electrode potential of gaseous electrodes is represented by the electron level $\alpha_{e(M/S/V)}$ (the real potential of electrons) in the electrode which, in the state of reaction equilibrium, equals the redox electron level $\alpha_{e(M/S/V)}$ (the real potential of redox electrons) of the gaseous electrode reaction.

Gas electrodes may also be regarded as electronic-and-ionic electrodes covered with a gaseous film in which the electron transfer occurs between the electrode and the gaseous film ($2\,H^+_{gas} + 2\,e_{(M)} \rightleftarrows H_{2(gas)}$); and the ion transfer occurs between the gaseous film and the electrolyte solution ($2\,H^+_{gas} \rightleftarrows 2\,H^+_{aq}$) as shown in Fig. 4–22. For the hydrogen electrode, the interfacial potential difference $\Delta\phi_{M/H_2}$ between the electrode metal and the hydrogen gas film is determined by the electron transfer equilibrium; and the interfacial potential difference $\Delta\phi_{H_2/S}$ between the hydrogen gas film and the aqueous solution is determined by the proton transfer equilibrium; both being, respectively, given by Eqn. 4–31:

$$e\,\Delta\phi_{M/H_2} = \mu_{e(M)} - \mu_{e(H^+/H_2,\,gas)}\,,\quad e\,\Delta\phi_{H_2/S} = \mu_{H^+(aq)} - \mu_{H^+(gas)}\,.\tag{4-31}$$

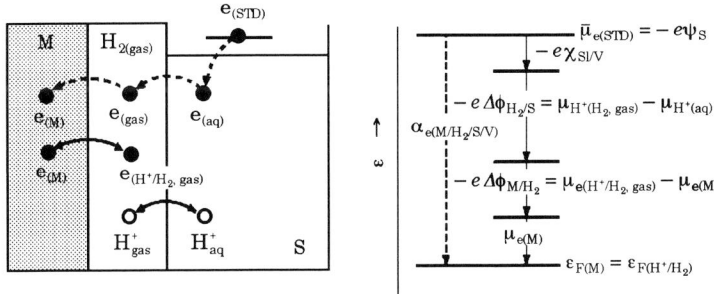

Fig. 4–22. Electron energy levels of the hydrogen electrode in the electron-and-ion transfer equilibrium: $H_{2\,(gas)}$ = gaseous hydrogen molecule on the electrode; $e_{(H^+/H_2,\,gas)}$ = gaseous redox electrons in equilibrium with the hydrogen reaction, $H_{2(gas)} = H^+_{gas} + 2\,e_{(H^+/H_2,\,gas)}$; H^+_{gas} = gaseous proton; H^+_{aq} = hydrated proton, $\varepsilon_{F(H^+/H_2)}$ = Fermi level of redox electrons in equilibrium with the hydrogen reaction (hydrated protons and redox electrons to form gaseous hydrogen molecules).

The chemical potential $\mu_{e(H^+/H_2,\,gas)}$ of the redox electron $e_{(H^+/H_2,\,gas)}$ in equilibrium with the hydrogen redox reaction ($2\,H^+_{gas} + 2\,e_{(H^+/H_2,\,gas)} = H_{2(gas)}$) is given by $\mu_{e(H^+/H_2,\,gas)} = (1/2)\,\mu_{H_2(gas)} - \mu_{H^+(gas)}$. The equilibrium electrode potential E_{eq} of the hydrogen redox reaction is, then, given by Eqn. 4–32:

$$E_{eq} = -\frac{\mu_{e(M)}}{e} + \Delta\phi_{M/G} + \Delta\phi_{G/S} + \chi_{S/V} = \frac{-\dfrac{\mu_{H_2(gas)}}{2} + \mu_{H^+(aq)}}{e} + \chi_{S/V}$$

$$= E_0 + \frac{k\,T}{e}\,\ln\frac{m_{H^+(aq)}}{\sqrt{p_{H_2(gas)}}}\,,\tag{4-32}$$

where $m_{\text{H}^+(\text{aq})}$ is the concentration (or the activity) of hydrated protons in aqueous solution, $p_{\text{H}_2(\text{gas})}$ is the pressure (or the fugacity) of gaseous hydrogen molecules, and E_0 is the standard equilibrium potential; $\mu_{\text{H}^+(\text{aq})}$ and $\mu_{\text{H}_2(\text{gas})}$ are represented by $\mu_{\text{H}^+(\text{aq})} = \mu^0_{\text{H}^+(\text{aq})} + kT \ln m_{\text{H}^+(\text{aq})}$ and $\mu_{\text{H}_2(\text{gas})} = \mu^0_{\text{H}_2(\text{gas})} + kT \ln p_{\text{H}_2(\text{gas})}$, respectively. This equilibrium electrode potential, E_{eq}, corresponds to the free enthalpy for the formation of gaseous hydrogen molecules from both hydrated protons and gaseous electrons in the standard state as shown in Eqn. 4–33:.

$$\text{H}^+_{\text{aq}} + e_{(\text{STD})} \rightarrow \tfrac{1}{2} \text{H}_{2(\text{gas})}, \qquad \Delta G = \alpha_{e(\text{H}^+_{\text{aq}}/\text{H}_2(\text{gas}))} = -e E_{\text{eq}}. \qquad (4\text{--}33)$$

As described in Sec. 2.11, the electron level in the normal hydrogen electrode (gaseous hydrogen molecules at unit fugacity and hydrated protons at unit activity) is -4.5 eV (or -4.44 eV in the IUPAC report [Trasatti, 1986]). Hence, the equilibrium potential of the normal hydrogen electrode E_{NHE} (= E_0 in Eqn. 4–32) is given by Eqn. 4–34:

$$E_{\text{NHE}} = 4.5 \text{ or } 4.44 \text{ V}. \qquad (4\text{--}34)$$

4.5 Measurement of Electrode Potentials

To measure the electrode potential of a test electrodes, M, we usually use an electrochemical cell consisting of test electrode M and reference electrode M^0 both of which are connected by a metal lead of A and A^0 of the same material to a potentiometer outside the cell as shown in Fig. 4–23. The difference of electrode potential, E, measured between the test electrode and the reference electrode, conventionally called the electromotive force, equals the difference in the Fermi level of electrons between the two electrodes: $E = -(\varepsilon_{\text{F(M)}} - \varepsilon_{\text{F(M}^0)})/e$.

In the cell used for measuring electrode potential, in which the two electrodes are immersed in a single phase of electrolyte solution, the outer potential, ψ_S, of the test electrode-solution should equal the outer potential, ψ^0_S, of the reference electrode-solution as shown in Fig. 4–24. Therefore, the difference in the Fermi level of electrons, $\varepsilon_{\text{F(M)}} - \varepsilon_{\text{F(M}^0)}$, between the test electrode M and the reference electrode M^0 is represented by the difference in the real potential of electrons, $\alpha_{e(\text{M/S/V})} - \alpha_{e(\text{M}^0/\text{S/V})}$, and hence by the difference in the electrode potential (absolute electrode potential), $\Delta E = E - E^0$, between the two electrodes. This difference also equals the difference in the work function, $\Phi_{\text{M}^0/\text{S/V}} - \Phi_{\text{M/S/V}}$, between the two electrodes. Thus, the potential E of the test electrode relative to the reference electrode is represented by the difference in the electrode potential (absolute electrode potential) between the two electrodes as indicated in Eqn. 4–35:.

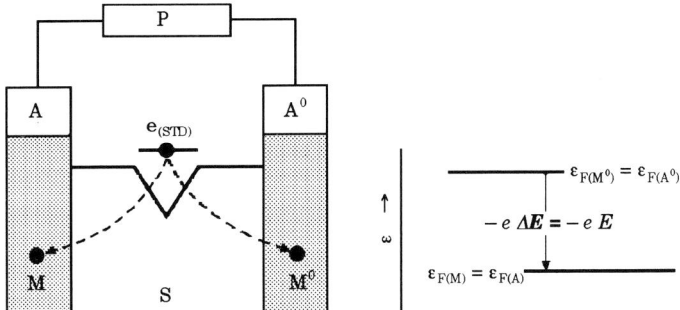

Fig. 4–23. Measurement of the relative electrode potential by a potentiometer: M = test electrode; M⁰ = reference electrode; P = potentiometer; $\varepsilon_{F(M)}$ = Fermi level of electrons in electrode M; $\varepsilon_{F(A)}$ = Fermi level of electrons in terminal A; E = relative electrode potential.

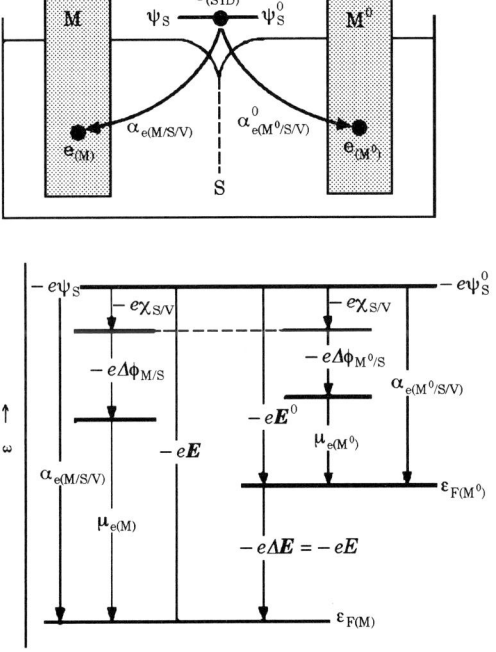

Fig. 4–24. Electron energy levels for the electrode potential relative to a reference electrode: E = electrode potential (absolute); E = relative electrode potential; ψ_S = outer potential of the electrolyte solution of test electrode; ψ_S^0 = outer potential of the electrolyte solution of reference electrode; $\psi_S = \psi_S^0$ in the potential measuring cell.

$$\varepsilon_{F(M)} - \varepsilon_{F(M^0)} = \alpha_{e(M/S/V)} - \alpha_{e(M^0/S/V)} = \Phi_{M^0/S/V} - \Phi_{M/S/V}$$

$$= -e(\boldsymbol{E} - \boldsymbol{E}^0) = -e\,\Delta \boldsymbol{E} = -e\,E\ . \tag{4-35}$$

The relative electrode potential E_{NHE} referred to the normal (or standard) hydrogen electrode (NHE) is used in general as a conventional scale of the electrode potential in electrochemistry. Since the electrode potential of the normal hydrogen electrode is 4.5 or 4.44 V, we obtain the relationship between the relative electrode potential, E_{NHE}, and the absolute electrode potential, $\boldsymbol{E}$, as shown in Eqn. 4–36:

$$\boldsymbol{E} = E_{\text{NHE}} + 4.5\ \text{eV}\ . \tag{4-36}$$

Fig. 4–25 compares the relative electrode potential, E_{NHE}, both with the (absolute) electrode potential, $\boldsymbol{E}$, and with the real potential, $\alpha_{e(M/S/V)}$, of electrons in the electrodes.

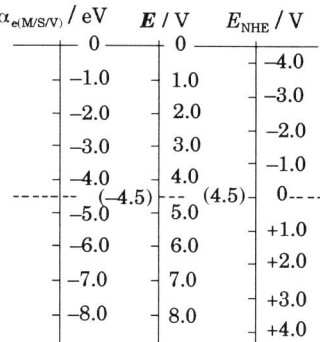

Fig. 4–25. Comparison between the real potential $\alpha_{e(M/S/V)}$ of electrons in an electrode, the absolute electrode potential $\boldsymbol{E}$, and the relative electrode potential E_{NHE} referred to the normal hydrogen electrode NHE: $\alpha_{e(M/S/V)} = -e\,\boldsymbol{E}$.

4.6 Potential of the Emersed Electrode

In recent investigations, it appears that the interfacial potential difference $\Delta\phi_{M/S}$ between the metal electrode and the aqueous solution somehow survives after the electrode is taken out of aqueous solution and into ultra high vacuum or an inactive gas phase [Wagner, 1993]. This circumstance is referred to as "emersion". As shown in Fig. 4–26, the electrode potential $\boldsymbol{E}_{\text{em}}$ of the *emersed* electrode is represented in the same way as Eqn. 4–16 by the real potential $\alpha_{e(M/S_{em}/V)}$ of electrons in the emersed electrode and, hence, by the negative work function $-\Phi_{M/S_{em}/V}$ $(= \alpha_{e(M/S_{em}/V)})$ as shown in Eqn. 4–37:

$$E_{em} = -\frac{\alpha_{e(M/S_{em}/V)}}{e} = -\frac{\mu_{e(M)}}{e} + \Delta\phi_{M/S_{em}} + \chi_{S_{em}/V} = \frac{\Phi_{M/S_{em}/V}}{e}, \qquad (4-37)$$

where S_{em} denotes a thin layer of aqueous solution adsorbed on the surface of the emersed electrode.

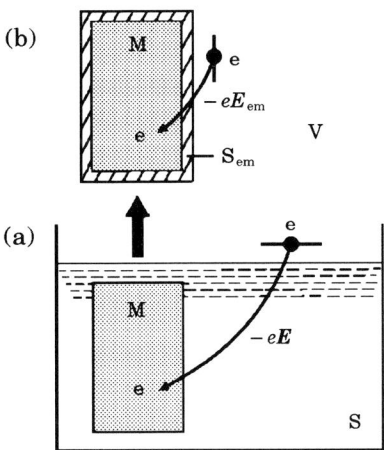

Fig. 4–26. (a) Electrode potential E of the "immersed" electrode in aqueous solution and (b) electrode potential E_{em} of the "emersed" electrode removed from aqueous solution: M = metal electrode; S = aqueous bulk solution; S_{em} = aqueous solution adsorbed on the emersed electrode; V = vacuum or gas phase.

4.6.1 Potential of emersed electrodes in vacuum

The work function of electrodes emersed from aqueous solutions in ultra high vacuum (UHV) can be measured by means of the ultra-violent photo-electron emission spectroscopy (UPS) [Kötz-Neff-Müller, 1986]. Fig. 4–27 shows the work function $\Phi_{M/S_{ex}/V}$ measured by UPS of the emersed metal electrodes of gold, silver and platinum in UHV as a function of the relative electrode potential, E, at which the electrodes have been maintained in perchloric acid solution before emersion. Obviously, the work function of emersed metal electrodes relates linearly to the electrode potential before emersion, irrespective of the nature of the type of metal of the electrode. It follows that there is a definite one-to-one relationship between the electrode potential E of metal electrodes in aqueous solutions and the electrode potential E_{em} of the metal electrodes emersed from aqueous solution in UHV as shown in Eqn. 4–38:

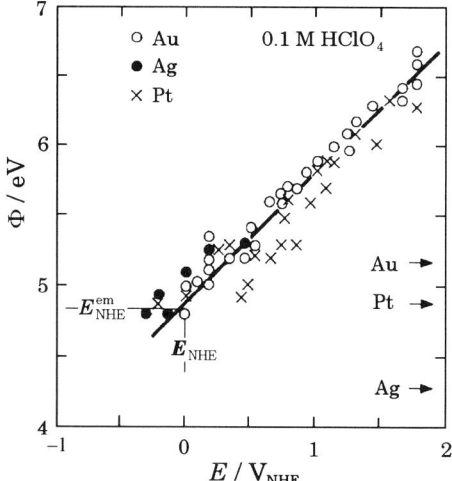

Fig. 4-27. Work function Φ of the "emersed electrodes" of gold, silver and platinum in ultra high vacuum (UHV) as a function of electrode potential E at which the electrodes have been maintained in 0.1 M perchloric acid solution before "emersion": E_{NHE} = normal hydrogen electrode potential = 4.5 or 4.44 V; E_{NHE}^{em} = emersed normal hydrogen electrode potential in UHV = 4.85 V; arrow = work function of the free clean surfaces of gold, platinum and silver in vacuum. [From Kötz-Neff-Müller, 1986.]

$$E_{em} = \frac{\Phi_{M/S_{em}/V}}{e} = E + \delta_{UHV} \, , \tag{4-38}$$

where δ_{UHV} is a shift in the electrode potential caused by the transfer (i.e. emersion) of metal electrodes from aqueous solution into UHV.

The results shown in Fig. 4-27 indicate that the potential E_{NHE}^{em} of the emersed normal hydrogen electrode is 4.85 V, while the normal hydrogen electrode potential E_{NHE} in aqueous solution is 4.5 (or 4.44) V. The difference between the two electrode potentials is thus $\delta_{UHV} = E_{NHE}^{em} - E_{NHE} = 0.4$ V; this difference may result from a change in the double layer structure caused by the emersion. The difference in electrode potential has been interpreted as the sum total of a change of -0.3 V due to the reorganization of water molecules in the interfacial aqueous solution layer and a change of $+0.7$ V due to the desorption of water molecules, at least those free from hydrated ions, after the emersion from solution into UHV [Kötz-Neff-Müller, 1986].

4.6.2 Potential of emersed electrodes in inactive gas

The relative work function and the relative electrode potential of electrodes in aqueous solutions and in inactive gases can be measured by a vibrating capacitor technique called Kelvin's method [Samec-Johnson-Doblhofer, 1992]. The Kelvin method estimates the difference of work function between the test electrode and the Kelvin probe (KP) by measuring the applied voltage V at which the difference in the outer potential $\psi_S - \psi_{KP}$ between the test electrode and the Kelvin probe becomes zero ($V = \psi_S - \psi_{KP}$) as shown in Fig. 4–28.

Since the difference in the outer potential $\psi_S - \psi_{KP}$ equals the difference in the real potential $-(\alpha_{e(M/S/V)} - \alpha_{e(KP)})/e$ of electrons and hence the difference in the work function $(\Phi_{M/S/V} - \Phi_{KP})/e$, Eqn. 4–39 can be obtained and represents the difference between the electrode potential E of an immersed electrode (M/S/V) and the potential E_{KP} of the Kelvin probe (Fig. 4–28a):

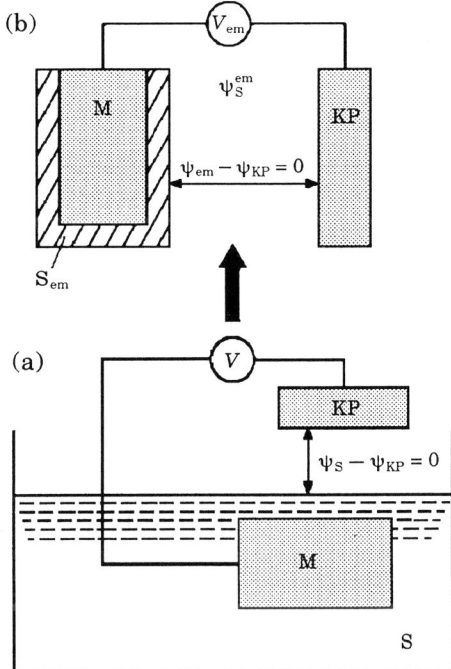

Fig. 4–28. Schematic layout of Kelvin's vibrating capacitor method to measure the relative electrode potential of (a) electrodes immersed in aqueous solution and of (b) electrodes emersed from solution: KP = Kelvin's probe; ψ_S = outer potential of aqueous solution; ψ_{KP} = outer potential of Kelvin's probe; V and V_{em} = applied voltages to cancel the difference of the outer potential.

$$E - E_{KP} = -\frac{\alpha_{e(M/S/V)} - \alpha_{e(KP)}}{e} = \frac{\Phi_{M/S/V} - \Phi_{KP}}{e} = \psi_{KP} - \psi_S = V. \quad (4\text{–}39)$$

This equation applies also to the electrodes emersed from aqueous solution into inactive gas (Fig. 4–28b) as shown in Eqn. 4–40, which gives the relationship between the electrode potential E_{em} of an emersed electrode and the potential E_{KP} of the Kelvin probe:

$$E_{em} - E_{KP} = -\frac{\alpha_{e(M/S_{em}/V)} - \alpha_{e(KP)}}{e} = \frac{\Phi^{em}_{M/S_{em}/V} - \Phi_{KP}}{e} = \psi_{KP} - \psi^{em}_S = V_{em}. \quad (4\text{–}40)$$

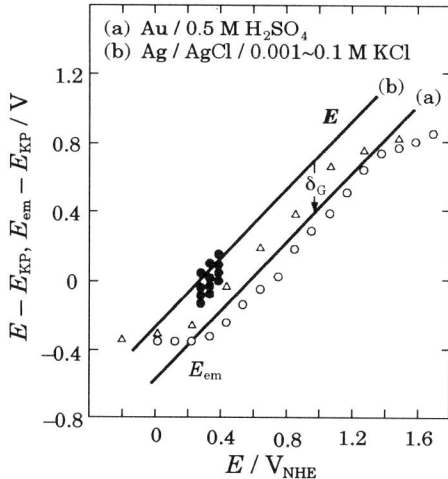

Fig. 4–29. Relationship (a) between the relative electrode potential E (referred to NHE) measured by the conventional method of a gold electrode immersed in a sulfuric acid solution and the relative electrode potential $E_{em} - E_{KP}$ measured by Kelvin's method of the same gold electrodes emersed in nitrogen gas; and relationship (b) between the relative electrode potential E measured by the conventional method and the relative electrode potential $E - E_{KP}$ measured by Kelvin's method of the equilibrium silver–silver chloride electrode in chloride solutions: gold electrode = Au / 0.05 M H_2SO_4; silver–silver chloride electrode = Ag/AgCl/0.001 ~ 0.1 M KCl; E_{KP} = electrode potential of the Kelvin probe (work function), δ_G = change in the electrode potential due to transferring the electrode from solution into nitrogen gas, solid circle = electrode potential of the equilibrium silver–silver chloride electrode, open circle = in ascending electrode potential, open triangle = in descending electrode potential. [From Samec-Johnson-Doblhofer, 1992.]

Fig. 4–29 illustrates the relationship for a gold electrode (a) between the electrode potential E measured in aqueous solution by the conventional method before the emersion and the electrode potential $E_{em} - E_{KP}$ measured in wet nitrogen gas by the Kelvin method after the emersion from aqueous sulfuric acid solution. Fig. 4–29 also shows the relationship for an equilibrium silver–silver chloride electrode (b) between the relative electrode potential E measured by the conventional method and the relative electrode potential $E - E_{KP}$ measured by the Kelvin method in chloride solutions of three different chloride concentrations; this procedure calibrates the Kelvin method versus the conventional method.

A linear relationship (straight line with unit slope) is observed in Fig. 4–29 between the electrode potential E measured by the conventional method and the electrode potential $E - E_{KP}$ measured by the Kelvin method for the equilibrium silver–silver chloride electrode: a linear relation ship is also observed between E and $E_{em} - E_{KP}$ for the gold electrode in sulfuric acid. The difference between the two straight lines (a) and (b) in Fig. 4–29 is $\delta_G = E_{em} - E = -0.32$ V, which represents the difference between the potential $E - E_{KP}$ of electrodes in aqueous solution and the potential $E_{em} - E_{KP}$ of electrodes in wet nitrogen gas. This potential difference $\delta_G = E_{em} - E$ may result from the reorganization of the interfacial dipoles of water molecules that occurs in transferring electrodes from the aqueous solution into the wet gaseous nitrogen.

As illustrated in Fig. 4–29, δ_G is constant and independent of the electrode potential at which the electrode has been maintained in aqueous solution before the emersion suggesting that the interfacial dipole of the water molecule exerts almost no effects on the change of electrode potential in aqueous solution. The potential difference δ_G, however, depends upon the nature of hydrated anions in the solution; δ_G is smaller as the affinity for anion adsorption on the electrode interface is greater. For instance, the potential difference has been estimated to be $\delta_G = -0.07$ V in an aqueous solution containing iodide anions which are strongly adsorbed on a gold electrode; whereas, it is $\delta_G \doteq -0.3$ V in sulfuric and perchloric acid solutions in which the anions are only weakly adsorbed on the gold electrode [Samec-Johnson-Doblhofer, 1992]. The fact that δ_G changes with different anions suggests that the adsorption of anions influences the electric double layer on the electrode interface.

References

[Kötz-Neff-Müller, 1986]: E. R. Kötz, H. Neff and K. Müller, *J. Electroanal. Chem.*, **215**, 331(1986).
[Samec-Johnson-Doblhofer, 1992]: Z. Samec, B. W. Johnson and K. Doblhofer, *Surface Science*, **264**, 440(1992).
[Sato, 1993]: N. Sato, *Electrode Chemistry*, **Vol. 1**, p. 80-93, Japan Technical Information Service, Tokyo, (1993).
[Sato, 1995]: N. Sato, *Russian J. Electrochem.*, (English Version), **31**, 906(1995); *Materials Science Forum*, **185-188**, 397(1995).
[Trasatti, 1986]: S. Trasatti, *J. Electroanal. Chem.*, **139**, 1(1986): *Pure and Appl. Chem.*, **58**, 956(1986).
[Trasatti, 1990]: S. Trasatti, *Electrochimica Acta*, **35**, 269(1990).
[Wagner, 1993]: F. T. Wagner, *Structure of Electrified Interfaces*, (Edited by J. Lipkowski and P. N. Ross), p. 309, VCH publishers, Inc., New York, (1993).

CHAPTER 5

ELECTRIC DOUBLE LAYER AT ELECTRODE INTERFACES

5.1 Solid Surface and Adsorption

5.1.1 Clean surface of solids

The energy of surface atoms is known to differ from the energy of interior atoms in solids. The lattice structure of surface atoms, therefore, may not be the same as that of interior atoms. Usually, *surface relaxation* takes place on the clean solid surface produced by cleavage. As a result of surface relaxation, the distance between the first atom layer and the second atom layer in the surface lattice becomes different from the atom–atom distance in the interior lattice of the solid. For metals, in general, the planes of close-packed surface lattice (e.g. (100) plane of the face centered cubic lattice) undergo either almost no surface relaxation or slight expansion of the layer–layer distance; whereas, surface lattice planes of low atomic density sustain surface relaxation leading to a shrinkage of the layer–layer distance, the magnitude of which may reach 10 to 15% [Van Hove, 1993]. For ionic crystals the surface relaxation of cationic and anionic surface planes also differs [Blakely, 1973]. The surface relaxation decays with increasing depth.

The clean surfaces of solids sustain not only surface relaxation but also *surface reconstruction* in which the displacement of surface atoms produces a two-dimensional superlattice overlapped with, but different from, the interior lattice structure. While the lattice planes in crystals are conventionally expressed in terms of Miller indices (e.g. (100) and (110) for low index planes in the face centered cubic lattice), but for the surfaces of solid crystals, we use an index of the form (1×1) to describe a two-dimensional surface lattice which is exactly the same as the interior lattice. An index (5×20) is used to express a surface plane in which a surface atom exactly overlaps an interior lattice atom at every five atomic distances in the x direction and at twenty atomic distances in the y direction.

The surface lattice plane of Pt (100)-(1×1), created by cleavage along the close-packed cubic lattice plane (100) of platinum crystals, transforms into the

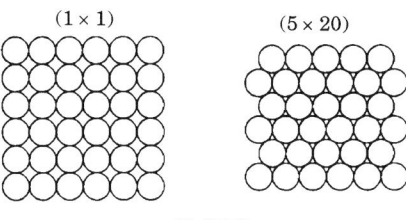

Fig. 5–1. Two-dimensional atomic structure on the (100) plane of platinum crystals: (1×1) = cubic close-packed surface plane identical with the (100) plane; (5×20) = hexagonal close-packed surface plane reconstructed from the original (100) plane. [From Kolb, 1993.]

close-packed hexagonal lattice plane of Pt (100)-(5×20) that is more stable than Pt (100)-(1×1) in vacuum as shown in Fig. 5–1. Such a surface reconstruction may be regarded as a two dimensional phase transformation of the surface lattice as indicated in Eqn. 5–1:

$$\text{Pt (100)-(1×1)} \rightleftarrows \text{Pt (100)-(5×20)} . \tag{5–1}$$

As the reconstructed surface (5×20) of platinum crystals contains as many atoms as 1.2 times the original surface (1×1) atoms, the transformation of surface lattice in the reverse direction from (5×20) to (1×1) forces the excess surface atoms to cohere in a striped pattern on the un-reconstructed (1×1) surface.

The reconstructed surface lattice on metal crystals depends on the interior lattice as well as on the nature of the metal, such as Au (100)-(5×20), Au (111)-(1×23) and Pt (110)-(1×2) [Kolb, 1993]. In general, the activation energy of surface reconstruction is relatively great (~ 1 eV) on clean metal surfaces so that the reconstruction is frequently suppressed at room temperature. Usually, the surface adsorption changes the activation energy that catalyzes or inhibits the surface reconstruction.

Besides metallic crystals, as mentioned in Sec. 2.7, covalent crystals also undergo surface reconstruction; and the dangling surface atoms reduces the number of their dangling bonds to stabilize the surface energy thereby forming a reconstructed surface lattice different from the interior lattice.

In the case of multi-component alloys and compounds, the surface composition may also change in addition to the surface relaxation and reconstruction that occur in pure solids. For instance, the first layer of the (100) plane on the surface of a nickel-aluminum alloy can enrich with aluminum of which the atomic size is larger than nickel. Such an enrichment of some constituents on the solid surface is called *surface segregation* [Van Hove, 1993]. It is also known that surface active minor impurities of oxygen, phosphorus and sulfur in metallic iron are segregated to the clean surface of iron [Nii-Yoshihara, 1980]. Further, the surface segregation of chloride ions, Cl⁻, has been reported for ionic crystals of NaCl and KCl [Tosi-Doyama, 1966]; also, the surface segregation of calcium ions, Ca^{2+}, has been reported on ionic crystals of NaCl to contain calcium impurities [Kummer-Young, 1963].

5.1.2 Adsorption

Adsorption of particles on solid surfaces may be classified into two modes: *physisorption* due to Van der Waals' force; and *chemisorption* due to the covalent, metallic, or ionic bonding between adsorbate particles and adsorbent solids. The chemisorption may also be divided into *atomic chemisorption* (adsorption in the form of atom), *ionic chemisorption* (in the form of ion), and *molecular chemisorption* (in the form of molecule). In this textbook, we deal only with adsorption in the form of simple atoms and ions.

In physisorption, such as the adsorption of inert gas atoms on metal surfaces, the affinity of adsorption is small; and the lifetime of adsorbed atoms remaining on the surface is short. The configuration of adsorbed atoms tends to the close-packed two-dimensional structure with increasing coverage almost independent of the surface lattice of the substrate metal. On the other hand, because of the strong adsorption affinity, chemisorption of atomic particles depends on the surface lattice of the adsorbent solid and affects the surface relaxation and reconstruction. In chemisorption, an adsorbate atom chooses its adsorption site preferentially at a hollow site of high coordination number in order to combine itself with as many surface atoms as possible. Therefore, as a result of chemisorption, the adsorbed atoms adjust their configuration in the adsorption plane to coincide in short or long intervals with the atoms in the first surface lattice plane of the adsorbent solid, thus producing a superlattice of adsorbed particles overlapping the adsorbent surface lattice.

The chemisorption of particles tends, in general, to suppress the relaxation and reconstruction of clean solid surfaces because the adsorption *stabilizes* the surface lattice energy by reducing the dangling surface bonds of solid substrates. In the case of metal surfaces, the chemisorption of electron-accepting particles, such as oxygen and sulfur, accelerates the *"un-reconstruction"* (the cancellation of reconstruction) of the previously reconstructed metal surface, giving rise to a surface lattice of the same structure as the interior lattice. For example, the adsorption of gaseous carbon monoxide makes the otherwise stable close-packed hexagonal superlattice of the reconstructed platinum surface, Pt (100)-(5×20), unstable and catalyzes the transformation of the reconstructed surface into the close-packed cubic superlattice of Pt (100)-(1×1) in Eqn. 5–1 as shown in Fig. 5–2.

In contrast, the adsorption of electron-donating particles, such as alkaline metal atoms on the surface (110) plane of nickel and silver, accelerates the

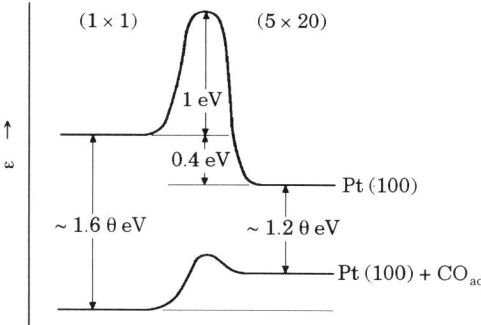

Fig. 5–2. Activation energy both for reconstruction of the surface (100) plane of platinum crystals in vacuum and for un-reconstruction of the reconstructed surface due to adsorption of CO: (1×1) ⇌ (5×20) is the surface lattice transformation (reconstruction and un-reconstruction). θ = adsorption coverage. [From Ertl, 1985.]

surface reconstruction from (1×1) to (1×2). Further, there are cases in which the chemisorption of particles produces a superlattice different from the original or reconstructed superlattice on the solid surface [Van Hove, 1993].

Coadsorption, in which two different kinds of particles are chemisorbed on the solid surface, may be classified into *cooperative adsorption* and *competitive adsorption*. Cooperative adsorption takes place with two different adsorbate particles of opposite characteristics, such as electron-donating particles and electron-accepting particles (e.g. Na and S), and the two adsorbate particles are adsorbed uniformly on the solid surface. On the other hand, competitive adsorption involves two different particles of similar characteristics, i.e. both being electron-donating or electron-accepting particles (e.g. O and S), which are adsorbed separately on the solid surface.

5.1.3 Electron level of adsorbed particles

We consider a simple case of the chemisorption of atoms on a clean surface of metals in vacuum. Fig. 5–3 shows the electron energy levels of both an isolated adsorbate atom and an adsorbent metal. Electrons in the metal are in the delocalized energy band, and their energy is represented by the Fermi level, ε_F, (negative work function, $-\Phi$). Electrons in the atom exist at localized discontinuous levels in which the *highest occupied molecular orbital level* (HOMO) corresponds to the negative ionization energy, $-I$, of the atom, and the *lowest unoccupied molecular orbital level* (LUMO) corresponds to the negative electron affinity, $-A$, of the atom.

As an isolated adsorbate particle approaches a metal surface, the eigenstate density of localized electrons in the particle increases due to its interaction with metal ions and electrons in the metal; and the electron level is *broadened* into a level band. With an increasing number of metal ions and electrons that interact with an approaching adsorbate particle, the width of the electron level band in the particle increases in proportion to $\exp(-\kappa d)$: where κ is the energy barrier for electrons in the particle (ionization energy I or electron affinity A) and d is the distance between the approaching adsorbate particle and the surface of adsorbent metal [Gurney, 1935].

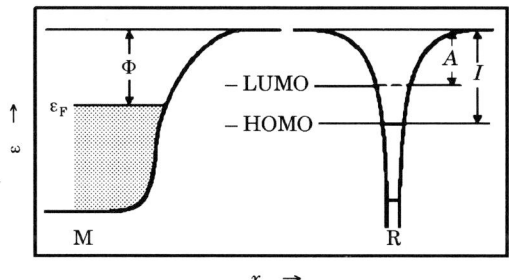

Fig. 5–3. Electron energy levels in an isolated adsorbate particle and an adsorbent solid metal: M = metal; R = isolated particle; LUMO = lowest unoccupied molecular orbital (lowest vacant electron level); HOMO = highest occupied molecular orbital (highest occupied electron level).

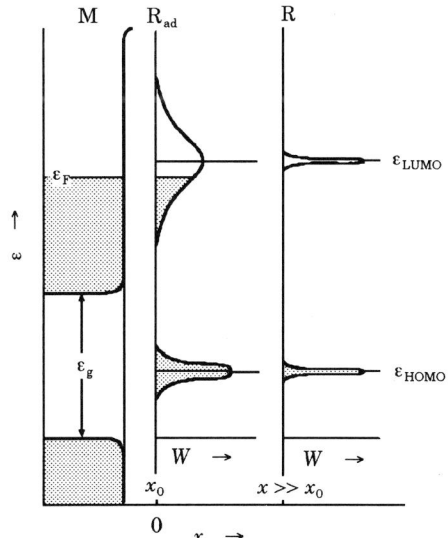

Fig. 5–4. Electron energy levels of an adsorbate particle broadened by the interaction with the adsorbent metal crystal: M = adsorbent metal; R = adsorbate atomic particle; R_{ad} = adsorbed particle; W = probability density of electron energy states; x = distance to adsorbate particle, x_0 = distance to adsorbed particle.

Fig. 5–4 illustrates the widening of the highest occupied electron level, ε_{HOMO}, and of the lowest unoccupied electron level, ε_{LUMO}, of an atomic adsorbate particle before and after its chemisorption on the adsorbent metal. In the case in which the electron level of the adsorbed atom is within the range of the conduction or valence band of the adsorbent metal, such as ε_{LUMO} shown in Fig. 5–4, the state density of metal electrons in the conduction or valence band which are interactive with the adsorbate atom is so great that a considerable widening of the electron level of the adsorbed atom occurs. On the other hand, the localized electron level of the adsorbate does not widen significantly if the electron level of the adsorbed atom is in the band gap of the adsorbent metal; this case is illustrated in Fig. 5–4 for ε_{HOMO} where the electron–electron interaction is slight.

Further, the electron level of adsorbed particles differs from that of isolated adsorbate particles in vacuum; as shown in Fig. 5–5, this electron level of the adsorbate particle shifts in the course of adsorption by a magnitude equivalent to the *adsorption energy* [Gomer-Swanson, 1963]. In the illustration of Fig. 5–5, the electron level of adsorbate particles is reduced in accordance with the potential energy curve of adsorption towards its lowest level at the plane of adsorption where the level width is broadened. In the case in which the allowed electron energy level of adsorbed particles, such as ε_{LUMO} and ε_{HOMO}, approaches the Fermi level, ε_F, of the adsorbent metal, an electron transfers between the adsorbent metal and the adsorbed particle which is then partially or fully ionized.

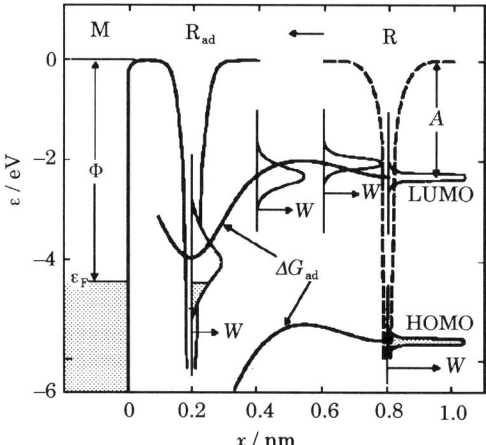

Fig. 5–5. Change in the electron level of an adsorbate particle approaching the adsorbent metal surface: ΔG_{ad} = adsorption energy curve of an adsorbate particle approaching the metal surface.

The adsorption of atoms on metals may be classified in four types [Benard, 1983] in terms of their electron levels relative to those of adsorbent metal as described in the following and illustrated in Fig. 5–6 and Fig. 5–7:

(a) *Cation adsorption*. The highest occupied electron level, ε_{HOMO}, of adsorbed atoms is higher than the Fermi level, ε_F, of the metal; hence, electrons in the adsorbed atoms are transferred into the absorbent metal to form adsorbed cations (e.g. cesium on metallic tungsten).

(b) *Metallic adsorption*: The occupied electron level, ε_{HOMO}, of adsorbed atoms is in the same level as the Fermi level ε_F of the metal, and electrons in the metal transfer freely into and out of adsorbed atoms to involve them in the metallic bonding of the adsorbent metal (e.g. lithium on metallic tungsten).

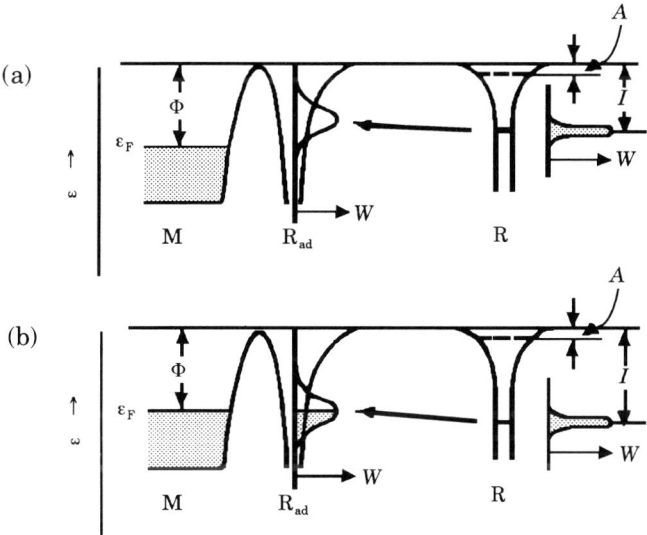

Fig. 5–6. Electron energy levels of an adsorbed particle on the adsorbent metal: (a) cationic adsorption, (b) metallic adsorption. A = electron affinity; I = ionization energy; Φ = work function. [From Benard, 1983.]

(c) *Localized covalent adsorption*. The unoccupied electron level, ε_{LUMO}, of adsorbed atoms is much higher than the Fermi level ε_F of the metal; and the occupied electron level, ε_{HOMO}, is in the band gap where no electrons, but a few surface metal atoms, interact with an adsorbed atom which is, thus, in the localized covalent bonding state (e.g. fluorine on metallic tungsten).

(d) *Anion adsorption.* The unoccupied electron level, ε_{LUMO}, of adsorbed atoms is lower than the Fermi level ε_F of the metal, thereby transferring electrons from the metal into adsorbed atoms to form adsorbed anions (e.g. oxygen on metallic tungsten).

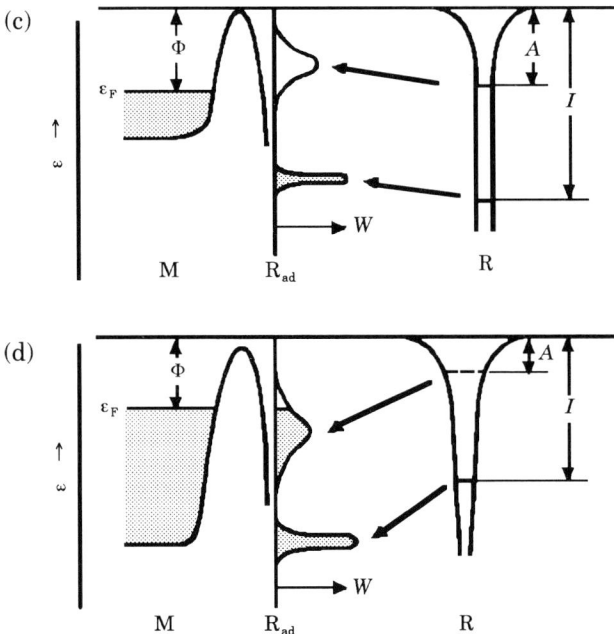

Fig. 5–7. Electron energy levels of an adsorbed particle on the adsorbent metal: (c) localized covalent adsorption, (d) anionic adsorption. [From Benard, 1983.]

Both cationic adsorption and anionic adsorption belong to what is called *ionic adsorption*. Covalent adsorption is due to the localized covalent bonding, and metallic adsorption is due to the delocalized covalent bonding. The distinction among these three modes of chemisorption, however, is not so definite that the transition from the covalent through the metallic to the ionic adsorption may not be discontinuous, but rather continuous, in the same way as the transition of the three dimensional solid compounds between the covalent, metallic, and ionic bonding.

5.2 Electric Double Layer at Solid/Aqueous Solution Interfaces

5.2.1 Electric double layer model

The difference of electrostatic potential between a solid and an aqueous solution can be explained in terms of a parallel plate condenser with a positive excess charge on one phase and a negative excess charge on the other. The interfacial charge on the solid (electronic conductor) is usually carried by mobile excess electrons and holes, while it is carried by mobile excess hydrated ions on the side of aqueous solution (ionic conductor).

An adsorbed layer of water molecules at the interface separates the hydrated ions from the solid surface. The interfacial electric double layer can be represented by a *condenser model* comprising three distinct layers: a diffuse charge layer in the ionic solution, a compact layer of adsorbed water molecules, and a diffuse charge layer in the solid as shown in Fig. 5–8. The interfacial excess charge on

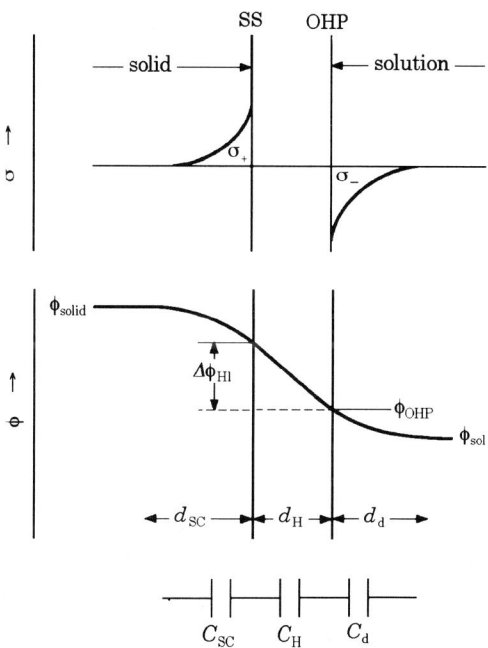

Fig. 5–8. An interfacial double layer model (triple-layer model): SS = solid surface; OHP = outer Helmholtz plane; ϕ = inner potential; σ = excess charge; d_H = distance from the solid surface to the closest approach of hydrated ions (Helmholtz layer thickness); C = electric capacity.

the aqueous solution side is distributed from the plane of the closest approach of hydrated ions towards the bulk solution, thereby forming a diffuse layer of excess ionic charge; similarly, the interfacial excess charge on the solid side is distributed from the solid surface toward the solid interior forming a diffuse layer of excess electrons or holes.

The *diffuse* layer of excess electrons and holes in solids is called the *space charge* layer; and the *diffuse* layer of excess hydrated ions in aqueous solution is simply called the diffuse layer and occasionally called the Gouy layer [Gouy, 1917]. The middle layer of adsorbed water molecules between the diffuse layer on the aqueous solution side and the space charge layer on the solid side is called the *compact* or the *inner* layer. This compact or inner layer is also called the Helmholtz layer [Helmholtz, 1879] or the Stern layer [Stern, 1924]; the plane of the closest approach of hydrated ions to the solid surface is called the outer Helmholtz plane (OHP) [Graham, 1947].

The thickness of the compact layer is nearly constant and independent of the nature of the solid, being 1.5 to 2.0 times the thickness of the mono-molecular water layer, i.e. d_H = 0.3 to 0.5 nm. On the other hand, the thicknesses of the diffuse and space charge layers depend on the concentrations of mobile charge carriers in the respective phases; the thicknesses decrease with increasing concentrations of the charge carriers. For example, the thickness of the diffuse layer is in the range of 10 to 100 nm in aqueous ionic solutions of low ion concentrations, and the thickness of the space charge layer is in the range of 100 to 1000 nm in semiconductors of low electron or hole concentrations. In concentrated ionic solutions of 1 M (kmol m^{-3}), however, the thickness of the diffuse layer, d_d, decreases to about the mono-molecular layer. Similarly, in the case of metals in which the electron state density is great ($\sim 10^{22}$ cm^{-3}V^{-1}), the space charge layer thickness, d_{SC}, is less than the thickness of a monatomic layer. Accordingly, the electric double layer at the interface between a solid metal and a concentrated ionic solution is, to a first approximation, represented by a simple parallel plate condenser consisting of the compact layer only in which the excess charge is accumulated two-dimensionally at the outer Helmholtz plane and at the metal surface.

Usually, there is a difference of electrostatic potential of the order of 1 V across the electric double layer at the interface between the metal and the aqueous solution; this difference produces an intense electric field of the order of 10^6 V cm^{-1} in a compact layer 0.3 to 0.5 nm thick. Such an intense electric field can not be realized in any dielectrics of macroscopic size, because of dielectric breakdown by the *electron avalanche*; but the intense electric field can be sustained in a layer of several atomic thickness where an electron avalanche can not occur.

In the *three layer model* shown in Fig. 5–8, the electric capacity C of an interfacial electric double layer is represented by a series connection of three capacities: the capacity of space charge layer C_{SC} on the solid side; the capacity of compact layer C_H at the interface, and the capacity of diffuse layer C_d in the aqueous solution as shown in Eqn. 5–2:

$$\frac{1}{C} = \frac{1}{C_{SC}} + \frac{1}{C_H} + \frac{1}{C_d} , \tag{5-2}$$

where the smallest capacity predominates in determining the overall capacity. Usually, the smallest capacity occurs in the compact layer (C_H) for metal electrodes and in the space charge layer (C_{SC}) for semiconductor electrodes.

5.2.2 Diffuse charge layer (Space charge layer)

The space charge layer in solids and the diffuse charge layer in aqueous solutions can both be described, in principal, using the same theoretical treatment. We consider first the ionic diffuse charge layer formed on the side of aqueous solution of metal electrodes. As shown in Fig. 5–9, the interfacial charge, σ_M, on the metal side is located on the metal surface without any space charge layer; but the charge, σ_S, on the solution side is distributed from the plane of the closest approach of hydrated ions (OHP) towards the solution interior. Both charges, σ_M and σ_S, are equal to each other in amount but opposite in sign.

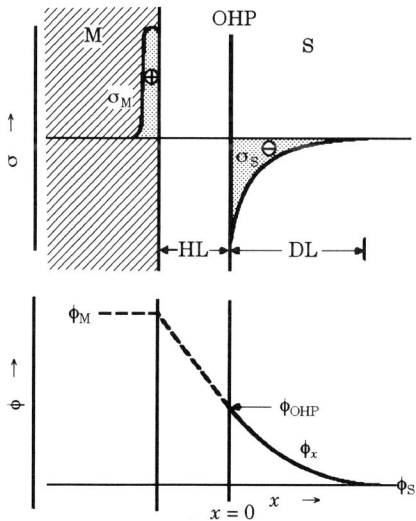

Fig. 5–9. Diffuse charge layer on the solution side of a metal electrode: M = electrode metal; S = aqueous solution; HL = compact layer (Helmholtz layer); DL = diffuse charge layer; x = distance from the outer Helmholtz plane (OHP).

The distribution of excess charge of hydrated ions in the diffuse layer can be derived by using Poisson's equation, $d^2\phi/dx^2 = -\sigma(x)/\varepsilon$, and Boltzmann's distribution equation, $c_i(x) = c_{i(S)} \exp\{-z_i e \phi(x)/kT\}$, to obtain the relationship in Eqn. 5–3 between the interfacial charge, σ_M, and the diffuse layer potential, ϕ_{OHP}:

$$\sigma_M = -\sigma_S = \pm \sqrt{2kT\varepsilon} \sqrt{\Sigma_i c_{i(S)} \left(\exp\frac{z_i e \phi_{OHP}}{kT} - 1\right)}, \qquad (5\text{–}3)$$

where ε is the permittivity (dimension F m^{-1}), z_i is the ionic valence, $c_i(x)$ is the concentration of ions at distance x from the interface, $c_{i(S)}$ is the concentration of ions in the bulk solution, and ϕ_{OHP} is the electrostatic inner potential at the outer Helmholtz plane referred to the electrostatic inner potential of the solution interior ($\phi_S = 0$) [Reeves, 1980].

It follows that, as the interfacial charge σ_M increases in positive or negative charge, the potential of the diffuse layer, ϕ_{OHP}, shifts in the positive or negative direction. The electrode potential at which the interfacial charge is zero ($\sigma_M = -\sigma_S = 0$; hence, $\phi_{OHP} = 0$) is called the *potential of zero charge*.

Differentiating Eqn. 5–3 with respect to the potential, we obtain the *differential electric capacity*, $C_d = (\partial \sigma_M / \partial \phi_{OHP})$, of the diffuse layer in the aqueous ionic solution of z-z valence as shown in Eqn. 5–4:

$$C_d = \sqrt{\frac{2 z_i^2 e^2 \varepsilon c_{i(S)}}{kT}} \cosh \frac{e \phi_{OHP}}{2kT}. \qquad (5\text{–}4)$$

To a first approximation, the capacity of the diffuse layer in Eqn. 5–4 may be represented by an inverse parabolic function of the diffuse layer potential ϕ_{OHP} with its minimum at the potential of zero charge, E_{pzc} ($\phi_{OHP} = 0$); the minimum capacity is given by Eqn. 5–5:

$$C_d = \sqrt{\frac{2 z_i^2 e^2 \varepsilon c_{i(S)}}{kT}} = \frac{\varepsilon}{L_D}, \quad \frac{1}{L_D} = \sqrt{\frac{2 z_i^2 e^2 c_{i(S)}}{\varepsilon kT}}, \qquad (5\text{–}5)$$

which corresponds to the capacity of a parallel plate condenser of thickness L_D. The length, L_D, called the *Debye length* after P. J. Debye (physicochemist, 1984–1966), is a measure of the atmospheric space of hydrated ions in aqueous solution. Fig. 5–10 illustrates the differential electric capacity of the interfacial double layer on a mercury electrode in dilute and concentrated sodium fluoride solutions. The capacity of diffuse layer is observed to be an inverse parabolic curve around the potential of zero charge in the dilute ionic solution as shown in Fig. 5–10 (a); but this same capacity of the diffuse layer is concealed in the concentrated solution as shown in Fig. 5–10 (b), where $1/C_d$ becomes much smaller than $1/C_H$.

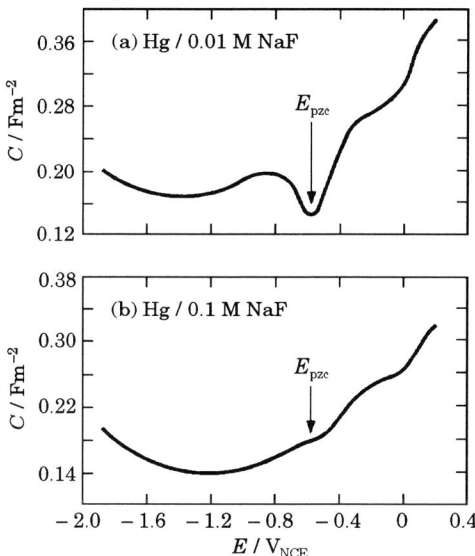

Fig. 5–10. Differential electric capacity, C, observed for mercury electrodes in dilute and concentrated solutions of sodium chloride: (a) 0.01 M NaF, (b) 0.1 M NaF. E_{pzc} =potential of zero charge; V_{NCE} = volt referred to the normal calomel electrode. [From Graham, 1954.]

The potential distribution in the diffuse layer can be derived from Eqn. 5–3 as an approximate equation given in Eqn. 5–6:

$$\phi_x = \phi_{OHP} \exp\left(\frac{-x}{L_D}\right), \tag{5-6}$$

where x is the distance from the outer Helmholtz plane. Eqn. 5–6 yields the thickness of the diffuse layer, $x = d_d$, at which the potential ϕ_x is almost zero ($\phi_x / \phi_{OHP} = 0.0001$); for 0.1 M and 0.0001 M concentrations of a 1-1 valence ionic solution $d_d \doteq 8.8$ nm and $d_d \doteq 280$ nm, respectively. In general, the potential difference of the diffuse layer, ϕ_{OHP}, is negligibly small in concentrated ionic solution; whereas, it constitutes a main part of the potential difference across the interfacial electric double layer in dilute solution [Reeves, 1980].

The concentration profile of excess ions in the diffuse layer may be derived from Eqn. 5–3 and Boltzmann's distribution equation as a function of interfacial charge, σ_M. Simple calculation gives the interfacial ionic concentration (at the OHP) to be $c_{x=0} = 1$ M for an interfacial charge $\sigma_M = 0.1$ C m^{-2} (corresponding to

ϕ_{OHP} = 0.3 V) in a dilute 0.001 M ionic solution. In other words, a shift of the electrode potential by 0.3 V from the potential of zero charge increases the interfacial ionic concentration about 1000 times more than the bulk ionic concentration in dilute solution. Such an increase in the interfacial ionic concentration is reduced with increasing ionic concentration in the bulk solution and is negligible in ionic solutions more concentrated than 1 M.

This same theoretical approach can be applied to the space charge layer formed in solid semiconductors. Instead of the concentration of ions in aqueous solution, however, the concentration of electrons or holes is used with the space charge layer in semiconductors. Then, the Debye length is given by Eqn. 5–7:

$$\frac{1}{L_D} = \sqrt{\frac{\Sigma_i e^2 n_{i(SC)}}{\varepsilon k T}}, \tag{5-7}$$

where $n_{i(SC)}$ is the concentration of electrons or holes in semiconductors. The Debye length, L_D, is in the range from 100 to 1000 nm in semiconductors of usual concentrations of impurities (10^{14} to 10^{15} cm^{-3}) and decreases to the range from 10 to 100 nm in semiconductors of high concentrations of impurities.

In the case of an interface between a metal and an aqueous solution, the space charge layer on the metal side is less than a monatomic layer; hence, the charge in the space charge layer is included in the interfacial charge of the compact layer although, as is discussed in Sec. 5.5, a charge distribution in the atomic dimension exists on the metal side of electric double layer.

5.3 The Potential of Zero Charge on Metal Electrodes

5.3.1 Classical model of the compact double layer at interfaces

Fig. 5–11 shows a simple model of the *compact double layer* on metal electrodes. The electrode interface adsorbs water molecules to form the first mono-molecular adsorption layer about 0.2 nm thick; next, the second adsorption layer is formed consisting of water molecules and hydrated ions; these two layers constitute a compact electric double layer about 0.3 to 0.5 nm thick. Since adsorbed water molecules in the compact layer are partially bound with the electrode interface, the permittivity of the compact layer becomes smaller than that of free water molecules in aqueous solution, being in the range from 5 to 6 compared with 80 of bulk water in the relative scale of dielectric constant. In general, water molecules are adsorbed as monomers on surfaces of metals on which the affinity for adsorption of water is large (e.g. d-metals); whereas, the water molecules are adsorbed as clusters in addition to monomers on surfaces of metals on which the affinity for adsorption of water is relatively small (e.g. sp-metals).

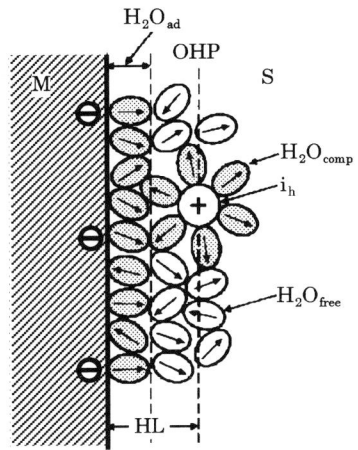

Fig. 5-11. A simple model of the interfacial compact double layer on metal electrodes: H_2O_{ad} = adsorbed water molecule; H_2O_{comp} = water molecule coordinated with ions; H_2O_{free} = free water molecule; i_h = hydrated ions. [From Bockris-Devanathan-Muller, 1963.]

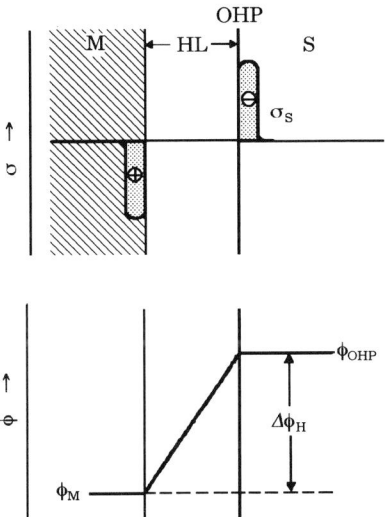

Fig. 5-12. Interfacial excess charge in the compact double layer comprising adsorbed water molecules and hydrated ions: $\Delta\phi_H$ = potential difference of the compact layer (HL).

A simple parallel plate condenser model (Fig. 5–12) gives the electric capacity C_H of the compact double layer as shown in Eqn. 5–8:

$$C_H = \frac{\varepsilon}{d_H}, \tag{5-8}$$

where ε and d_H are the permittivity and the thickness of the compact layer thickness, respectively. In the range of potential where ε and d_H are constant, the compact layer capacity, C_H, remains constant. In fact, a compact layer of constant capacity (C_H = 16 to 18 mF cm^{-1}) has been found on metal electrodes at relatively negative potentials over a range of several hundreds of millivolts.

The potential difference, $\Delta\phi_H$, of the compact layer is the sum of the potential differences g_{ion} generated by the interfacial charge and g_{dip} generated by the interfacial dipoles; g_{dip} is composed of $g_{M,dip}$ due the dipole on the metal side and $g_{S,dip}$ due the dipole on the solution side as shown in Eqn. 5–9:

$$\Delta\phi_H = g_{ion} + g_{dip} = g_{ion} + g_{M,dip} - g_{S,dip} . \tag{5-9}$$

The potential differences of the interfacial dipoles $g_{M,dip}$ and $g_{S,dip}$ differ, respectively, from the potential differences of the free surface dipoles $\chi_{M,dip}$ and $\chi_{S,dip}$ of the metal and solution as shown in Eqn. 5–10 and in Fig. 5–13:

$$g_{M,dip} = \chi_{M,dip} + \delta\chi_{M,dip} , \quad g_{S,dip} = \chi_{S,dip} + \delta\chi_{S,dip} , \tag{5-10}$$

where $\delta\chi_{M,dip}$ and $\delta\chi_{S,dip}$ are the changes in the surface dipoles that occur at the metal/aqueous solution interface as a result of the interfacial contact.

Partially differentiating $\Delta\phi_H$ with respect to the interfacial charge, σ, we obtain the inverse capacity, $1/C_H$, of the compact layer shown in Eqn. 5–11:

$$\frac{\partial \Delta\phi_H}{\partial \sigma} = \frac{\partial g_{ion}}{\partial \sigma} + \frac{\partial g_{M,dip}}{\partial \sigma} - \frac{\partial g_{S,dip}}{\partial \sigma} , \quad \frac{1}{C_H} = \frac{1}{C_{ion}} + \frac{1}{C_{M,dip}} - \frac{1}{C_{S,dip}} . \tag{5-11}$$

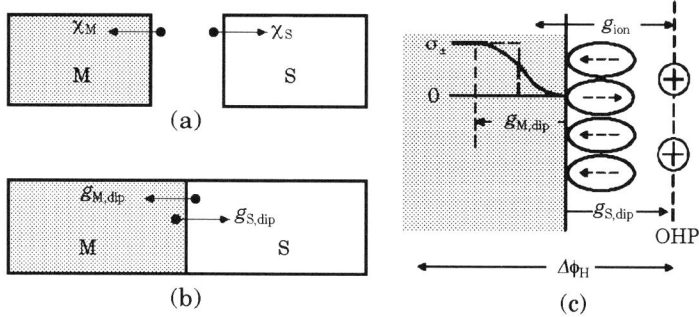

Fig. 5-13. Potential difference created at the contact interface between metal M and aqueous solution S: (a) before contact, (b) after contact, (c) charge-induced and dipole-induced potential differences. χ = surface potential difference at free surfaces; g_{dip} = potential difference due to the interfacial dipole; g_{ion} = potential difference due to the interfacial charge; $\Delta\phi_H$ = potential difference across the interfacial compact layer.

The interfacial charge-related capacity, C_{ion}, due to the internal polarization of adsorbed water molecules remains constant in the potential range where the reorganization of water molecule does not occur in the adsorbed water molecules. On the other hand, the interfacial capacity related to water dipoles, $C_{S,dip}$, on the aqueous solution side depends on the orientation of adsorbed water molecules which changes with the interfacial charge and, hence, with the electrode potential. Further, the dipole capacity, $C_{M,dip}$, on the metal side appears to slightly depend upon the interfacial charge and electrode potential. Therefore, Eqn. 5–11 yields Eqn. 5–12:

$$\Delta\phi_H = \frac{\sigma}{C_{ion}} + g_{M,dip} - g_{S,dip} \ . \tag{5-12}$$

5.3.2 The potential of zero charge

The potential E of metal electrodes at which the interfacial charge σ is zero (hence, $g_{ion} = 0$) is the *potential of zero charge* (or the *zero charge potential*), E_{pzc}. It follows from Eqn. 5–12 that the potential difference, $\Delta\phi_{pzc}$, across the compact layer at the potential of zero charge is composed of $g_{M,dip}$ and $g_{S,dip}$ as shown in Eqn. 5–13:

$$\Delta\phi_{pzc} = g_{M,dip(pzc)} - g_{S,dip(pzc)} \ . \tag{5-13}$$

Obviously, $\Delta\phi_{pzc}$ is not zero but maintains a potential difference due to the interfacial dipoles. Table 5–1 shows the potential of zero charge for various metals in aqueous solution. Also, the potential of zero charge appears to depend to some extent on the crystal plane of metal surfaces.

Table 5–1. The potential of zero charge of metallic electrodes in aqueous solution. [From Trasatti, 1971, 1975.]

Metal	E_{pzc}/V_{NHE}	Metal	E_{pzc}/V_{NHE}	Metal	E_{pzc}/V_{NHE}
Ag	– 0.44	Fe	– 0.35	Pt	0.02
Al	– 0.52	Hg	– 0.19	Sb	– 0.14
Au	0.18	In	– 0.65	Sn	– 0.43
Cd	– 0.72	Nb	– 0.79	Ta	– 0.85
Co	– 0.45	Ni	– 0.30	Ti	– 1.05
Cr	– 0.45	Pb	– 0.62	Tl	– 0.75
Cu	0.09	Pd	0.00	Zn	– 0.63

Since the dipole potential difference, $\chi_{M,dip}$, at the free surface of metals of high electron density is in the range of 1 to 5 V (Fig. 2–11), the dipole potential difference, $g_{M,dip}$, at the electrode interface on the metal side is assumed to be in the order of 1 V. The interfacial potential, $g_{S,dip}$, due to the dipole of adsorbed water molecules on the solution side has been reported, at its maximum, to be 0.4 V [Trasatti, 1979] and 1.1 V [Schmickler, 1983]. In general, hydrophilic or hydrophobic nature of the electrode metal affects significantly the dipole potential difference due to adsorbed water molecules at the electrode interface. For example, for strongly hydrophilic metals (e.g. d-metals), that strongly adsorb water molecules with preferential dipole orientation, the dipole potential difference, $g_{S,dip}$, at the potential of zero charge is greater than the dipole potential difference, $\chi_{S,dip}$ (= 0.13 V), at the free water surface being about $g_{S,dip(pzc)} = 0.4$ V; for weakly hydrophilic metals such as tin and lead the dipole potential difference is close to the dipole potential at free surfaces of liquid water being $g_{S,dip(pzc)} = 0.16$ V and $g_{S,dip(pzc)} = 0.11$ V, respectively; for hydrophobic metals (sp metals) such as gold and mercury the dipole potential difference is smaller than the free surface dipole potential of liquid water being $g_{S,dip(pzc)} = 0.0$ V and $g_{S,dip(pzc)} = 0.07$ V, respectively [Trasatti, 1979].

At the potential of zero charge the difference of electrostatic inner potential, $\Delta\phi_{pzc}$, across the electrode interface is related to the difference of outer potential, $\Delta\psi_{pzc}$, between the free surface of the metal electrode and the free surface of the aqueous solution as formulated in Eqn. 5–14 and shown in Fig. 5–14:

$$\Delta\phi_{pzc} = (\psi_M + \chi_{M,dip}) - (\psi_S + \chi_{S,dip})$$

$$= g_{M,dip(pzc)} - g_{S,dip(pzc)} = \chi_{M,dip} - \chi_{S,dip} + \Delta\psi_{pzc} \ . \tag{5–14}$$

Then, from Eqns. 5–10 and 5–14 we obtain Eqn. 5–15:

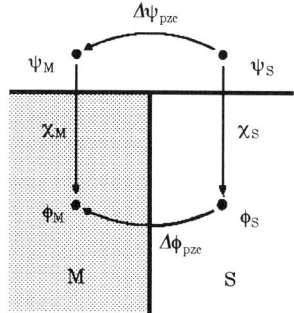

Fig. 5–14. The inner and the outer potential differences, $\Delta\phi_{pzc}$ and $\Delta\psi_{pzc}$, at the zero charge interface between metal M and aqueous solution S.

$$\Delta\psi_{pzc} = \delta\chi_{M,dip} - \delta\chi_{S,dip} , \qquad (5-15)$$

where $\delta\chi_{S,dip}$ increases as the hydrophilicity of metal increases. For the interface between metallic mercury and an aqueous solution, it has been reported that $\Delta\psi_{pzc} = -0.26$ V, $\delta\chi_{M,dip} = -0.32$ V, and $\delta\chi_{S,dip} = -0.06$ V [Trasatti, 1979].

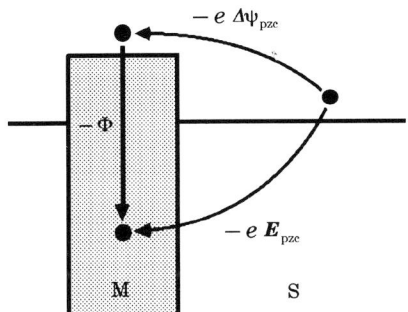

Fig. 5–15. Work function, Φ, of a metal and the potential of zero charge, E_{pzc}, of a metal electrode: $E_{pzc} = -\alpha_{e(M/S/V)}/e$; $\Phi = -\alpha_{e(M/V)}$; $\alpha_{e(M/S/V)}$ = real potential of electrons in a metal electrode; $\alpha_{e(M/V)}$ = real potential of electrons in an isolated metal.

As illustrated in Fig. 5–15, the relationship is obtained among the electrode potential of zero charge E_{pzc} ($= -\alpha_{e(M/S/V)}/e$), the work function Φ of the electrode metal, and the difference of outer potential $\Delta\psi_{pzc}$ between the free surface of the electrode metal and the free surface of the aqueous solution. Thus, taking Eqns. 5–14 and 5–15 into account, we obtain Eqn. 5–16:

$$E_{pzc} = \frac{\Phi}{e} + \delta\chi_{M,dip} - \delta\chi_{S,dip} , \qquad (5-16)$$

where $\alpha_{e(M/S/V)}$ and $\alpha_{e(M/V)}$ are the real potentials of electrons in the metal electrode and in the isolated electrode metal, respectively. The magnitude of $\delta\chi_{S,dip}$ appears to be constant for strongly hydrophilic metals (e.g. d-metals) on which adsorbed water molecules are in the state of saturated dipole orientation ($g_{S,dip(pzc)} = 0.4$); the constancy of $\delta\chi_{S,dip}$ applies also to weekly hydrophilic metals (e.g. sp-metals) on which little chemisorption of water molecules occurs ($g_{S,dip(pzc)} = 0.13$ V). If we assume that $\delta\chi_{M,dip}$ is nearly constant or extremely small, Eqn. 5–16 would give a linear relationship between E_{pzc} and Φ. Such a linear relationship has been proved experimentally with d-metals and sp-metals as shown in Fig. 5–16. A potential gap of 0.3 V has been found between a series of d-metals and that of sp-metals; this gap may be explained as a difference between the dipole potential of adsorbed water molecules on d-metals ($g_{S,dip(pzc)} = 0.4$ V) and that on sp-metals ($g_{S,dip(pzc)} = 0.13$ V equivalent to the dipole potential $\chi_{S,dip}$ of the free water surface).

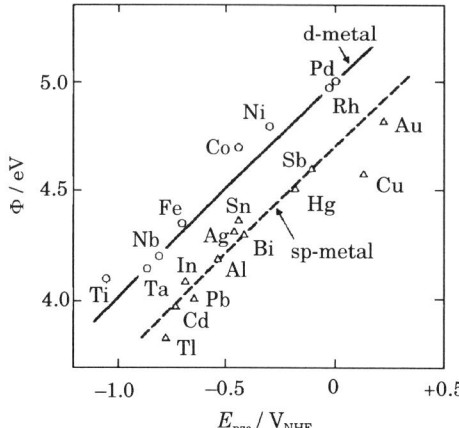

Fig. 5-16. Relationship observed between the work function, Φ, and the potential of zero charge, E_{pzc}, of sp- and d-metals in aqueous solution. [From Reeves, 1974; Trasatti, 1979.]

The *potential of zero charge*, which is defined not in terms of thermodynamics but in terms of electrostatics, is characteristic of the nature of metal electrodes and, hence, is used as a *reference potential* for individual metal electrodes. In this section we have dealt with metal electrodes on which no contact adsorption of ions takes place. As is discussed in Sec. 5.4, the contact adsorption of ions at the electrode interface may alter the potential of zero charge.

5.4 Thermodynamics of Adsorption on Metal Electrodes

5.4.1 Gibbs' adsorption equation

Adsorption equilibrium of hydrated ions at the interfaces of metal electrodes is represented by the Gibbs adsorption equation as in Eqn. 5–17:

$$-d\gamma = \Sigma_i \, \Gamma_i \, d\bar{\mu}_i \,, \tag{5-17}$$

where γ is the interfacial tension, $\bar{\mu}_i$ is the electrochemical potential of adsorbate particle i, and Γ_i is the interfacial excess of adsorbed particles of i. This excess is the number of adsorbed particles given by $n_{i(ad)}$ ($n_{i(ad)} = n_{i(T)} - n_{i(M)} - n_{i(S)}$); here $n_{i(T)}$ is the total number of particles of i, $n_{i(M)}$ is the number of the particles of i on the metal side, and $n_{i(S)}$ is the number of the particles of i on the aqueous solution side. Then, $\Gamma_i = n_{i(ad)} / A$, where A is the interfacial area.

In the case of adsorption of chloride ions on the interface of metal electrodes in aqueous potassium chloride solution, the Gibbs adsorption equation is written as in Eqn. 5–18:

$$-d\gamma = \sigma_M dE + \Gamma_{Cl^-,w} d\mu_{KCl(S)} , \qquad (5\text{--}18)$$

where σ_M is the interfacial charge on the metal side, E is the electrode potential, $\mu_{KCl(S)}$ is the chemical potential of KCl in aqueous solution, and $\Gamma_{Cl^-,w}$ is the relative interfacial excess of adsorbed Cl^- ions; this relative excess is given by $\Gamma_{Cl^-,w} = \Gamma_{Cl^-} - (x_{KCl}/x_W)\Gamma_W$, where Γ_W is the interfacial excess of adsorbed water molecules, x_{KCl} and x_W are the molar fractions of KCl and H_2O in aqueous solution, respectively. From Eqn. 5–18 we obtain $\sigma_M = -(\partial\gamma/\partial E)$ for the interfacial charge and $\Gamma_{Cl^-,w} = -(\partial\gamma/\partial\mu_{KCl(S)})$ for the relative interfacial excess of adsorbed Cl^- ions.

5.4.2 Ion adsorption on mercury electrodes

The potential of zero charge, E_{pzc}, can be obtained from the condition at which $\sigma_M = -(\partial\gamma/\partial E) = 0$. This is the potential at which the interfacial tension is maximum in the electrocapillary curve (γ vs. E) and is called the *electrocapillary maximum*. Fig. 5–17 illustrates the electrocapillary curves observed for the liquid mercury electrode in aqueous solutions of various anions. It is found that the greater the adsorption affinity of the anions ($Cl^- < Br^- < I^-$) on mercury, the more negative is the potential of zero charge (the potential of electrocapillary maximum) of the liquid mercury electrode.

The interfacial charge $\sigma_M = -\sigma_S$ is given by the interfacial excess of adsorbed anions Γ_- and cations Γ_+ in Eqn. 5–19:

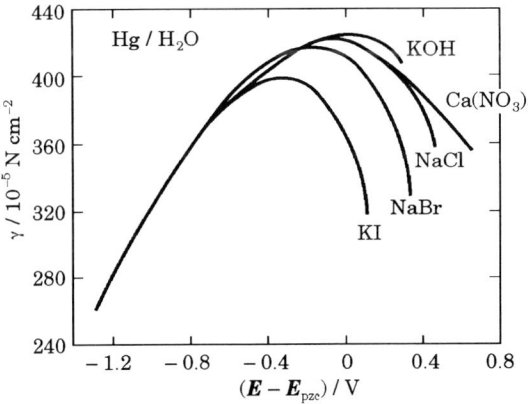

Fig. 5–17. Interfacial tension γ of the mercury electrode observed in aqueous solutions of various anions as a function of electrode potential: E_{pzc} = potential of zero charge in sodium fluoride solution in which no contact adsorption occurs. [From Grahame, 1947.]

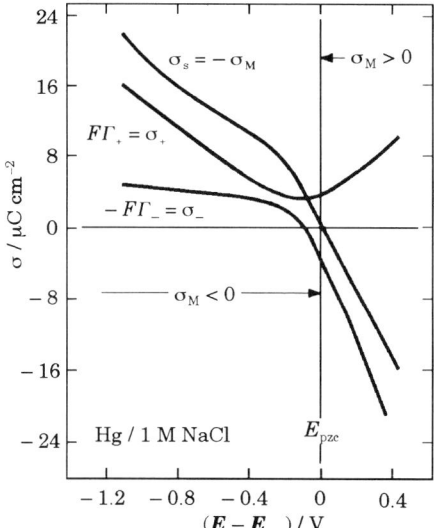

Fig. 5–18. Parts of interfacial charge carried by excess cations and anions on the solution side of the electric double layer as a function of the electrode potential of the mercury electrode in a sodium chloride solution: σ_M = interfacial charge on the side of metal electrode; σ_S = interfacial charge on the solution side; σ_+ = excess positive charge carried by cations; σ_- = excess negative charge carried by anions; Γ_+ = interfacial cation excess; Γ_- = interfacial anion excess. [From Grahame, 1947.]

$$-\sigma_M = \sigma_S = \sigma_+ + \sigma_- = z_+ e\, \Gamma_+ + z_- e\, \Gamma_- \,, \tag{5–19}$$

where z_+ and z_- are the valences of the cations and anions, respectively. Fig. 5–18 illustrates the interfacial charge versus potential curve ($\sigma_\pm$ vs. E) observed for the mercury electrode in a 1 M sodium chloride solution. At negative potentials (cathodic potentials), the charge on the metal side of the electrode is negative; the positive charge on the solution side is carried by excess hydrated cations as shown in Fig. 5–19 (a). At positive potentials (anodic potentials), the positive charge on the metal side is balanced with the negative charge carried by excess hydrated anions on the solution side as shown in Fig. 5–19 (b).

When anions are strongly adsorbed on metal electrodes, their negative charge is frequently greater than the positive charge on the metal side so that an additional positive charge of hydrated cations is accumulated on the solution side to balance the overall charge as shown in Fig. 5–19 (c). This strong adsorption is due to the chemisorption of dehydrated anions which predominates over the weaker electrostatic physical adsorption of hydrated ions.

Chemisorption of anions at an electrode interface involves the dehydration of hydrated anions followed by the adsorption of dehydrated anions which penetrate into the compact double layer to contact the interface directly; this result is called *contact adsorption* or *specific adsorption*. The plane of dehydrated anions of contact adsorption (the plane of contact adsorption) is occasionally called the

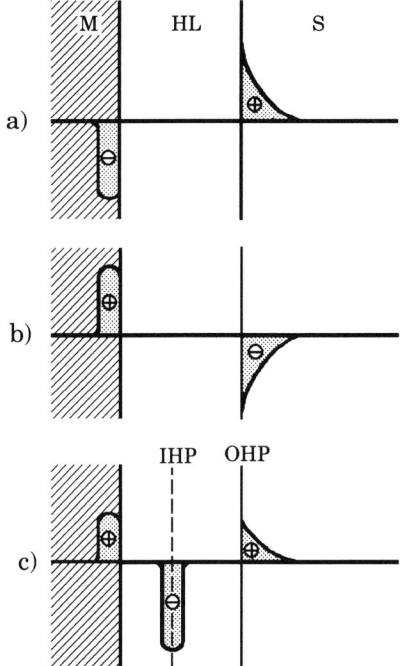

Fig. 5–19. Interfacial charge at the electric double layer on metal electrodes: (a) negative charge on the electrode side without contact ion adsorption, (b) positive charge on the electrode side without contact ion adsorption, (c) positive charge on the electrode side with contact anion adsorption.

inner Helmholtz plane (IHP). The layer between the electrode interface and the IHP is called the inner Helmholtz layer, and the layer between the IHP and the OHP is called the outer Helmholtz layer. In general, the interfacial chemisorption of dehydrated ions takes place at the IHP, and the physisorption of hydrated ions takes place at the OHP.

Many anions are chemisorbed at the interface of the metal electrode; whereas, most cations are not prone to contact adsorption because the energy of hydration–dehydration of cations is usually greater than that of anions; cations are usually smaller in size and greater in the hydration energy than anions.

The shift of the potential of zero charge toward the negative direction induced by the contact adsorption of chloride ions has been found not only with the liquid mercury electrode but also with solid metal electrodes such as gold [Jiang-Seo-Sato, 1990].

5.4.3 Contact adsorption of ions

For the equilibrium of ion adsorption on electrode interfaces, the electrochemical potential $\bar{\mu}_{i,aq}$ of hydrated adsorbate ions in aqueous solution equals the electrochemical potential $\bar{\mu}_{i,ad}$ of adsorbed ions as shown in Eqn. 5–20:

$$\bar{\mu}^0_{i,aq} + kT \ln a_i = \bar{\mu}^0_{i,ad} + kT \ln f(\theta_i) , \qquad (5\text{–}20)$$

where a_i is the activity of hydrated adsorbate ions, $f(\theta_i)$ is the activity of adsorbed ions in terms of the coverage θ_i, $\bar{\mu}^0_{i,aq}$ and $\bar{\mu}^0_{i,ad}$ are the standard electrochemical potentials of hydrated adsorbate ions and dehydrated adsorbed ions, respectively. Eqn. 5–20, then, yields Eqn. 5–21:

$$f(\theta_i) = a_i \exp\left(\frac{-\Delta \bar{G}^0_{ad}}{kT} \right) = \beta_E a_i , \qquad (5\text{–}21)$$

where $\Delta \bar{G}^0_{ad} = \bar{\mu}^0_{i,ad} - \bar{\mu}^0_{i,aq}$ is the standard electrochemical free enthalpy of the ion adsorption. $\Delta \bar{G}^0_{ad}$ is composed of the standard chemical free enthalpy ΔG^0_{ad} and the potential-dependent electrostatic energy $z_i e \Delta\phi_I$ as shown in Eqn. 5–22:

$$\Delta \bar{G}^0_{ad} = \Delta G^0_{ad} + z_i e \Delta\phi_I , \qquad (5\text{–}22)$$

where z_i is the ionic valence and $\Delta\phi_I$ is the potential difference between the plane of adsorption (IHP) and the interior of the bulk solution.

Table 5–2 shows the standard chemical free enthalpy for the adsorption of anions observed on several metal electrodes in aqueous solutions of various anions. The magnitude of ΔG^0_{ad} in the range of 0.1 to 0.7 eV, illustrated in Table 5–2, is of the same order of magnitude as the potential difference of 0.1 to 1.0 V usually present across interfaces of metal electrodes. On electrodes such as Hg/F$^-$ and

Table 5–2. The standard chemical free enthalpy, ΔG^0_{ad}, of anion adsorption on metal electrodes in aqueous solution at 25°C. [From Bode, 1972.]

Electrode	ΔG^0_{ad} / eV	Electrode	ΔG^0_{ad} / eV
Hg/Br$^-$	– 0.516	Ag/F$^-$	– 0.104
Hg/Cl$^-$	– 0.447	Ag/OH$^-$	– 0.148
Hg/F$^-$	+ 0.113	Au/Br$^-$	– 0.608
Hg/OH$^-$	+ 0.113	Au/Cl$^-$	– 0.508
Ag/Br$^-$	– 0.751	Au/F$^-$	– 0.117
Ag/Cl$^-$	– 0.651	Au/OH$^-$	– 0.161

Hg/OH⁻ of which the standard electrochemical free enthalpy of anion adsorption is positive ($\Delta \bar{G}^0_{ad} > 0$), only the physical adsorption of hydrated anions occurs at the OHP; whereas, the chemisorption of anions takes place at the IHP on electrodes of which the standard electrochemical free enthalpy of anion adsorption is negative ($\Delta \bar{G}^0_{ad} < 0$).

Since contact adsorption requires dehydration of ions, it decreases as the hydration energy increases. In general, the hydration energy decreases as ion size increases. Therefore, anions predominate in contact adsorption over cations of which the size is usually smaller than anions in aqueous solution. In the same series of halogen ions, the contact-adsorbability increases with increasing anion size (hence, with decreasing hydration energy) in the order of $F^- < Cl^- < Br^- < I^-$ [Anderson-Bockris, 1964].

The relationship between the activity of adsorbed ions $f(\theta_i)$ in Eqn. 5–21 and the adsorption coverage θ_i is known as an *adsorption isotherm*. Eqns. 5–23 and 5–24 show simple adsorption isotherms:

$$\beta_E a_i = \frac{\theta}{1-\theta}, \qquad (5\text{–}23)$$

which is called the Langmuir isotherm [Langmuir, 1918]; and

$$\beta_E a_i = \frac{\theta}{1-\theta} \exp(-2\alpha\theta), \qquad (5\text{–}24)$$

which is called the Frumkin isotherm [Frumkin, 1925], where α is a parameter of the adsorbed ion–ion interaction. From Eqns. 5–21 and 5–22, it follows that β_E is a function of the electrode potential.

5.5 Electric Double Layer at Metal Electrodes

5.5.1 Interfacial electric capacity (Electrode capacity)

Conventionally, for the range of potential in which the interfacial charge is relatively small, the reciprocal of the interfacial electric capacity, C, of metal electrodes has been represented by the Laurent series with respect to the Debye length L_D of aqueous solution as shown in Eqn. 5–25 [Schmickler, 1993]:

$$\frac{1}{C} = a(L_D)^1 + b(L_D)^0 + c(L_D)^{-1} + \cdots, \qquad (5\text{–}25)$$

where the first term in the right hand side corresponds to the diffuse layer on the solution side, the second term corresponds to the compact double layer (Helmholtz layer), and the succeeding terms arise from the nonideal property of concentrated solution. The second term, $1/C_H$, for the electric capacity of the compact layer may be divided into two terms: $1/C_M$ for the capacity on the metal side and the $1/C_S$ for the capacity on the solution side:

$$\frac{1}{C_H} = \frac{1}{C_M} + \frac{1}{C_S} .\tag{5-26}$$

5.5.2 The effective image plane on metal surfaces

According to the jellium model of metals described in Sec. 2.2, metal electrons diffuse away from the metal surface to form surface dipoles (Fig. 2–10). As the metal is charged with an image charge of approaching ions, the diffuse tailing of metal electrons changes with the interfacial charge $\sigma_m(x)$ carried by an excess (negative charge) or deficit (positive charge) of metal electrons which distribute around the jellium metal surface, i.e. the jellium metal edge ($x = 0$) one-half lattice constant away from the metal surface.

The surface potential difference, χ_M, due to the interfacial dipole of the electron tailing away from the metal surface is given as a function of the excess or deficit of metal electrons $\sigma_m(x)$ in Eqn. 5–27:

$$\chi_M(\sigma_M) = -\frac{1}{\varepsilon}\int_{-\infty}^{+\infty} \sigma_m(x)\, x\, dx = \chi_M(\sigma_M = 0) + \delta\chi_M(\sigma_M) ,\tag{5-27}$$

where ε is the permittivity (dimension F m^{-1}) of the compact layer, σ_M is the total excess charge on the metal side, and $\delta\chi_M(\sigma_M)$ is the change of χ_M due to the image charge σ_M induced on the metal surface.

The plane of the center of mass of induced image charge, $\sigma_m(x)$, is called the *effective image plane* [Lang-Kohn, 1973], and its position, x_{im}, is given by Eqn. 5–28:

$$x_{im} = \int_{-\infty}^{+\infty} \frac{[\sigma_m(x, \sigma_M + \delta\sigma_M) - \sigma_m(x, \sigma_M)]}{\delta\sigma_M}\, x\, dx .\tag{5-28}$$

As shown in Fig. 5–20, the effective image plane is located close to but away from the jellium metal edge. Combining Eqns. 5–27 and 5–28 yields the surface potential difference χ_M as a function of σ_M in Eqn. 5–29:

$$\chi_M(\sigma_M) = -\frac{1}{\varepsilon}\int_0^{\sigma_M} x_{im}\, d\sigma_M .\tag{5-29}$$

In calculating the distribution of image charge $\sigma_m(x)$, we use an improved jellium model in which the quasi-potential due to the lattice ions is taken into account instead of using uniformly distributed positive charge. The effective image plane, x_{im}, obtained from Eqn. 5–28 by using the improved jellium model with the distribution of image charge $\sigma_m(x)$, shifts outwards with increasing electron density in the metal as shown in Table 5–3 (a metal/vacuum interface). The position of the image plane is also dependent on the crystal plane of the metal surface, being the closest to the (111) crystal plane and a little more away from the (110) crystal plane on aluminum crystals as shown in Table 5–4. It is also known that the effective image plane shifts more outwards, the more is the excess negative charge on the metal side [Gies-Gerhardt, 1985] as schematically shown in Fig. 5–22.

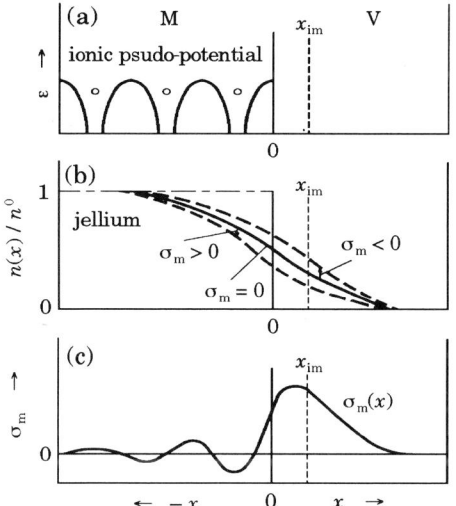

Fig. 5–20. Charge distribution profile across the interface between metal and vacuum (M/V): (a) ionic pseudo-potential in the metal, (b) diffuse electron tailing away from the jellium metal edge, (c) excess charge profile. $n(x)$ = electron density at distance x; n^0 = electron density in the metal; x_{im} = effective image plane; σ_m = differential excess charge; $\sigma_m = 0$ corresponds to the zero charge interface.

Table 5–3. The effective image plane position of a metal in vacuum estimated as a function of the valence electron density in metals: x_{im} = distance at the effective image plane from the jellium metal edge; r_{WS} = Wigner-Seitz radius (a sphere containing one electron) which is related to the electron density n_e in metals ($1 / n_e = 4 \pi r_{WS}^3 / 3$); au = atomic unit (0.529 Å). [From Schmickler, 1993.]

r_{WS} / au	x_{im} / nm
2	0.085
3	0.075
4	0.065

Table 5–4. The effective image plane position, x_{im}, estimated on three crystal surface planes of metallic aluminum in vacuum. The electron density increases in order from the (111), (100) to (110). [From Schmickler, 1993.]

Crystal Plane	x_{im} / nm
111	0.053
100	0.066
110	0.127

In the range of potential where the interfacial charge is relatively small, the electric capacity, C'_M, due to the electron tailing from the metal side is, to a first approximation, represented by the capacity of a parallel plate condenser of thickness x_{im} as shown in Eqn. 5–30:

$$\frac{1}{C'_M} = -\frac{x_{im}(\sigma_M)}{\varepsilon} \quad . \tag{5–30}$$

The fact that the effective image plane is located slightly away from the metal surface means that the thickness of the compact double layer becomes thinner with the capacity to be greater than the ideal parallel plate condenser; therefore, we place a negative sign on the right hand side of Eqn. 5–30. Simple calculations have shown that the effective image plane becomes closer to the metal surface with increasing excess positive charge on the metal side [Russier-Rosinberg, 1988] as shown in Fig. 5–22.

5.5.3 The closest approach of water molecules to electrode interfaces

The interfacial solution layer contains hydrated ions and dipoles of water molecules. According to the hard sphere model or the mean sphere approximation of aqueous solution, the plane of the center of mass of the excess ionic charge, $\sigma_s(x)$, is given at the distance x_s from the jellium metal edge in Eqn. 5–31:

$$x_s = \int_{-\infty}^{+\infty} \frac{[\sigma_s(x, \sigma_S + \delta\sigma_S) - \sigma_s(x, \sigma_S)]}{\delta\sigma_S} x \, dx \quad . \tag{5–31}$$

This plane of the center of mass of the excess ionic charge $\sigma_s(x)$ is the *effective excess charge plane* on the solution side which may be compared with the effective image plane on the metal side. In simple cases, the effective excess charge plane coincides with the outer Helmholtz plane (the plane of closest approach of hydrated ions) as shown in Fig. 5–21.

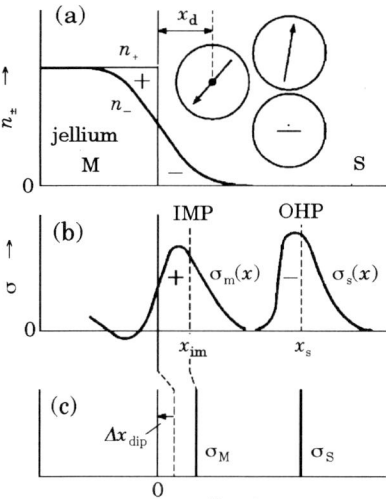

Fig. 5–21. Charge distribution profile across the metal/aqueous solution interface (M/S): (a) the hard sphere model of aqueous solution and the jellium model of metal (the jellium-sphere model), (b) the effective image plane x_{im} (IMP) and the effective excess charge plane x_s (OHP in case of no diffuse layer), (c) the reduction in distance Δx_{dip} to the closest approach of water molecules due to electrostatic pressure. σ_s = differential excess charge on the solution side; σ_S = total excess charge on the solution side; σ_M = total excess charge on the metal side.

Next, we discuss the plane of the closest approach ($x = x_{dip}$) of water molecules to the jellium metal edge ($x = 0$). At the zero charge interface, this plane of closest approach of water molecules is separated by a distance equal to the radius of water molecules from the metal surface. As the interfacial excess charge increases, the electrostatic pressure (electrostriction pressure) reduces the distance of x_{dip} in proportion to the square of the interfacial charge, σ ($= \sigma_M = -\sigma_S$); the electrostatic force in the compact layer is proportional to $\sigma_M \times \sigma_S$. The change in x_{dip} due to the interfacial charge is given by Eqn. 5–32:

$$\Delta x_{dip}(\sigma) = \int_0^\sigma \frac{\partial x_{dip}}{\partial \sigma} d\sigma . \tag{5-32}$$

As shown in Fig. 5–21, the reduction of x_{dip} reduces the thickness of the compact layer, the magnitude of which increases as an inverse-parabolic function of the interfacial charge (Fig. 5–22).

Such a change of the thickness of the compact layer (thickness reduction, $\Delta x_{dip} < 0$) due to the electrostatic pressure may be represented in terms of the conventional electric capacity C'_{es} given by Eqn. 5–33:

$$\frac{1}{C'_{es}} = \frac{\Delta x_{dip}(\sigma)}{\varepsilon} . \tag{5-33}$$

5.5.4 Electric capacity of the compact layer

The interfacial electric double layer is represented by a combination of the improved jellium model on the metal side and the hard sphere model of ions and dipoles on the aqueous solution side. Then, the electric capacity, C_H, of the compact layer is given by Eqn. 5–34:

$$\frac{1}{C_H} = \frac{x_s(\sigma) + \Delta x_{dip}(\sigma) - x_{im}(\sigma)}{\varepsilon} - \frac{\partial}{\partial \sigma} \int_{-\infty}^{+\infty} \frac{P(x)\,dx}{\varepsilon} , \qquad (5\text{–}34)$$

where ε is the permittivity (F m^{-1}) of the layer, and $P(x)$ is the polarization of adsorbed water molecules consisting of the internal polarization $P^{pol}(x)$ and the orientational polarization $P^{or}(x)$.

Capacity C_H in Eqn. 5–34 may be conventionally divided into capacity C_M on the metal side and capacity C_S on the aqueous solution side as shown in Eqns. 5–35 and 5–36:

$$\frac{1}{C_M} = \frac{\Delta x_{dip}(\sigma) - x_{im}(\sigma)}{\varepsilon} , \qquad (5\text{–}35)$$

$$\frac{1}{C_S} = \frac{x_s(\sigma)}{\varepsilon} - \frac{\partial}{\partial \sigma} \int_{-\infty}^{+\infty} \frac{P^{pol}(x)}{\varepsilon}\,dx - \frac{\partial}{\partial \sigma} \int_{-\infty}^{+\infty} \frac{P^{or}(x)}{\varepsilon}\,dx . \qquad (5\text{–}36)$$

The third term of the right hand side of Eqn. 5–36 represents the inverse capacity of $1/C_{S,dip}$ due to the orientational polarization of adsorbed water molecules. Usually, C_M and C_S are of the same order of magnitude near the potential of zero charge.

From the molecular dynamics based on the jellium-sphere model, Eqn. 5–37 has been derived for the capacity of the compact layer formed near the potential of zero charge [Schmickler, 1993]:

$$\frac{1}{C_H} = \frac{1}{2\varepsilon}\left(r_i + \frac{\kappa-1}{\lambda} r_d\right) + \frac{\Delta x_{dip} - x_{im}}{\varepsilon} , \qquad (5\text{–}37)$$

where r_i is the radius of the hydrated ions, r_d is the radius of water dipoles, κ is the relative dielectric constant, and λ is a parameter related to κ, the magnitude of which is of the order of $\lambda = 2$ to 3: $\lambda^2(1+\lambda)^4 = 16\kappa$. In Eqn. 5–37, the first term on the right hand side represents the inverse capacity $1/C_S$ on the solution side and the second term represents the inverse capacity $1/C_M$ on the metal side. Since $(\kappa-1)/\lambda$ in the first term on the right hand side of Eqn. 5–37 is greater than unity, it follows that the contribution of dipoles predominates, over that of ions, to the inverse capacity $1/C_S$ of the interfacial compact layer on the solution side.

According to observations with metal electrodes in aqueous solutions [Amokrane-Badiali, 1992], the interfacial charge versus capacity curve is described for the inverse capacity, $1/C_M$, as a parabolic curve with its maximum near the zero charge; the capacity, C_S, is described as a parabolic curve with its maximum near the zero charge. Both of their dependences are shown schematically in Fig. 5–22. Usually, therefore, the total capacity C_H of the compact layer is represented by a parabolic curve with its maximum near the zero charge ($\sigma = 0$) of the interface. Fig. 5–23 shows the capacity of the compact layer observed for silver and mercury electrodes in aqueous solution as a function of the interfacial charge.

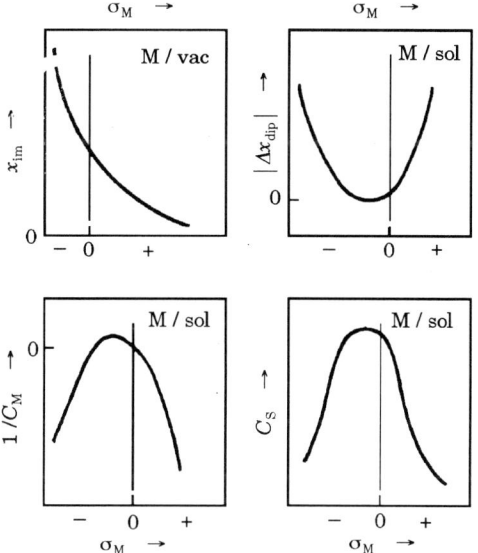

Fig. 5–22. Effect of interfacial excess charge, σ_M, on: the effective image plane, x_{im}; the shift of the plane of closest approach of water molecules, Δx_{dip}; the inverse interfacial capacity on the metal side, $1/C_M$; and the interfacial capacity on the solution side, C_S. M/vac = metal/vacuum interface; M/sol = metal/ solution interface.

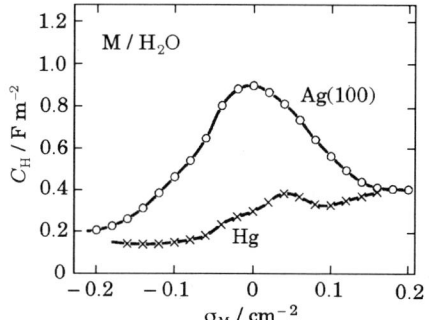

Fig. 5–23. Differential capacity C_H of the compact layer observed as a function of the interfacial charge σ_M on a mercury electrode and on the (100) surface of silver electrode in aqueous solution. [From Schmickler, 1993.]

As described in this section, the distance x_{im} to the image plane increases with increasing electron density in the electrode metals. Correspondingly, as shown in Fig. 5–24, the capacity of the compact layer, C_H, of sp metal electrodes in aqueous solution increases with the increasing electron density of the metal. It appears that the interfacial electric double layer is affected significantly by the nature (electron density) of electrode metals.

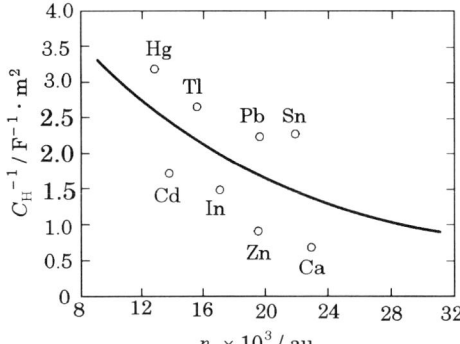

Fig. 5–24. Inverse differential capacity $1/C_H$ of the compact layer as a function of electron density n_e of polycrystalline sp-metal electrodes in aqueous solution: open circles = observed; solid line = calculated; au = atomic unit. [From Schmickler, 1993.]

5.5.5 Potential difference across the compact double layer

Have now developed the jellium-sphere model it is useful to compare it with the classical model as described in this section. From Eqns. 5–34, 5–35 and 5–36, the potential difference, $\Delta\phi_H$, across the compact layer is given by Eqn. 5–38:

$$\Delta\phi_H = \int_0^\sigma \frac{\sigma}{C_H} d\sigma = \int_0^\sigma \frac{(\Delta x_{dip} - x_{im})\sigma}{\varepsilon} d\sigma$$

$$+ \int_0^\sigma \frac{x_s \sigma}{\varepsilon} d\sigma - \int_{-\infty}^{+\infty} \frac{P^{pol}(x)}{\varepsilon} dx - \int_{-\infty}^{+\infty} \frac{P^{or}(x)}{\varepsilon} dx \ . \tag{5-38}$$

The first term on the right hand side of Eqn. 5–38 can be expressed by $-z_{im}$, the second term by $+z_s$, and the integral constant of the two terms by $\chi_m(\sigma = 0)$; then, we obtain Eqn. 5–39:

$$\Delta\phi_H = \chi_m(\sigma = 0) + \sigma(z_s - z_{im}) - \int_{-\infty}^{+\infty} \frac{P^{pol}(x)}{\varepsilon} dx - \int_{-\infty}^{+\infty} \frac{P^{or}(x)}{\varepsilon} dx \ . \tag{5-39}$$

In comparing Eqn. 5–39 with Eqn. 5–9 ($\Delta\phi_H = g_{ion} + g_{M,dip} - g_{S,dip}$), which is based on the classical double layer model, it appears that: the sum of the first, second and third terms on the right hand side of Eqn. 5–39 corresponds to the sum of g_{ion} due to the interfacial charge and $g_{M,dip}$ due to the interfacial dipole on the metal side; and the fourth term corresponds to $g_{S,dip}$ due to the interfacial dipole of adsorbed water molecules on the solution side. These equivalences give Eqns. 5–40 and 5–41:

$$g_{M,dip} + g_{ion} = \chi_m(\sigma = 0) + \sigma(z_s - z_{im}) - \int_{-\infty}^{+\infty} \frac{P^{pol}(x)}{\varepsilon} dx, \quad (5–40)$$

$$g_{S,dip} = \int_{-\infty}^{+\infty} \frac{P^{or}(x)}{\varepsilon} dx. \quad (5–41)$$

These equations amplify the meaning of the classical model in terms of the improved jellium-sphere model.

5.6 Contact Adsorption and Electric Double Layer

5.6.1 Contact adsorption and work function

In general, adsorption of particles on the metal surface in vacuum changes the work function Φ of the metal as shown in Eqn. 5–42:

$$\Phi = \Phi_0 - \frac{\Gamma_i \mu}{\varepsilon}, \quad (5–42)$$

where Φ_0 is the work function without adsorption, Γ_i is the interfacial excess of adsorbed particle i, and μ is the dipole moment of adsorbed particles. Fig. 5–25 shows the change in the work function observed as a function of the adsorption coverage of bromine and water molecules on the surface of a single crystal of silver metal. Generally, adsorbed water molecules tend to align with their oxygen on the metal side (positive dipole moment) as shown in Fig. 5–26 (a); this alignment reduces the work function of the metal.

The dipole moment of the adsorbed water molecules is estimated to be μ = 0.22 D (unit of D = 3.36 × 10^{-30} C m) from the slope of the observed curves shown in Fig. 5–25. Since this dipole moment is nearly one tenth of the dipole moment of the gaseous water molecules (μ = 1.84 D), the dipole of the adsorbed water molecules on the silver surface is suggested to be aligned almost in parallel to the metal surface by forming hydrogen-bonded two-dimensional clusters of water molecules. On the other hand, bromine molecules are in the state of dissociative adsorption on the silver surface producing adsorbed bromine atoms which receive

152   ELECTRIC DOUBLE LAYER AT ELECTRODE INTERFACE   CHAP. 5

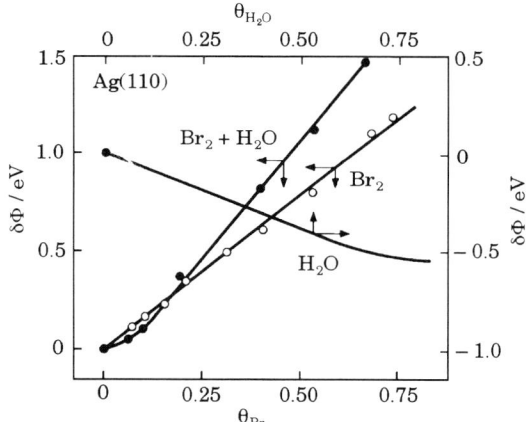

Fig. 5–25. Relative change in the work function, δΦ, of the (100) surface of single crystal of silver as a function of adsorption coverage θ of bromine atoms and water molecules on the clean surface, and of bromine atoms on the surface with pre-adsorbed water molecules. [From Bange-Straehler-Sass-Parsons, 1987.]

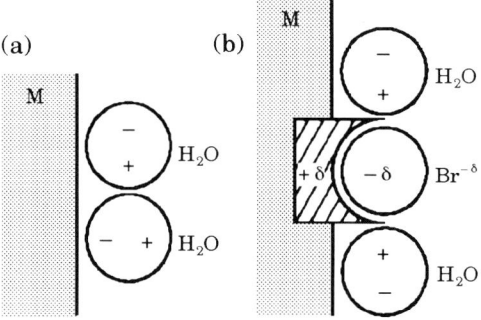

Fig. 5–26. For the hard sphere model on metal electrodes: (a) interfacial dipoles induced by adsorbed water molecules and (b) interfacial dipoles induced by the contact adsorption of partially ionized bromine atoms. $-\delta$ = charge number of adsorbed particle (z_{ad}). [From Schmickler, 1993.]

electrons from the silver metal to form interfacial electric dipoles as shown in Fig. 5–26 (b). This bromine-induced dipole moment is estimated to be $\mu = -0.5$ D, which is less than the dipole moment of silver bromides ($\mu = -9.36$ D), suggesting that only a slight electron transfer occurs from the silver metal to the adsorbed bromine atoms.

Fig. 5–25 illustrates also the change in the work function of silver metal observed in the course of substitutional adsorption of bromine atoms in place of adsorbed water molecules on the silver surface. As this substitutional adsorption progresses on the silver surface covered with a mono-molecular layer of adsorbed water molecules, the work function of the silver metal increases linearly with the bromine coverage; from the rate of this increase with the bromine coverage, we may estimate the interfacial dipole moment induced by the adsorbed bromine atom to be $\mu = -0.35$ D at low coverages and $\mu = -0.73$ D at high coverages. The fact that the dipole moment ($\mu = -0.35$ D) at low coverages is smaller than that ($\mu = -0.5$ D) of adsorbed bromine alone may be explained by the reorganization of coadsorbed water molecules on the silver surface. As the coverage of the adsorbed bromine atom increases, the coverage of the adsorbed water molecule decreases and the dipole moment due to the bromine–water combination predominates over that of the adsorbed water molecule.

5.6.2 Interfacial dipole moment induced by contact adsorption

In the course of contact adsorption on the interface of metal electrodes, hydrated ions are first dehydrated and then adsorbed at the inner Helmholtz plane in the compact layer as shown in Fig. 5–27 and as described in Sec. 5.6.1. In the interfacial double layer containing adsorbed ions, the charge balance may be expressed by the combined charge of the metal σ_M and adsorbed ions $\sigma_{ad} = z_{ad}\, e\, \Gamma$

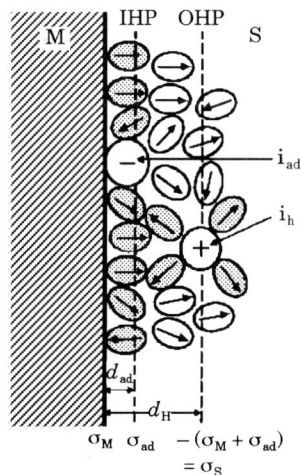

Fig. 5–27. Contact ion adsorption on metal electrodes in aqueous solution: IHP = inner Helmholtz plane; OHP = outer Helmholtz plane; i_{ad} = adsorbed ion; i_h = hydrated ion; σ_M = charge on the metal electrode; σ_{ad} = charge of adsorbed ions; σ_{ad} = charge of excess hydrated ions in solution. [From Bockris-Devanathan-Muller, 1963.]

on the metal side which equals the charge of excess hydrated ions σ_{OHP} on the solution side: $\sigma_{OHP} = -(\sigma_M + \sigma_{ad})$; here z_{ad} is the ionic valence of adsorbed ions and Γ is the interfacial excess of adsorbed ions.

The potential difference $\Delta\phi_H$ across the compact layer is a function of the interfacial charges σ_M and $\sigma_{ad} = z_{ad} e \Gamma$ as shown in Eqn. 5–43:

$$d\Delta\phi_H = \left(\frac{\partial \Delta\phi_H}{\partial \sigma_M}\right)_\Gamma d\sigma_M + \left(\frac{\partial \Delta\phi_H}{\partial \Gamma}\right)_{\sigma_M} d\Gamma . \tag{5–43}$$

In the case that σ_{OHP} is constant, Eqn. 5–43 yields Eqn. 5–44:

$$\left(\frac{\partial \Delta\phi_H}{\partial \Gamma}\right)_{\sigma_{OHP}} = \left(\frac{\partial \Delta\phi_H}{\partial \sigma_M}\right)_\Gamma \left(\frac{\partial \sigma_M}{\partial \Gamma}\right)_{\sigma_{OHP}} + \left(\frac{\partial \Delta\phi_H}{\partial \Gamma}\right)_{\sigma_M} , \tag{5–44}$$

with $(\partial \sigma_M / \partial \Gamma)_{\sigma_{OHP}} = -z_{ad} e$. Consequently, we further obtain Eqn. 5–45:

$$\left(\frac{\partial \Delta\phi_H}{\partial \Gamma}\right)_{\sigma_{OHP}} = -z_{ad} e \left(\frac{\partial \Delta\phi_H}{\partial \sigma_M}\right)_\Gamma + \left(\frac{\partial \Delta\phi_H}{\partial \Gamma}\right)_{\sigma_M} = z_{ad} e \left(\frac{1}{C_{\sigma_{OHP}}} - \frac{1}{C_{\sigma_M}}\right) , \tag{5–45}$$

where $C_{\sigma_M} = (\partial \sigma_M / \partial \Delta\phi_H)_\Gamma$ is the capacity of the compact layer at constant interfacial excess of adsorbed ions, and $C_{\sigma_{OHP}} = (\partial z_{ad} e \Gamma / \partial \Delta\phi_H)_{\sigma_M}$ is the capacity of the compact layer at constant metal charge.

Assuming that the capacity of the compact layer is constant at potentials near the potential of zero charge, we obtain, to a first approximation, Eqn. 5–46:

$$C_{\sigma_{OHP}} = \frac{\varepsilon}{d_H - d_{ad}} , \quad C_{\sigma_M} = \frac{\varepsilon}{d_H} , \tag{5–46}$$

where d_H is the thickness of the compact layer, and d_{ad} is the thickness of the inner Helmholtz layer for contact adsorption. From this base, we obtain the interfacial dipole moment, μ_{ad}, induced by the contact adsorption of ions near the potential of zero charge as shown in Eqn. 5–47:

$$\left(\frac{\partial \Delta\phi_H}{\partial \Gamma}\right)_{\sigma_{OHP}} = -\frac{z_{ad} e d_{ad}}{\varepsilon} = \mu_{ad} . \tag{5–47}$$

The charge number, z_{ad}, of adsorbed ions is not always the same as the ionic valence z of the hydrated ions; the difference between z_{ad} and z is called the *charge transfer coefficient*, δz, in the contact adsorption of ions as identified in Eqn. 5–48:

$$\delta z = z - z_{ad} . \tag{5–48}$$

Usually, δz ranges from |0.1| to |2.0| with a negative sign for anions (the negative charge number of adsorbed anions is less than that of hydrated anions) and with a positive sign for cations. Such a partial electron transfer has also been found to take place in the adsorption of gaseous ions on metals, and the charge transfer coefficient has been found to decrease with increasing hydration of adsorbate ions [Kornyshev-Schmickler, 1985].

In addition to Eqn. 5–47, the adsorption-induced dipole moment has been derived based on the molecular dynamics [Schmickler, 1993] as shown in Eqn. 5–49:

$$\mu_{ad} = z_{ad} e \left[\frac{r_i}{2\kappa} + \frac{r_i r_d (\kappa - 1)}{2\kappa (r_d + \lambda r_i)} - x_{im} \right], \qquad (5\text{–}49)$$

where the symbols are the same as in Eqn. 5–37. From Eqn. 5–49 we estimate the adsorption-induced dipole moment to be in the range of 0.2 to 1.5 D.

5.6.3 Interfacial potential difference affected by contact adsorption

As a result of the contact adsorption of ions on metal electrodes that induces an interfacial dipole, the potential difference, $\Delta\phi_H$, across the compact layer is altered. Fig. 5–28 shows the change of the potential difference, $\delta\Delta\phi_H$ ($= E - E_{pzc}$), observed as a function of the coverage θ_{Br} of adsorbed bromide ions on the silver electrode interface in aqueous solution; this change indicates that $\Delta\phi_H$ increases with increasing θ_{Br}. Also plotted in Fig. 5–28 is the change in the work function (hence, in the surface potential difference, χ_M) as a function of θ_{Br} for the coadsorption of water molecules and dissociated bromine atoms on the silver surface in vacuum; this is equivalent to the change of work function, $\delta\Phi$, shown

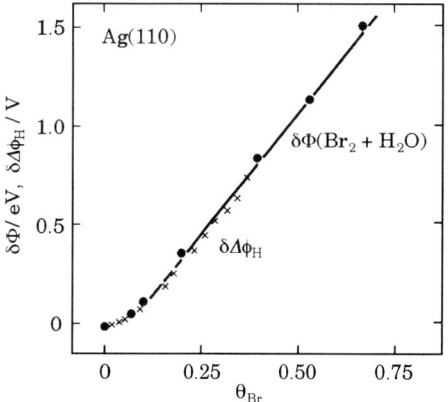

Fig. 5-28. Potential difference $\Delta\phi_H$ across the compact layer observed in aqueous solution and work function Φ observed in vacuum as a function of the adsorption coverage of bromine atoms on the (100) interface of single crystal silver electrode: $\delta\Phi$ = relative change of the work function in vacuum; $\delta\Delta\phi_H$ = relative change in the potential difference across the compact layer; work function data corresponds to Fig. 5-25. [From Bange-Straehler-Sass-Parsons, 1987.]

in Fig. 5–25. There appears a good coincidence between the change in $\Delta\phi_H$ and the change in Φ. Hence, the interfacial dipole induced by the contact adsorption of ions plays an important role in determining the potential difference at both the metal/aqueous solution and the metal/gas interfaces.

The contact adsorption of ions on the interface of metal electrodes alters the potential profile across the compact layer at constant electrode potential. If anions are adsorbed on the metal electrode at positive potentials, the adsorption-induced dipole generates a potential difference $\Delta\phi_{IHL}$ across the inner Helmholtz layer (IHL) as illustrated in Fig. 5–29. As a result, the field in the outer part (OHL) of the compact layer becomes different from and frequently opposite to that in the inner part (IHL) of the compact layer.

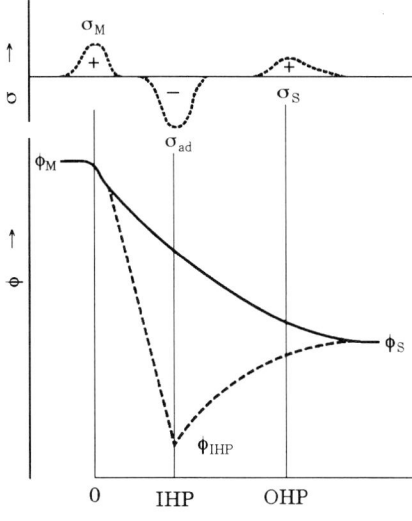

Fig. 5–29. Change in potential profile across the compact layer due to contact adsorption of anions at constant potential on a metal electrode: solid line = without contact anion adsorption; broken line = with contact anion adsorption; ϕ_M = inner potential of the metal electrode, ϕ_S = inner potential of the solution; ϕ_{IHP} = inner potential at the IHP.

In general, the contact adsorption of anions creates an electric field of intensity E_{in} in the inner Helmholtz layer, which may be greater than the average field intensity of $E_{av} = (E_{in} + E_{out})/2$, where E_{out} is the field intensity in the outer Helmholtz layer. The rate of field increase η may be derived from electrostatics as shown in Eqn. 5–50 [Liu, 1983]:

$$\eta = \frac{E_{in}}{E_{av}} = \frac{\sigma_M - 0.5\,(\kappa - 1)\,\sigma_{ad}}{\sigma_M + 0.5\,\sigma_{ad}}, \qquad (5\text{--}50)$$

where κ is the relative dielectric constant ($\kappa = 5$ to 10). As the adsorbed charge σ_{ad} changes from zero to $-\sigma_M$, the rate of field increase η increases from 1 to ($\kappa - 1$); hence, the field intensity in the inner part becomes greater than the average field intensity in the compact layer. Accordingly, the differential capacity C_H of the compact layer with the contact adsorption is η times greater than that without the contact adsorption as shown in Eqn. 5–51 [Liu, 1983]:

$$C_H = \frac{\sigma_M}{E_{av}\,d_H} = \frac{\eta\,\varepsilon}{d_H}. \qquad (5\text{--}51)$$

As the field intensity E_{in} in the inner Helmholtz layer becomes extremely high, the field intensity E_{out} in the outer Helmholtz layer is reversed as shown in Fig. 5–29. Fig. 5–30 illustrates the potential profile across the interfacial double layer of a mercury electrode in an aqueous chloride solution; this result was obtained by calculations at various electrode potentials ranging from negative (cathodic) to positive (anodic) potentials.

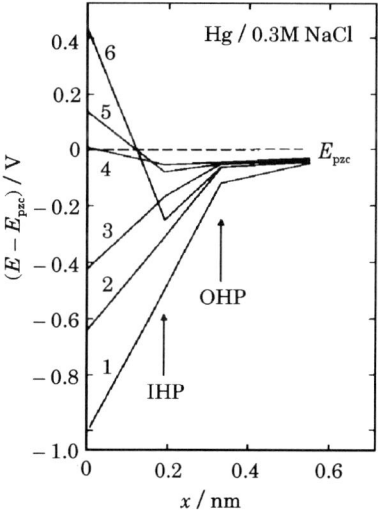

Fig. 5–30. Potential profile across the compact layer estimated by calculations at various electrode potentials for a mercury electrode in a 0.3 M sodium chloride solution: the electrode potential changes from No. 1 (a cathodic potential) to No. 6 (an anodic potential), and the contact adsorption of chloride ions takes place at anodic potentials. E = electrode potential; E_{pzc} = zero charge potential; x = distance from the interface. [From Graham, 1947.]

5.7 Particle Adsorption on Metal Electrodes

5.7.1 Adsorption of water molecules

In the adsorption of water molecules on metal electrodes in aqueous solutions, the unpaired electrons in the frontier orbital of oxygen atoms in water molecules form covalent bonds with the metal surface atoms. Then, the adsorbate water molecules act as a *Lewis base* (covalent-electron providers); and the adsorbent surface metal atoms act as a *Lewis acid* (covalent-electron receivers). Since the bond energy (0.4 to 0.7 eV) of water molecules with the surface metal atoms is close to the energy of hydrogen bond (0.2 to 0.4 eV) between water molecules, the adsorbed water molecule is combined not only with the surface metal atoms but also with the adjacent water molecules to form a *bi-molecular layer* rather than a monomer layer as shown in Fig. 5–31.

This bilayer, which is also called the "puckered" layer, consists of a first layer of adsorbed water molecules combined directly with the metal surface atoms and a second layer of water molecules hydrogen-bonded with the first layer of adsorbed

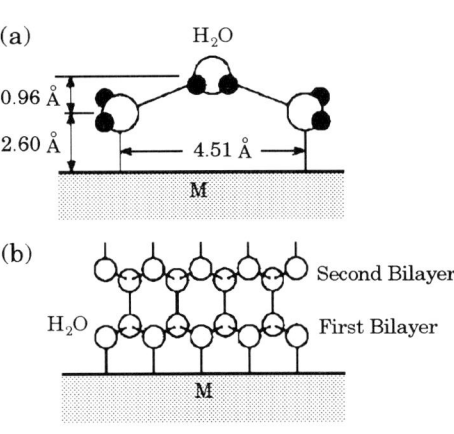

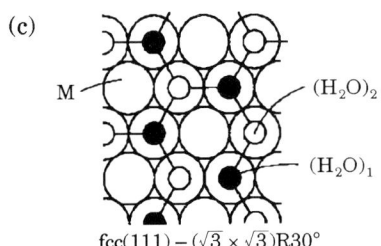

Fig. 5–31. Coordination structure of adsorbed water molecules on the interface of metal electrodes: (a) hydrogen-bonded clusters, (b) bilayer clusters of adsorbed water molecules, (c) the superficial ($\sqrt{3} \times \sqrt{3}$) R30° lattice of adsorbed water molecules on the (111) surface plane of face-centered cubic metals. $(H_2O)_1$ = first layer of adsorbed water molecules. [From Thiel-Madey, 1987.]

water molecules; this array forms a *cluster-like network* structure similar to the basal plane of a hexagonal lattice of ice crystals. On the (111) plane of a face-centered cubic metal surface, the bilayer lattice of the adsorbed water molecules is expressed by ($\sqrt{3} \times \sqrt{3}$) R30°, in which an adsorbed water molecule is overlapped with a metal surface atom at the position $\sqrt{3}$ times the lattice constant rotated by 30° on the surface plane [Soriaga, 1993]. The adsorbed water molecules are not in the frozen state but are in an exchange equilibrium with the water molecules in the aqueous bulk solution.

Fig. 5–32 shows the normalized density profile of oxygen and hydrogen atoms in the adsorption layer of water molecules on the (100) plane of the platinum electrode interface in aqueous solution. The oxygen and hydrogen atoms in the first layer of adsorbed water molecules manifest their maximum density at almost the same position 0.25 nm away from the platinum electrode interface, indicating that the dipole of the adsorbed water molecule is nearly parallel to the electrode interface. This also suggests that, even in the first layer, the water–water interaction predominates over the water–platinum interaction in determining the coordination of adsorbed water molecules. In Fig. 5–32, the normalized density of oxygen atoms is 1.30 times as great as that of hydrogen atoms, indicating that an excess negative charge of water dipoles exists in the first adsorption layer; this excess gradually disappears in the second and succeeding layers of water molecules. Observations have also shown that application of anodic or cathodic potentials to the electrode alters the atomic density profile and separates slightly the position of maximum density of oxygen atoms from that of hydrogen atoms [Heinzinger, 1993].

There are some cases in which the adsorbed water molecules are not in the molecule state but in the state of dissociation forming hydroxyl radicals (-OH)

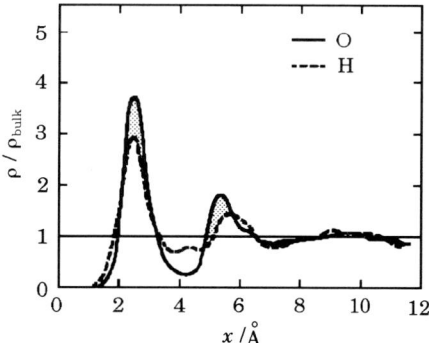

Fig. 5–32. Normalized density profile of oxygen atom and hydrogen atom on the interface (111) plane of platinum electrode in aqueous solution: ρ = atomic density; ρ_{bulk} = mean atomic density in aqueous solution; x = distance normal to the interface. [From Heinzinger, 1993.]

and atomic oxygen radicals (-O). In most of the cases of dissociative adsorption, water molecules are dissociated into hydroxyl radicals and hydrogen atoms (partial dissociation) rather than into oxygen radicals and hydrogen atoms (complete dissociation). For its great electronegativity, the hydroxyl radical tends to receive electrons in its partially occupied frontier orbital (1p) to form an hydroxide ion (OH^-). Hence, the adsorbed hydroxyl radical generates an interfacial dipole; this is in agreement with observations that the work function of metallic palladium increases with the adsorption of hydroxyl radicals [Nyberg-Uvdal, 1986]. In contrast to the adsorbed water molecules which are covalent-electron providers (Lewis base), the adsorbed hydroxyl radicals are covalent-electron receivers (Lewis acid) on metal electrodes.

Whether water molecules are adsorbed in the state of molecules or in the state of partially dissociated hydroxyl radicals is determined by the adsorption energy as shown in Fig. 5–33. In general, sp-metals such as silver, platinum, gold and mercury, for which the adsorption energy of water molecules is relatively small, are apt to adsorb water molecules in the molecule state rather than in the radical state; whereas, the adsorption of hydroxyl radicals predominates on the surface of d-metals such as titanium, chromium and iron on which the adsorption energy of water molecules is relatively large. Cobalt, nickel and copper, which are in an intermediate group for the adsorption energy, adsorb water molecules in the molecule state or in the state of partially dissociated hydroxyl radicals depending upon the crystal plane of metal surfaces.

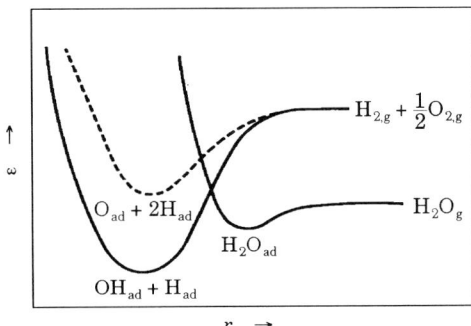

Fig. 5–33. Potential energy curves for water adsorption on metal surfaces in the states of molecules and hydroxyl radicals: ε = energy; r = reaction coordinate; solid curve = adsorption as water molecules and as partially dissociated hydroxyl and hydrogen radicals; broken curve = adsorption of completely dissociated oxygen and hydrogen radicals.

5.7.2 Coadsorption of water molecules and third-particles

Adsorption of third particles other than water molecules on metal electrodes influences the microstructure and the electrochemical activity of electrode interfaces. For example, the interface of a metal electrode usually acts as a Lewis acid in the adsorption of water molecules, but its Lewis acid–base property is altered by the adsorption of third particles. Electronegative particles such as oxygen molecules, if adsorbed, increase the local *Lewis acidity* of interfacial metal atoms around the adsorption sites; whereas, electropositive particles such as sodium atoms, if adsorbed, increase the local *Lewis basicity* around their adsorption sites. Furthermore, the adsorption energy of water molecules is altered by the coadsorption of third particles on metal electrodes.

The adsorption of third particles influences the electronic configuration as well as the coordination structure of coadsorbed water molecules, either accelerating or inhibiting the dissociation of coadsorbed water molecules to hydroxyl radicals. Coadsorbed third particles of which the interaction affinity with water molecules is relatively weak alters the dipole orientation of adsorbed water molecules, thereby changing the work function of the metal surface. As the interaction affinity increases between the third particles and the water molecules, the coadsorption accelerates the dissociation of adsorbed water molecules to hydroxyl radicals, thereby further changing the work function. Fig. 5–34 shows the adsorption coverage of hydroxyl radicals on, and the work function of, the (111) platinum surface observed as a function of the coverage of coadsorbed

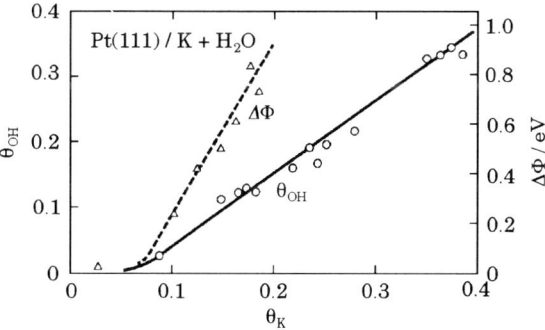

Fig. 5–34. Adsorption coverage of hydroxyl radicals on, and work function of, the platinum (111) surface plane observed as functions of the coverage of potassium atoms coadsorbed with water molecules: the adsorption of water vapor takes place on the potassium-adsorbed surface of platinum at 305 K. θ_K = coverage of adsorbed potassium atoms; θ_{OH} = coverage of hydroxyl radicals adsorbed by partial dissociation of water molecules; $\Delta\Phi$ = change in the work function. [From Bonzel-Pirug-Ritke, 1991; Kiskinova-Pirug-Bonzel, 1985.]

potassium atoms. Here, the dissociative adsorption of hydroxyl radicals increases with increasing coadsorption of electropositive potassium atoms on the platinum surface. It has also been reported that the coadsorption of electronegative oxygen molecules accelerates the adsorption of hydroxyl radicals on the surface of copper, silver and nickel [Thiel-Madey, 1987].

In electrochemistry, the effect of acidic anions on the electrode interface is important. The interaction of the adsorbed water molecules is relatively weak with perchlorate ions (ClO_4^-) and fluoride ions (F^-), intermediate with chloride ions (Cl^-) and sulfate ions (SO_4^{2-}), and relatively strong with phosphate ions (PO_4^{3-}) and bromide ions (Br^-). In a series of halogen anions, fluoride ions (F^-), which interact weakly with the metal surface, are adsorbed as hydrated ions ($H_3O^+ \cdot F^-$); chloride and bromide ions (Cl^- and Br^-), which interact strongly with the metal surface, are adsorbed as dehydrated ions ($Cl^- \cdot M$ and $Br^- \cdot M$).

In general, the contact adsorption of dehydrated anions changes the interfacial lattice structure of adsorbed water molecules, thereby changing the interfacial property. For example, the clean surfaces of metallic gold and silver, which are hydrophobic, become hydrophilic with the contact adsorption of dehydrated halogen anions.

For polyatomic anions, such as ClO_4^-, BF_4^- and PF_6^-, the adsorption layer frequently comprises polyhedral clathrate hydrates forming a cluster network structure of host–guest inclusions, in which a guest polyanion is included in the host cluster network structure of hydrogen-bonded water molecules. With increasing concentration of anions at the interface, the adsorption layer of clathrate hydrates grows into a solid-like film (e.g. $HClO_4 \cdot 5.5\,H_2O$) in which a space charge layer of protons is eventually formed [Borkowska-Stimming, 1993].

5.7.3 Surface lattice transformation due to contact adsorption

The clean surface of metals in vacuum sustains the surface lattice transformation, as described in Sec. 5.1. Similarly, the transformation of interfacial lattice takes place also on metal electrodes in aqueous solutions. In general, the interfacial lattice transformation of metal electrodes is affected by the electrode potential and the contact adsorption of ions.

For example, the clean surfaces of single crystals of metallic gold with the (100) plane sustain the transformation of surface lattice in vacuum as shown in Eqn. 5–52:

$$Au\,(100): (1\times1) \rightleftarrows (5\times20). \tag{5–52}$$

The superficial (5×20) lattice has a greater atomic density and is more stable than the superficial (1×1) lattice on the (100) plane of gold surfaces in vacuum.

Anodic polarization (shifting the potential in the positive direction) of the interfacial (5×20) lattice plane of the gold electrode produces no change in the interfacial lattice structure at potentials less positive than a certain critical potential, E_T; but at potentials more positive than E_T, anodic polarization causes the transformation from the interfacial (5×20) plane to the interfacial (1×1) plane as shown in Fig. 5–35. Further, the potential of zero charge is seen to shift by about 0.22 V between the (5×20) interface and the (1×1) interface. Fig. 5–36 shows that the anodic critical potential E_T for the interfacial lattice transformation depends on the nature and the concentration of hydrated anions, shifting in the cathodic (negative) direction with increasing concentration of anions. On the analogy of the platinum surface in vacuum with and without CO adsorption shown in Fig. 5–2, it appears evident that the stable (5×20) interface of gold electrodes becomes unstable and transforms to the (1×1) interface as a result of anion adsorption. Thus, it is suggested that the critical potential of the interfacial lattice transformation corresponds to the onset potential of the contact adsorption

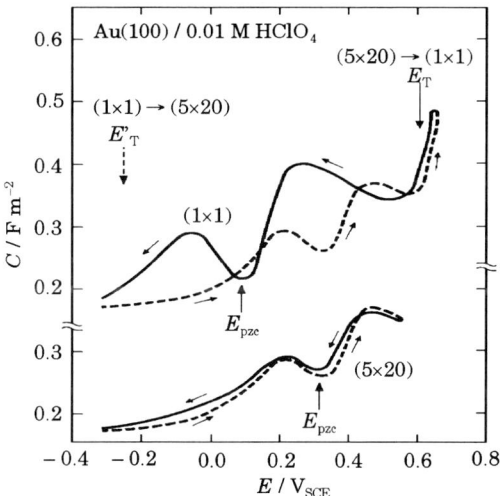

Fig. 5–35. Change in differential capacity of the interfacial double layer leading or not leading to the transformation of interfacial lattice in anodic and cathodic potential sweeps for a gold electrode surface (100) in perchloric acid solution: E_T = critical potential beyond which the interfacial lattice transforms from (5×20) to (1×1); E'_T = critical potential below which the interfacial lattice transforms from (1×1) to (5×20); E_{pzc} = potential of zero charge; V_{SCE} = volt referred to the saturated calomel electrode. [From Kolb-Schneider, 1985.]

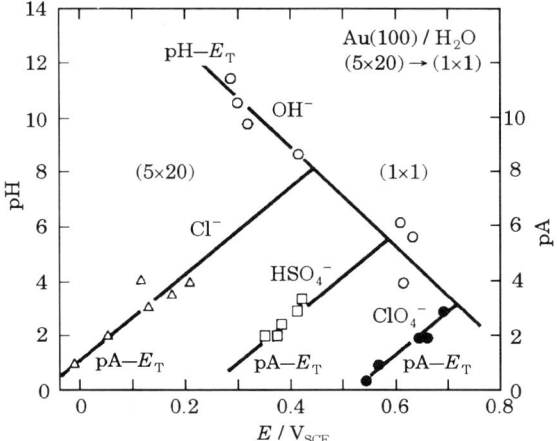

Fig. 5–36. Effects of the concentration of hydrated anions on the critical potential, E_T, beyond which the transformation of interfacial lattice (5×20) → (1×1) occurs on gold electrodes with the (100) plane in aqueous solution: the interfacial (5×20) and (1×1) lattices are stable at potentials more cathodic and more anodic than pA–E_T and pH–E_T curves, respectively. pA = – log (c_i / M); c_i / M = concentration of hydrated anion in molarity M. [From Kolb, 1993.]

of anions. The transformation of interfacial lattice that occurs at the anodic critical potential E_T, therefore, is regarded as an anion-induced transformation of the interfacial lattice.

Cathodically polarizing the anion-adsorbed interface (1×1) of the gold electrode again transforms the (1×1) interface to the (5×20) interface at potentials more cathodic than another critical potential, E'_T, as shown in Fig. 5–35. Fig. 5–37 illustrates the degree of the transformation of interfacial lattice from the (1×1) plane to the (5×20) plane on the gold electrode observed as a function of the electrode potential in acid solutions. Since this reverse transformation is little affected by the nature of anions, it is considered to be a potential-induced or charge-induced transformation of an interfacial lattice. A few cases may arise in which the cation adsorption induces the transformation of interfacial lattice. This potential-induced transformation of interfacial lattice can be explained in terms of an increased electron density at the interface of cathodically polarized electrodes; this increased electron density reinforces the atom–atom bond at the metallic interface to produce an interfacial lattice (5×20) of high atomic density from the interfacial lattice (1×1) of lower atomic density.

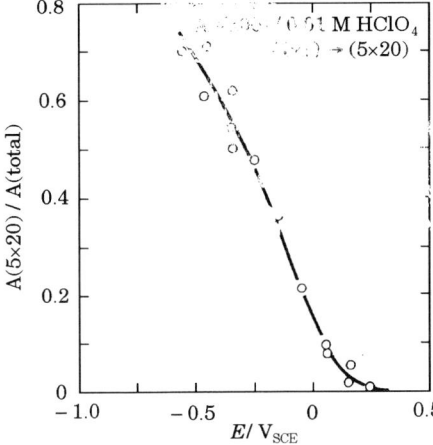

Fig. 5-37. Degree of the transformation of interfacial lattice (1×1) → (5×20) observed as a function of cathodic potential for a gold electrode with the plane (100) in perchloric acid solution: A(total) = total interface area; A(5×20) = interface area of (5×20) supper lattice. [From Kolb-Schneider, 1988.]

The transformation of interfacial lattice has been observed not only on gold metal electrodes but also on other sp-metal electrodes [Kolb, 1993]. In general, on the electrode interface of sp-metals the interfacial lattice of high atomic density vanishes with anion adsorption in anodic polarization and reappears with increasing electron density at the interface in cathodic polarization as shown in Fig. 5-35.

5.7.4 Electron energy levels of adsorbed particles

The frontier electron level of adsorbed particles is split into the occupied level (donor level) in the reduced state (reductant, RED) and the vacant level (acceptor level) in the oxidized state (oxidant, OX), because the reduced and oxidized particles differ from each other both in their respective adsorption energies on the interface of metal electrodes and in the interaction energy with molecules of adsorbed water. The most probable electron levels, ε_{RED} and ε_{OX}, of the adsorbed reductant and oxidant particles are separated from each other by a magnitude equivalent to the reorganization energy $2\lambda_{ad}$ in the same way as occurs with hydrated redox particles described in Sec. 2.10.

We consider adsorbed redox particles comprising protons and hydrogen atoms as shown in Eqn. 5–53:,

$$H^+_{ad} + e_{H^+/H} = H_{ad} \ , \tag{5–53}$$

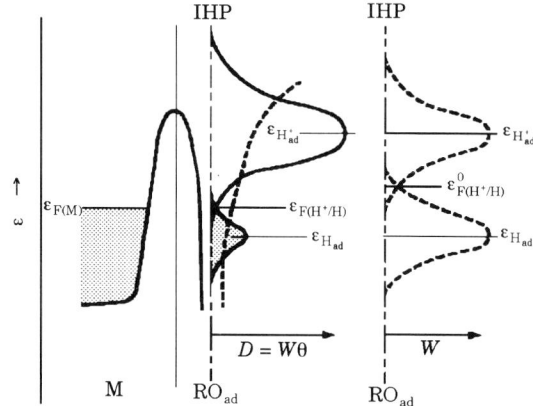

Fig. 5–38. Electron levels of dehydrated redox particles, $H^+_{ad} + e_{H^+/H} = H_{ad}$, adsorbed on the interface of metal electrodes: D = state density (electron level density); θ = adsorption coverage; $\varepsilon_{H^+_{ad}}$ = most probable vacant electron level of adsorbed protons (oxidants); $\varepsilon_{H_{ad}}$ = most probable occupied electron level of adsorbed hydrogen atoms (reductants); RO_{ad} = adsorbed redox particles.

where $e_{H^+/H}$ denotes the interfacial redox electron for the redox equilibrium. Fig. 5–38 shows the electron level for the adsorbed protons and hydrogen atoms at the interface of metal electrodes. The most probable level $\varepsilon_{H^+_{ad}}$ vacant of electrons in the 1s orbital of adsorbed protons (OX) is higher than the most probable level $\varepsilon_{H_{ad}}$ occupied by electrons in the 1s orbital of adsorbed hydrogen atoms (RED); the energy gap nearly equals the difference in their respective adsorption energies on the electrode interface. Owing to the thermal fluctuations the state density (level density) D is distributed around the most probable levels presumably in accordance with the Gaussian distribution function as shown in Eqn. 5–54:

$$D_{H^+} = W_{H^+}\, \theta_{H^+_{ad}}, \qquad D_H = W_H\, \theta_{H_{ad}}, \tag{5–54}$$

where W is the probability density of Gaussian distribution and θ is the coverage of absorbed particles.

In the adsorption equilibrium, the Fermi level $\varepsilon_{F(M)}$ of electrons in the metal electrode equals the Fermi level $\varepsilon_{F(H^+_{ad}/H_{ad})}$ of redox electrons in the adsorbed redox particles; the state density of the occupied electron level equals the state density of the vacant electron level at the Fermi level ($D_{H^+} = D_H$). Assuming the Langmuir adsorption isotherm at low adsorption coverages and the Gaussian distribution for the state density, we obtain Eqn. 5–55 for the Fermi level:

$$\varepsilon_{F(M)} = \varepsilon_{F(H_{ad}^+/H_{ad})} = \varepsilon^0_{F(H_{ad}^+/H_{ad})} + k\,T\ln\left(\frac{\theta_{H_{ad}}}{\theta_{H_{ad}^+}}\right), \qquad (5\text{--}55)$$

where $\varepsilon^0_{F(H_{ad}^+/H_{ad})}$ is the standard Fermi level of redox electrons in the adsorbed redox particles.

The mean charge number $z_{ad} = z_{H^+}\{\theta_{H_{ad}^+}/(\theta_{H_{ad}^+} + \theta_{H_{ad}})\}$ of the adsorbed proton is given in Eqn. 5–56:

$$z_{ad} = \frac{z_{H^+}}{\left(1 + \dfrac{\theta_{H_{ad}}}{\theta_{H_{ad}^+}}\right)} = \frac{z_{H^+}}{\left\{1 + \exp\left(\dfrac{\varepsilon_{F(M)} - \varepsilon^0_{F(H_{ad}^+/H_{ad})}}{k\,T}\right)\right\}}, \qquad (5\text{--}56)$$

where $z_{H^+} = +1$ is the charge number of hydrated protons. The charge transfer coefficient, $\delta z = z - z_{ad}$, introduced in Eqn. 5–48 is, then, obtained in Eqn. 5–57:

$$\delta z = z_{H^+} - z_{ad} = z_{H^+}\left(\frac{\theta_{H_{ad}}}{\theta_{H_{ad}^+} + \theta_{H_{ad}}}\right), \qquad (5\text{--}57)$$

Since the electron transfer of the interfacial redox reaction, $H_{ad}^+ + e_M = H_{ad}$, on electrodes takes place between the inner Helmholtz plane (adsorption plane at distance d_{ad}) and the electrode metal, the ratio of adsorption coverages $\theta_{H_{ad}}/\theta_{H_{ad}^+}$ in the electron transfer equilibrium (hence, the charge transfer coefficient, δz) is given in Eqn. 5–58 as a function of the potential drop $\eta\,(d_{ad}/d_H)\,\Delta\phi_H$ across the inner Helmholtz layer:

$$\left(\frac{\theta_{H_{ad}}}{\theta_{H_{ad}^+}}\right) = K\exp\left(\frac{-z_{H^+}\,e\,\Delta\phi_H\,\eta\left(\dfrac{d_{ad}}{d_H}\right)}{k\,T}\right), \qquad (5\text{--}58)$$

where η is the rate of field increase given in Eqn. 5–50. Consequently, the charge transfer coefficient, δz, changes depending on the electrode potential (increasing with cathodic polarization), except for the case in which the potential drop across the inner Helmholtz layer is constant and independent of the electrode potential.

The same approach may be applied to the adsorption of redox particles other than the adsorption of proton–hydrogen atoms on metal electrodes. To understand electrosorption phenomena, various concepts have been proposed such as the charge transfer coefficient and the adsorption valence [Vetter-Schultze, 1972]. The concept of the redox electron level in adsorbed particles introduced in this textbook is useful in dealing with adsorption at electrodes.

5.8 Electric Double Layer at Semiconductor Electrodes

5.8.1 Electric double layer model

The interfaces of semiconductor electrodes, as has been shown in Fig. 5–7, consists of three layers: the space charge layer on the semiconductor side; the compact layer at the interface; and the diffuse layer on the side of aqueous solution. Fig. 5–39 illustrates the *three-layer model* in the presence of the surface states and adsorbed ions. The interfacial charge consists of: the charge σ_{SC} in the space charge layer; the charge σ_{ss} in the surface states; the charge σ_{ad} of adsorbed ions in the compact layer, and the ionic excess charge σ_S carried by hydrated ions in the diffuse layer of the aqueous solution. Electroneutrality at the interface yields Eqn. 5–59:

$$\sigma_{SC} + \sigma_{ss} + \sigma_{ad} + \sigma_S = 0 \ . \tag{5-59}$$

Further, the total potential difference $\Delta\phi$ across the electrode interface is given by the sum of the potential differences across the space charge layer, the compact layer, and the diffuse layer as shown in Eqn. 5–60:

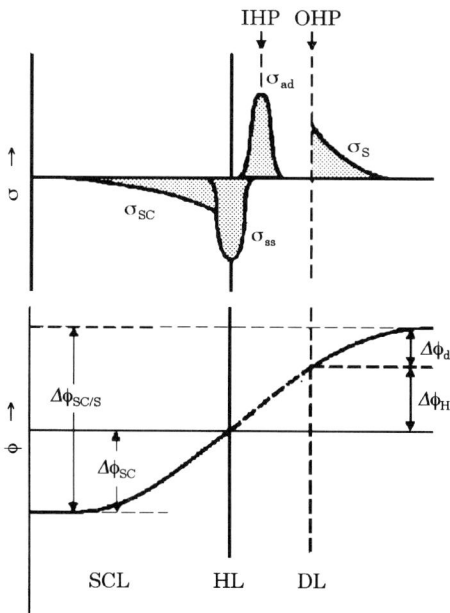

Fig. 5–39. Interfacial electric double layer on semiconductor electrodes: σ_{ss} = charge of the surface state; σ_{ad} = interfacial charge of adsorbed ions; IHP = inner Helmholtz plane.

$$\Delta\phi = \Delta\phi_{SC} + \Delta\phi_H + \Delta\phi_d ,\tag{5-60}$$

where $\Delta\phi_d$ is the potential difference of the diffuse layer referred to the solution interior, $\Delta\phi_H$ is the potential difference of the compact layer referred to the outer Helmholtz plane, and $\Delta\phi_{SC}$ is the potential difference of the space charge layer referred to the semiconductor surface ($\Delta\phi_{SC}$ is often referred to the semiconductor interior in texts of semiconductor physics).

The thickness of these layers is in the range of $d_H = 0.4$ to 0.6 nm for the compact layer, $d_{SC} = 10$ to 1000 nm for the space charge layer, and $d_d = 1$ to 10 nm for the diffuse layer. The thickness of space charge layer, d_{SC}, and the thicknesses of diffuse layer, d_d, depend on the concentrations of mobile charge carries n_i in the semiconductors and in the aqueous solutions, respectively. The Debye length, L_D, may be used as a measure of the thickness of these two respective layers as shown in Eqn. 5–61:

$$L_D = \sqrt{\frac{\varepsilon k T}{e^2 \sum n_i z_i^2}} .\tag{5-61}$$

Further, the thicknesses of the diffuse and space charge layers depend on the potential differences $\Delta\phi_d$ and $\Delta\phi_{SC}$ across the layers; for the space charge layer the thickness, d_{SC}, is expressed, to a first approximation, by $d_{SC} = 2 L_D \times \sqrt{(e \Delta\phi_{SC}/kT) - 1}$ [Memming, 1983]. The Debye length, L_D, is about 100 nm in usual semiconductors with impurity concentrations in the order of 10^{15} cm^{-3} and is about 10 nm in dilute 0.01 M ionic solutions.

The differential electric capacity C of the interface of semiconductor electrodes can be represented by a series connection of three capacities as shown in Eqn. 5–62:

$$\frac{1}{C} = \frac{1}{C_{SC}} + \frac{1}{C_H} + \frac{1}{C_d} ,\tag{5-62}$$

where C_{SC}, C_H, and C_d are the capacities of the space charge layer, the compact layer, and the diffuse layer, respectively. For germanium electrodes that are intrinsic semiconductors $C_{SC} \approx 0.01$ mF, $C_H \approx 10$ mF, and $C_d \approx 0.1$ mF. Since the total capacity is predominantly determined by the smallest capacity, the interfacial capacity of ordinary semiconductor electrodes may be approximated by the capacity of the space charge layer, C_{SC}.

5.8.2 Potential distribution across the electrode interface

The total potential difference $\Delta\phi$ across the electrode interface may be expressed approximately by using the thickness of space charge layer approximated by the Debye length $L_{D,SC}$, the thickness of compact layer d_H, and the thickness of diffuse layer approximated by the Debye length $L_{D,S}$ as shown in Eqn. 5–63:

$$\Delta\phi = \Delta\phi_{SC} + \Delta\phi_H + \Delta\phi_d = E_{SC} L_{D,SC} + E_H d_H + E_S L_{D,S} , \tag{5–63}$$

where E is the field strength in the layers. Further, in the absence of both the surface state charge ($\sigma_{ss} = 0$) and the charge of absorbed ions ($\sigma_{ad} = 0$) electrostatics gives Eqn. 5–6:

$$E_{SC} \varepsilon_{SC} = E_H \varepsilon_H = E_S \varepsilon_S , \tag{5–64}$$

where ε is the permittivity ($\varepsilon_{SC}/\varepsilon_0 \doteq 10$ to 20, $\varepsilon_H/\varepsilon_0 \doteq 5$ to 10, $\varepsilon_S/\varepsilon_0 \doteq 80$ referred to the permittivity ε_0 of the free space).

Therefore, the potential distribution is approximately given by Eqn. 5–65:

$$\frac{\delta\Delta\phi_H}{\delta\Delta\phi_{SC}} \doteq \frac{d_H \varepsilon_{SC}}{L_{D,SC} \varepsilon_H} \ll 1 , \quad \frac{\delta\Delta\phi_d}{\delta\Delta\phi_{SC}} \doteq \frac{L_{D,S} \varepsilon_{SC}}{L_{D,SC} \varepsilon_S} \ll 1 . \tag{5–65}$$

In Eqn. 5–65 it appears that most of the change of electrode potential occurs in the space charge layer with little change of potential both in the compact layer and in the diffuse layer. This pattern may be regarded as characteristic of semiconductor electrodes.

In the case in which the charge of the surface state exists at the electrode interface, electrostatics gives Eqn. 5–66 in place of Eqn. 5–64:

$$E_H \varepsilon_H = E_{SC} \varepsilon_{SC} + \sigma_{ss} \doteq \sigma_{ss} . \tag{5–66}$$

If the charge of the surface state is relatively great, the potential difference of the compact layer, $\Delta\phi_H$, is given by Eqn. 5–67:

$$\Delta\phi_H = E_H d_H = \frac{\sigma_{ss} d_H}{\varepsilon_H} . \tag{5–67}$$

The potential difference $\Delta\phi_{SC}$ of the space charge layer can also be derived as a function of the charge σ_{ss} in the surface state (the surface state density D_{ss} multiplied by the Fermi function). The relationship between of σ_{ss} and $\Delta\phi_{SC}$ thus derived can be compared with the relationship between σ_{ss} and $\Delta\phi_H$ (Eqn. 5–67) to obtain, to a first approximation, Eqn. 5–68 for the distribution of the electrode potential in the space charge layer and in the compact layer [Myamlin-Pleskov, 1967; Sato, 1993]:

$$\frac{\delta\Delta\phi_H}{\delta\Delta\phi_{SC}} \doteq \frac{e d_H \sigma_{ss}}{kT \varepsilon_H} . \tag{5–68}$$

Simple calculation gives a comparable distribution of the electrode potential in the two layers, $(\delta\Delta\phi_H / \delta\Delta\phi_{SC}) \doteq 1$ for a surface state density of about 10^{13} cm^{-3} that is about one percent of the surface atoms of semiconductors. Fig. 5–40 shows the distribution of the electrode potential in the two layers as a function of the surface state density. At a surface state density greater than one percent of the surface atom density, almost all the change of electrode potential occurs in the compact layer, $(\delta\Delta\phi_H / \delta\Delta\phi_{SC}) > 1$, in the same way as occurs with metal electrodes. Such a state of the semiconductor electrode is called the *quasi-metallic state* or *quasi-metallization* of the interface of semiconductor electrodes, which is described in Sec. 5.9 as *Fermi level pinning* at the surface state of semiconductor electrodes.

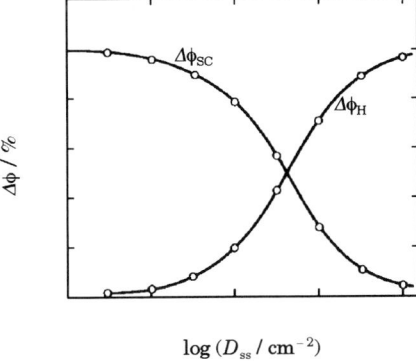

Fig. 5–40. Interfacial potential difference, $\Delta\phi$, distributed to $\Delta\phi_{SC}$ in the space charge layer and to $\Delta\phi_H$ in the compact layer as a function of the concentration of surface states, D_{ss}. [From Chandrasekaran-Kainthla-Bockris, 1988.]

5.9 Band Edge Level Pinning and Fermi Level Pinning

As described in Sec. 5.8, the total difference of potential across the interface of semiconductor electrodes is distributed to the space charge layer in the semiconductor and to the compact layer at the electrode interface if no diffuse layer exists in aqueous solutions. The state of the semiconductor electrode in which the entire change of electrode potential occurs in the space charge layer of the semiconductor is called *band edge level pinning* of the semiconductor electrode. In the state of band edge level pinning, the levels of the conduction and valence band edges at the electrode interface are constant referred to the zero electron level (at the outer potential of the electrode-solution system); hence, the potential difference $\Delta\phi_H$ across the compact layer remains constant and independent of the electrode potential. Therefore, all the change in the electrode potential is distributed in the potential difference $\Delta\phi_{SC}$ across the space charge layer.

On the other hand, the state in which the Fermi level at the interface of semiconductor electrodes is pinned at a specific surface state level of high state density (or at the band edge levels of the electrode interface) is called *Fermi level pinning* of the semiconductor electrode. In such a state of Fermi level pinning, the potential difference $\Delta\phi_{SC}$ of the space charge layer remains constant; and all the change in the electrode potential is distributed in the potential difference $\Delta\phi_H$ across the compact layer. As mentioned in the foregoing, Fermi level pinning also occurs when the Fermi level at the interface is located in the conduction or valence band of semiconductor electrodes.

Fig. 5–41 illustrates the profile of the electron level across the interfacial double layer of a semiconductor electrode: (A) in the state of band edge level

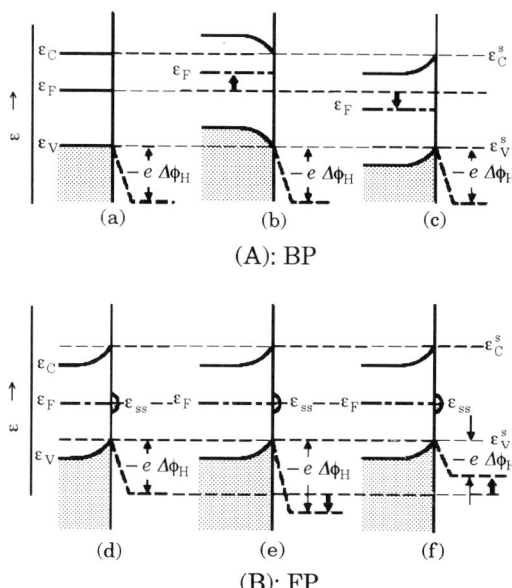

Fig. 5–41. Band edge levels and Fermi level of semiconductor electrodes: (A) the band edge level pinning, (a) the flat band electrode, (b) under cathodic polarization, (c) under anodic polarization; (B) the Fermi level pinning, (d) the initial electrode, (e) under cathodic polarization, (f) under anodic polarization. ε_F = Fermi level; ε_C^s = conduction band edge level at the interface; ε_V^s = valence band edge level at the interface; ε_{ss} = surface state level; $\Delta\phi_H$ = potential difference across the compact layer.

pinning; and (B) in the state of Fermi level pinning. In Fig. 5–41 the cathodic polarization (shifting the electrode potential in the negative direction) raises the Fermi level of the electrode as shown in (b) and (e), while the anodic polarization lowers the Fermi level of the electrode as shown in (c) and (f). Here, the change of electrode potential caused by the anodic or cathodic polarization occurs only across the space charge layer in the state of band edge level pinning, but it occurs only across the compact layer in the state of Fermi level pinning.

Fermi level pinning occurs when the Fermi level at the interface approaches the surface state level ε_{ss} of high state density or the band edge levels ε_C^s and ε_V^s at the interface; whereas, the band edge level pinning occurs when the Fermi level at the interface is located in the band gap away from both the surface state level and the band edge levels. Fig. 5–42 shows the distribution of electrode potential to the space charge layer, $\Delta\phi_{SC}$, and to the compact layer, $\Delta\phi_H$, as a function of the potential difference across the interface, $\Delta\phi_{SC/S}$, (the electrode potential).

As long as the Fermi level at the interface is located away from the surface state in the band gap, the interfacial charge remains constant and produces no

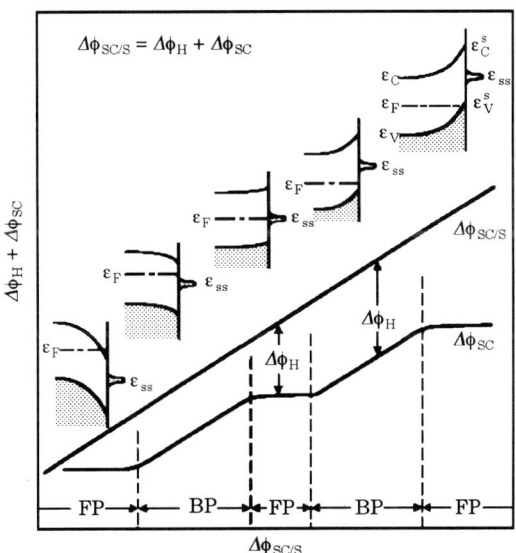

Fig. 5–42. Potential difference, $\Delta\phi_{SC/S}$, across the interface of semiconductor electrodes distributed to the space charge layer, $\Delta\phi_{SC}$, and to the compact layer, $\Delta\phi_H$, as a function of the potential difference, $\Delta\phi_{SC/S}$, over a wide range of electrode potential: BP = band edge level pinning; FP = Fermi level pinning. [From Sato, 1989, 1990, 1993.]

change in the potential difference across the compact layer; and all the change of the electrode potential occurs in the space charge layer. As the Fermi level at the interface approaches the surface state level, the Fermi level is pinned at the surface state until the surface state is fully charged or discharged by accepting or donating electrons thereby altering the interfacial charge and, hence, causing the potential across the compact layer $\Delta\phi_H$ to change.

In the state of Fermi level pinning, the Fermi level at the interface is at the surface state level both where the level density is high and where the electron level is in the state of *degeneracy* similar to the allowed band level for electrons in metals. Therefore, Fermi level pinning is regarded as quasi-metallization of the interface of semiconductor electrodes, making semiconductor electrodes behave like metal electrodes at which all the change of electrode potential occurs in the compact layer.

Such an interfacial degeneracy of electron energy levels (quasi-metallization) at semiconductor electrodes also takes place when the Fermi level at the interface is polarized into either the conduction band or the valence band as shown in Fig. 5–42 (Refer to Sec. 2.7.3.); namely, quasi-metallization of the electrode interface eventually results when semiconductor electrodes are polarized to a great extent in either the anodic or the cathodic direction. This quasi-metallization of electrode interfaces is important in dealing with semiconductor electrode kinetics, as is discussed in Chap. 8. It is worth noting that the interfacial quasi-metallization requires the electron transfer to be in the state of equilibrium between the interface and the interior of semiconductors; this may not be realized with wide band gap semiconductors.

5.10 The Space Charge Layer of Semiconductor Electrodes

5.10.1 Space charge layers

Space charge layers can be classified into four types: (a) the *accumulation layer* in which the mobile majority charge carriers (electrons in the n-type and holes in the p-type) are accumulated; (b) the *depletion layer* in which the mobile majority charge carriers are depleted and the excess charge is carried by immobile ionized donors or acceptors; (c) the *inversion layer* in which the mobile minority charge carriers (holes in the n-type and electrons in the p-type) are accumulated; (d) the *deep depletion layer* in which the mobile minority charge carriers are prevented from accumulating in otherwise the inversion layer, giving rise to an extension of the depletion layer.

Fig. 5–43 and Fig. 5–44 illustrate the band bending and the concentration profile of the charge carriers in these four types of the space charge layers.

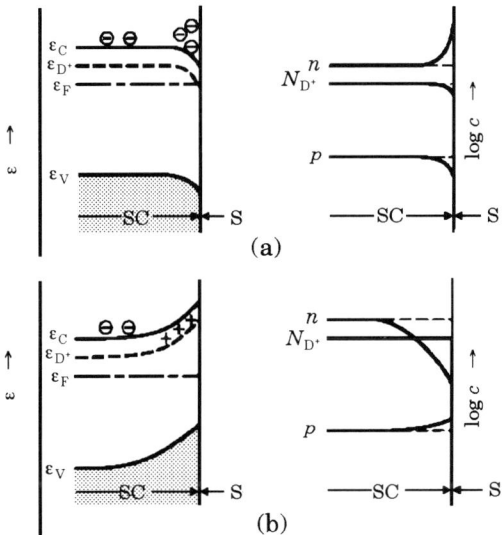

Fig. 5–43. Space charge layers of n-type semiconductor electrodes: (a) the accumulation layer, (b) the depletion layer. ε_D = donor level; c = concentration of particles; n = concentration of electrons in the conduction band; p = concentration of holes in the valence band; N_{D^+} = concentration of ionized donors.

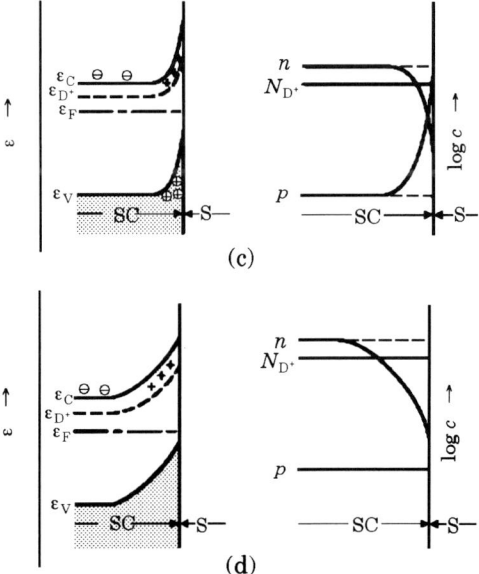

Fig. 5–44. Space charge layers of n-type semiconductor electrodes: (c) the inversion layer, (d) the deep depletion layer.

5.10.2 Differential electric capacity of space charge layers

In the same way as is described in Sec. 5.2 for the diffuse layer in aqueous solution, the differential electric capacity, C_{SC}, of the space charge layer of semiconductors can be derived from the Poisson's equation and the Fermi distribution function (or approximated by the Boltzmann distribution) to obtain Eqn. 5–69 for intrinsic semiconductor electrodes [Gerischer, 1961; Myamlin-Pleskov, 1967; Memming, 1983]:

$$C_{SC} = \frac{\varepsilon}{L_D} \cosh\left(\frac{e\,\Delta\phi_{SC}}{2\,k\,T}\right), \quad L_D = \sqrt{\frac{\varepsilon\,k\,T}{2\,n_i\,e^2}}, \tag{5-69}$$

where L_D is the Debye length, n_i is the concentration of electrons or holes, and $\Delta\phi_{SC}$ is the difference of potential across the space charge layer.

Fig. 5–45 shows the differential capacity for an intrinsic semiconductor electrode of germanium estimated by calculation as a function of the electrode potential. Here, the capacity is minimum at the flat band potential, E_{fb}, where $\Delta\phi_{SC}$ is zero. As the electrode potential shifts so far away from E_{fb} that the Fermi level at the interface may be close to the band edge levels, the Fermi level pinning is realized both with $\Delta\phi_{SC}$ remaining constant and with C_{SC} being constant and independent of the electrode potential.

Further, the capacity of the space charge layer of n-type or p-type semiconductor electrodes, in which all the donors or acceptors are ionized, has been derived as shown in Eqn. 5–70 [Gerischer, 1961]:

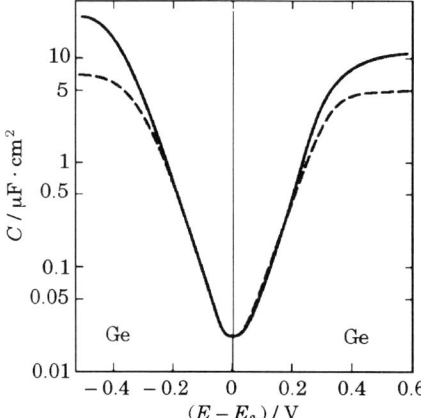

Fig. 5–45. Differential capacity estimated for an electrode of intrinsic semiconductor of germanium by calculation as a function of the electrode potential: C = electrode capacity; solid curve = capacity of the space charge; broken curve = capacity of a series connection of the space charge layer and the compact layer. [From Gerischer, 1961.]

$$C_{SC} = \frac{\varepsilon}{L_{D,\text{eff}}} \frac{1 - \exp\left(\frac{\pm e\, \Delta\phi_{SC}}{kT}\right)}{\sqrt{\left|1 + \frac{\pm e\, \Delta\phi_{SC}}{kT} - \exp\left(\frac{\pm e\, \Delta\phi_{SC}}{kT}\right)\right|}}, \tag{5–70}$$

where $L_{D,\text{eff}} (= \sqrt{\varepsilon kT / 2Ne^2})$ is the effective Debye length, N is the concentration of donors or acceptors, and $\pm$ denotes minus for the n-type and plus for the p-type. Eqn. 5–70 is a general equation for the capacity of the space charge layer from which the capacities of the accumulation, depletion, inversion, and deep depletion layers can be derived in the following:

(a) For the *accumulation layer*, the following approximate equation, Eqn. 5–71, is obtained if the relationship of $|e\, \Delta\phi_{SC}| > 3kT$ holds with $\Delta\phi_{SC} < 0$ in the n-type and with $\Delta\phi_{SC} > 0$ in the p-type:

$$C_{SC} \doteq \frac{\varepsilon}{2 L_{D,\text{eff}}} \sqrt{\exp\left(\frac{|\varepsilon\, \Delta\phi_{SC}|}{kT}\right) - 1} \doteq \frac{\varepsilon}{2 L_{D,\text{eff}}} \exp\left(\frac{|e\, \Delta\phi_{SC}|}{2kT}\right), \tag{5–71}$$

where Boltzmann distribution, instead of Fermi distribution, is used for the electron occupation assuming that the Fermi level is located away from the band edge levels. As the Fermi level at the interface approaches the band edge levels, the capacity of the space charge layer becomes independent of the electrode potential because the Fermi level is pinned in the conduction or valence band.

Fig. 5–46 shows the capacity observed for an n-type semiconductor electrode of zinc oxide in which the accumulation layer is formed at potentials more cathodic

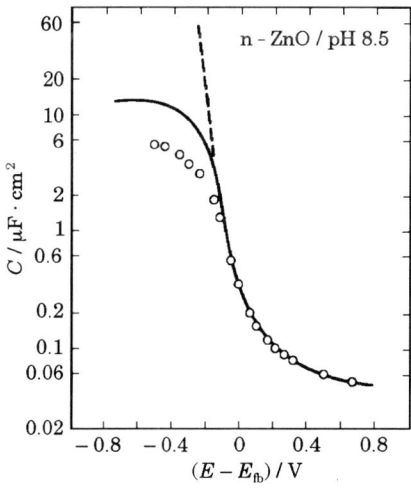

Fig. 5–46. Differential capacity observed and computed for an n-type semiconductor electrode of zinc oxide (conductivity 0.59 S cm^{-1}) in an aqueous solution of 1 M KCl at pH 8.5 as a function of the electrode potential: solid curve = calculated capacity on the Fermi distribution function; dotted curve = calculated capacity on the Boltzmann distribution function. [From Dewald, 1960.]

(negative) and the depletion layer is formed at potentials more anodic (positive) than the flat band potential.

(b) For the *depletion layer*, the following approximate equation, Eqn. 5–72, for C_{SC} is obtained if the relationship of $|e\,\Delta\phi_{SC}| > 3\,kT$ holds with $\Delta\phi_{SC} > 0$ in the n-type and with $\Delta\phi_{SC} < 0$ in the p-type:

$$C_{SC} \doteq \frac{\varepsilon}{2\,L_{D,\text{eff}}} \frac{1}{\sqrt{\left(\dfrac{|e\,\Delta\phi_{SC}|}{kT}\right) - 1}}\,. \tag{5-72}$$

Eqn. (5–72) yields Eqn. 5–73:

$$\frac{1}{C_{SC}^{2}} = \left(\frac{2\,L_{D,\text{eff}}}{\varepsilon}\right)^{2}\left(\frac{e\,\Delta\phi_{SC}}{kT} - 1\right) = \left(\frac{2\,L_{D,\text{eff}}}{\varepsilon}\right)^{2}\frac{e}{kT}\left(E - E_{fb} - \frac{kT}{e}\right), \tag{5-73}$$

where E is the electrode potential and E_{fb} denotes the flat band potential ($\Delta\phi_{SC} = E - E_{fb}$). Eqn. 5–73 is called the Mott-Schottky equation and the plot of $1/C_{SC}^{2}$ versus E is the *Mott-Schottky plot*.

Fig. 5–47 shows the Mott-Schottky plot of n-type and p-type semiconductor electrodes of gallium phosphide in an acidic solution. The Mott-Schottky plot can be used to estimate the flat band potential E_{fb} and the effective Debye length $L_{D,\text{eff}}$. The flat band potential of the p-type electrode is more anodic (positive) than that of the n-type electrode; this difference in the flat band potential between the two types of the same semiconductor electrode is nearly equivalent to the band gap (2.3 eV) of the semiconductor (gallium phosphide).

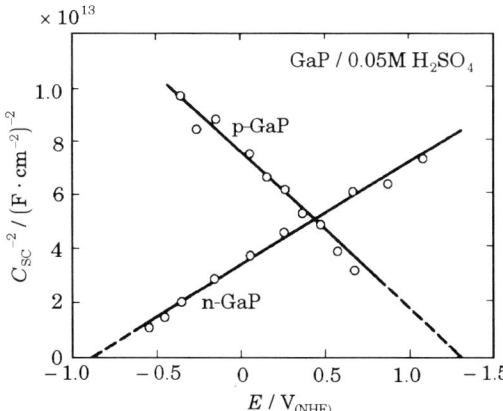

Fig. 5–47. Mott-Schottky plot of the electrode capacity observed for n-type and p-type semiconductor electrodes of gallium phosphide in a 0.05 M sulfuric acid solution. [From Memming, 1969.]

(c) The *inversion layer* is formed, when the Fermi level at the electrode interface approaches the band edge level of the mobile minority charge carriers with increasing potential difference $\Delta\phi_{SC}$ across the depletion layer. Since the minority charge carriers are accumulated in the inversion layer, the interfacial property changes from the n-type to the p-type property or from the p-type to the n-type property as a result of forming the inversion layer. Because the excess charge in the inversion layer is carried by mobile electrons or holes, the capacity C_{SC} of the inversion layer is derived similarly to the capacity of the accumulation layer as shown in Eqn. 5–74:

$$C_{SC} \doteq \frac{\varepsilon}{2 L_{D,\min}} \exp\left(\frac{|e \Delta\phi_{SC}|}{2kT} \right), \quad L_{D,\min} = \sqrt{\frac{\varepsilon k T}{2 n_{\min} e^2}}, \quad (5\text{--}74)$$

where $L_{D,\min}$ and $n_{\min}$ are the Debye length and the concentration of minority charge carriers, respectively. Eqn. 5–74 derived for the inversion layer, in which $\Delta\phi_{SC} > 0$ (n-type), is symmetric with Eqn. 5–71 derived for the accumulation layer, in which $\Delta\phi_{SC} < 0$ (n-type).

As the potential difference $\Delta\phi_{SC}$ of the inversion layer increases and as the Fermi level at the electrode interface coincides with the band edge level, the electrode interface is in the state of degeneracy (Fermi level pinning); and both the capacity C_{SC} and the potential difference $\Delta\phi_{SC}$ are maintained constant. Fig. 5–48 shows schematically the capacity of the space charge layer as a function of the electrode potential. As the electrode potential shifts in the anodic (positive) direction from a cathodic (negative) potential, the accumulation, depletion, and inversion layers are successively formed; here, the capacity of the space charge layer first decreases to a minimum and then increases to a steady value.

(d) The *deep depletion layer* is formed as an extension of the depletion layer not leading to the inversion layer but producing an insulating layer in which no minority charge carriers are accumulated because the minority charge carriers is prevented from transferring between the interface and the interior of the semiconductor. The deep depletion layer predominates over the inversion layer in the case that the band gap of the semiconductor is large or the minority charge carriers are annihilated in the interfacial reaction at the rate greater than the rate of their formation in the semiconductor. In both depletion and deep depletion layers the charge is carried by immobile ionized donors and acceptors. Accordingly, the Mott-Schottky relation given in Eqn. 5–73 applies not only to the depletion layer but also to the deep depletion layer as shown in Fig. 5–48 where the capacity of the deep depletion layer occurs on the extension (dotted curve) from the capacity of the depletion layer.

Fig. 5–49 illustrates the Mott-Schottky plot observed for n-type semiconductor electrodes of zinc oxide in a potential range in which the depletion and deep

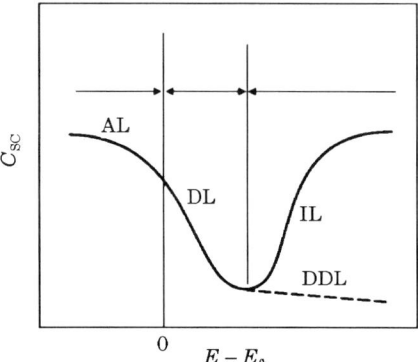

Fig. 5–48. Differential capacity of the space charge layer of an n-type semiconductor electrode as a function of electrode potential: solid curve = electronic equilibrium established in the semiconductor electrode; dotted curve = electronic equilibrium prevented to be established in the semiconductor electrode; AL = accumulation layer; DL = depletion layer; IL = inversion layer; DDL = deep depletion layer.

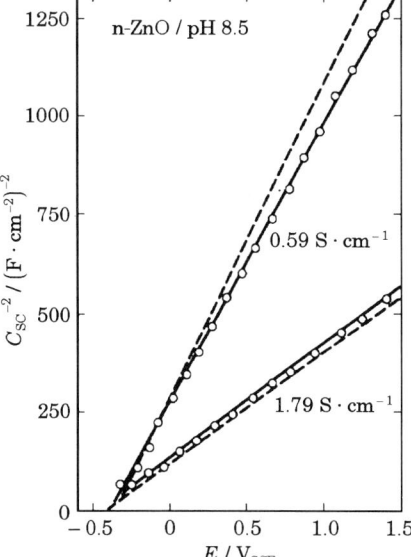

Fig. 5–49. Mott-Schottky plot of the electrode capacity observed for two n-type semiconductor electrodes of zinc oxide (conductivity 0.59 S cm^{-1} and 1.79 S cm^{-1}) in 1 M KCl at pH 8.5: solid curve = observed; dotted curve = calculated. [From Dewald, 1960.]

depletion layers are formed. At these zinc oxide electrodes the rate of annihilation of holes at the interface due to the anodic oxidation of zinc oxide is much greater than the rate of generation of holes in the electrode interior; thereby, forming the deep depletion layer instead of the inversion layer at more anodic potentials. Thus, the capacity of the space charge layer of the zinc oxide electrode is in agreement with the Mott-Schottky relation over a wide range of anodic potentials.

5.10.3 Schottky barrier

The energy barrier of the depletion layer (the potential difference across the depletion layer $|\Delta\phi_{SC}|$) is called the *Schottky barrier* in semiconductor physics. Assuming that all the impurity donors or acceptors are ionized to form a fixed space charge in the depletion layer, we obtain the following approximate equation, Eqn. 5–75, for the thickness of depletion layer, d_{SC}, [Memming, 1983]:

$$d_{SC} \doteq 2 L_{D,\text{eff}} \sqrt{\frac{e\,\Delta\phi_{SC}}{kT} - 1} \doteq 2 L_{D,\text{eff}} \sqrt{\frac{e\,\Delta\phi_{SC}}{kT}} \,. \tag{5–75}$$

Further, the total excess charge σ_{SC} in the depletion layer may be given approximately by the product of the impurity concentration N and the layer thickness d_{SC} in Eqn. 5–76:

$$\sigma_{SC} \doteq e\,N\,d_{SC} \,. \tag{5–76}$$

The thickness of the depletion and deep depletion layers may be approximated by the effective Debye length, $L_{D,\text{eff}}$, given in Eqn. 5–70; $L_{D,\text{eff}}$ is inversely proportional to the square root of the impurity concentration, $\sqrt{N}$. In ordinary semiconductors $L_{D,\text{eff}}$ is in the order of 100 nm. In contrast, the thickness of the accumulation and inversion layers, in which the mobile charge carriers (electrons or holes) are concentrated, is in the order of 5 to 10 nm and is thinner than the thickness of the depletion layer.

5.11 The Compact Layer at Semiconductor Electrodes

5.11.1 Hydroxylation of electrode interfaces

The surface of soluble and insoluble semiconductors is known to sustain *hydroxylation* in the presence of adsorbed water molecules, thereby forming two types of adsorbed hydroxyl groups: the *acid-type* and the *base-type* [Schindler-Stumm, 1987] as shown in Fig. 5–50. The ratio of concentrations of the acid-type to the base-type of surface hydroxyl groups has been considered to be unity (both types in the same concentration) [Boem, 1971]. The acid-type of hydroxyl group -OH(a) donates a proton as an acid and the base-type of hydroxyl group -OH(b) accepts a proton as a base, thereby forming on the interface the *acid site* and the *base site*, respectively. Since the acid site is negatively charged and the base site is positively charged, a hydrated cation K^+_{aq} is adsorbed at the acid site and a hydrated anion A^-_{aq} is absorbed at the base site, creating a compact double layer at the interface of semiconductor electrodes in aqueous solutions as indicated in Eqns. 5–77 and 5–78:

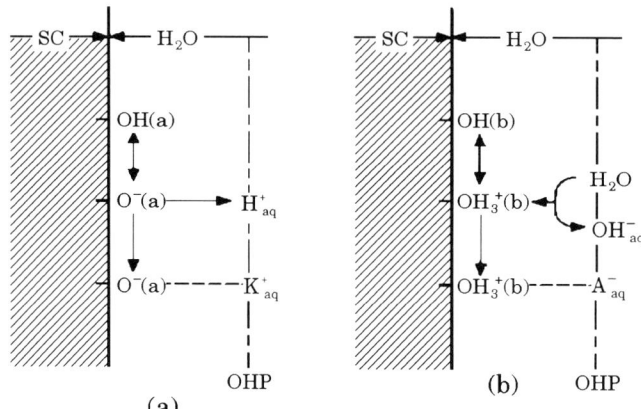

Fig. 5–50. Interfacial hydroxyl groups and their dissociation on semiconductor electrodes: (a) dissociation of the hydroxyl group of the acid-type, (b) dissociation of the hydroxyl group of the base-type. K^+_{aq} = hydrated cation; A^-_{aq} = hydrated anion. [From Tamura-Furuichi, 1991.]

$$-OH(a) + K^+_{aq} \rightleftarrows -O^-(a) \cdot K^+ + H^+_{aq} \,, \qquad \Delta \bar{G}^0_a \qquad (5\text{--}77)$$

$$-OH(b) + H_2O + A^-_{aq} \rightleftarrows -OH_2^+(b) \cdot A^-_{aq} + OH^-_{aq} \,, \qquad \Delta \bar{G}^0_b \qquad (5\text{--}78)$$

The surface acid or base sites in combination with adsorbed hydrated cations or anions, represented by $-O^-(a) \cdot K^+$ or $-OH_2^+(b) \cdot A^-_{aq}$, are called the interfacial ion pair [Tamura-Furuichi, 1991]. In some cases the surface acid or base site is covalent-bonded with dehydrated cations or anions to form an *interfacial complex* [Stumm, 1992].

The electrochemical standard free enthalpy, $\Delta \bar{G}^0$, of the dissociation of the surface acid or base sites consists of the chemical standard free enthalpy, ΔG^0, the electrostatic energy, $\pm e \Delta \phi_H$, and an interaction energy, $m \, \theta$, for the adsorption coverage in the Frumkin adsorption model; $\Delta \phi_H$ is the potential difference across the compact layer, θ is the adsorption coverage, and m is the Frumkin parameter [Frumkin, 1925]:

$$\Delta \bar{G}^0 = \Delta G^0 \pm e \, \Delta \phi_H + m \, \theta \,. \qquad (5\text{--}79)$$

The equilibria of dissociation of the surface acid and base sites are, respectively, represented by Eqn. 5–80 and Eqn. 5–81:

$$\frac{[-O^-(a)][H^+_{aq}]}{[-OH(a)][K^+_{aq}]} = \exp\left(\frac{-\Delta \bar{G}_a}{kT}\right)$$

$$= \exp\left(\frac{-\Delta G_a}{kT}\right)\exp\left(\frac{-e\,\Delta\phi_H}{kT}\right)\exp\left(\frac{-m_a\theta_a}{kT}\right), \tag{5-80}$$

$$\frac{[-OH_2^+(b)][OH^-_{aq}]}{[-OH(b)][H_2O_{aq}][A^-_{aq}]} = \exp\left(\frac{-\Delta \bar{G}_b}{kT}\right)$$

$$= \exp\left(\frac{-\Delta G_b}{kT}\right)\exp\left(\frac{-e\,\Delta\phi_H}{kT}\right)\exp\left(\frac{-m_b\theta_b}{kT}\right), \tag{5-81}$$

where the bracket [ ] denotes the activity of particles on the interface and in the aqueous solution.

In general, the associated acid sites of -OH(a) and the dissociated (ionized) base sites of -OH$_2^+$(b) predominate in acidic solutions, while the associated base sites of -OH(b) and the dissociated acid sites of -O$^-$(a) predominate in basic solutions. Therefore, it follows that the hydroxylated solid interface of semiconductor electrodes is positively charged in acidic solutions and negatively charged in basic solutions.

Assuming that the activities of the interfacial acidic and basic sites remain nearly constant in the range of the coverage from $\theta = 0.1$ to 0.9 and that the relation of $[K^+_{aq}] \doteq [A^-_{aq}]$ holds in aqueous solutions, we obtain Eqn. 5–82 from Eqns. 5–80 and 5–81:

$$\Delta\phi_H = \text{const} + kT\ln[H^+_{aq}] = \text{const} - 2.3\,kT\,(\text{pH}). \tag{5-82}$$

Eqn. 5–82 indicates that the potential difference $\Delta\phi_H$ across the compact layer depends linearly on solution pH. This potential difference $\Delta\phi_H$ includes both the potential difference $\Delta\phi_\sigma$ due to the interfacial charge σ_H ($= [-OH_2^+(b)] - [-O^-(a)]$) and the potential difference $\Delta\phi_{dip}$ due to the interfacial dipole, $\Delta\phi_H = \Delta\phi_\sigma + \Delta\phi_{dip}$, as shown in Fig. 5–51.

The pH at which the interfacial charge of σ_H is zero ($[O^-(a)]/[OH_2^+(b)] = 1$, $\theta_a/\theta_b = 1$) is called the *"iso-electric point pH"*. From Eqns. 5–80 and 5–81 the iso-electric point is obtained in terms of the concentration of hydrated protons, $[H^+_{aq}]_{iep}$, as shown in Eqn. 5–83:

$$[H^+_{aq}]_{iep} = \sqrt{\frac{[-OH(a)]\,K_W}{[-OH(b)]}}\,\exp\left(\frac{\Delta G_b - \Delta G_a + 2\,\Delta\phi_{dip} + m_b q_b - m_a q_a}{kT}\right), \tag{5-83}$$

where K_W is the ionic dissociation constant of water molecules. The iso-electric point pH$_{iep}$ is affected by the contact adsorption of ions; it is increased by the contact adsorption of cations and decreased by the contact adsorption of anions.

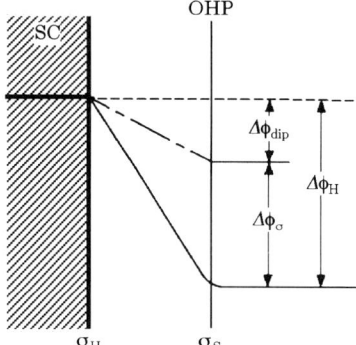

Fig. 5–51. Potential difference $\Delta\phi_H$ across the compact layer without adsorbed ions on semiconductor electrodes: σ_H = interfacial charge of the dissociated hydroxyl group; σ_S = excess charge at OHP on the solution side; $\Delta\phi_{dip}$ = potential difference of the compact layer due to the interfacial dipole; $\Delta\phi_\sigma$ = potential difference of the compact layer due to the interfacial charge.

In terms of the iso-electric point pH_{iep}, Eqn. 5–82 is expressed by Eqn. 5–84:

$$\Delta\phi_H \doteq \Delta\phi_{dip} - 2.3\, k\, T\, (\, pH - pH_{iep}\,)\,, \tag{5-84}$$

which is illustrated in Fig. 5–52.

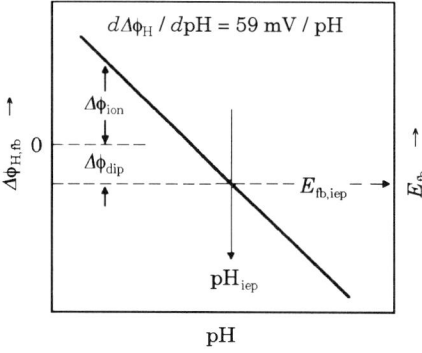

Fig. 5–52. Potential difference $\Delta\phi_{H,fb}$ across the compact layer at the flat band potential as a function of solution pH: pH_{iep} = iso-electric point pH; $E_{fb,iep}$ = flat band potential at the iso-electric point pH_{iep}.

5.11.2 The compact layer

The interfacial excess charge of semiconductor electrodes consists of: the space charge σ_{SC} on the semiconductor side; the charge of the surface states σ_{ss}; the charge σ_H of the interfacial hydroxyl groups; the charge σ_{ad} of adsorbed ions in the compact layer; and the charge σ_S of excess hydrated ions on the solution side as shown in Fig. 5–53. Charge balance at the interfacial electric double layer yields Eqn. 5–85:

$$\sigma_{SC} + \sigma_{ss} + \sigma_H + \sigma_{ad} + \sigma_S = 0 \ . \tag{5-85}$$

At the flat band potential of the semiconductor electrode at which the space charge σ_{SC} is zero, the charge balance is expressed in Eqn. 5–86:

$$\sigma_{ss} + \sigma_H + \sigma_{ad} + \sigma_S = 0 \ . \tag{5-86}$$

For simple semiconductor electrodes on which the charge of surface states σ_{ss} and the charge of adsorbed ions σ_{ad} are zero or remain constant, the flat band potential is obtained from Eqn. 5–84 to give Eqn. 5–87:

$$E_{fb} \doteq E_{fb,\,iep} - 2.3\, k\, T\, (\,\mathrm{pH} - \mathrm{pH}_{iep}\,) \ , \tag{5-87}$$

where $E_{fb,\,iep}$ is the flat band potential at the iso-electric point pH_{iep}.

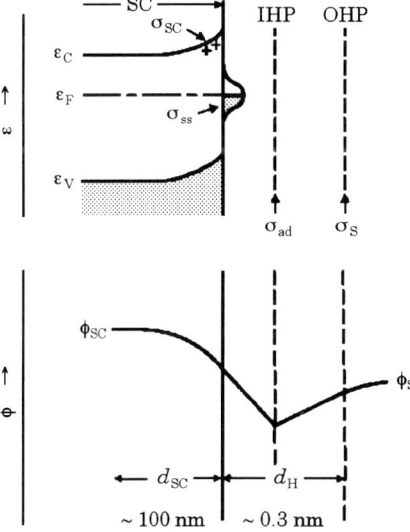

Fig. 5–53. Interfacial charges, electron levels and electrostatic potential profile across the electric double layer with the contact adsorption of dehydrated ions on semiconductor electrodes: σ_{SC} = space charge; σ_{ss} = charge of the surface state; σ_{ad} = ionic charge due to contact adsorption; d_{SC} = thickness of the space charge layer; d_H = thickness of the compact layer.

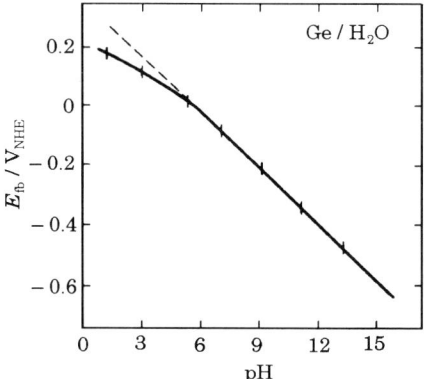

Fig. 5–54. The flat band potential for an intrinsic semiconductor electrode of germanium in aqueous solutions as a function of pH. [From Hoffman-Gerischer, 1960.]

Fig. 5–54 shows the flat band potential observed for an intrinsic semiconductor electrode of germanium in aqueous solutions as a function of pH. The flat band potential, E_{fb}, shifts in the cathodic (negative) direction with increasing pH at the rate of 59 mV/pH in agreement with Eqn. 5–87 at room temperature. The iso-electric point of the germanium electrode is found to be about $\mathrm{pH}_{iep} \doteq 2.5$ which is close to the iso-electric point of germanium oxide.

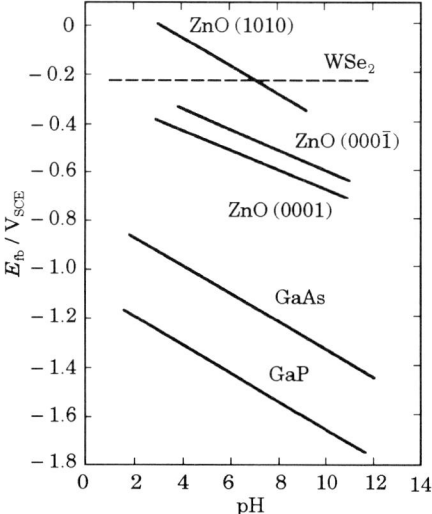

Fig. 5–55. The flat band potential for several compound semiconductor electrodes in aqueous solutions: ZnO with the surface lattice planes of (1010), (000$\bar{1}$) and (0001). [From Gerischer, 1990.]

Such a linear relationship between E_{fb} and pH has also been observed with compound semiconductors [Pleskov, 1980; Morrison, 1980]. Fig. 5–55 shows experimental observations for several compound semiconductors in which the flat band potential depends linearly on pH, except for WSe_2 of which the flat band potential does not depend on pH but depends on the concentration of hydrated selenide ions in aqueous solution.

The ion that determines the potential difference of the compact layer is called the *potential-determining ion*. In cases in which the potential difference of the compact layer is determined by the dissociation reaction of the adsorbed hydroxyl groups, the potential-determining ions are hydrated protons or hydroxide ions. For cadmium sulfide electrodes, the potential-determining ions are not hydrated protons but hydrated sulfide ions; the iso-electric point is at the sulfide ion concentration of 4×10^{-4} M [Ginley-Butler, 1978].

5.11.3 Differential electric capacity of electrode interfaces

The differential capacity of semiconductor electrodes depends on the electrode potential. In general, as the potential of semiconductor electrodes is changed, the capacity of the space charge layer changes more predominantly than the capacity of the compact layer. Fig. 5–56 shows the capacities of the compact and space charge layers obtained by calculation as a function of the electrode potential for an n-type semiconductor. The total capacity C_T is determined mainly by the compact layer ($C_H < C_{SC}$) at cathodic potentials where the semiconductor surface is in the state of degeneracy; whereas, it is determined by the space charge layer ($C_{SC} < C_H$) at anodic potentials where the depletion layer of small electric capacity is formed. Usually, the total capacity C_T of semiconductor electrodes at the flat band potential represents the capacity C_{SC} of the space charge layer.

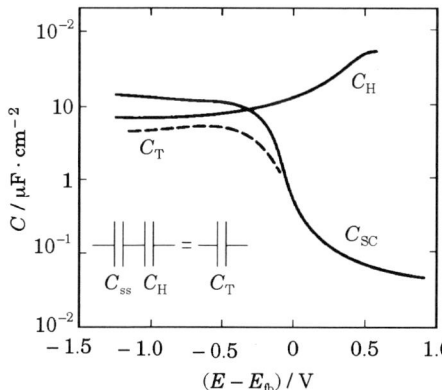

Fig. 5–56. Capacity C_{SC} of the space charge layer and capacity C_H of the compact layer calculated for an n-type semiconductor electrode as a function of the electrode potential: C_T = total capacity of the interfacial double layer ($1/C_T = 1/C_{SC} + 1/C_H$). [From Gerischer, 1990.]

5.12 The Surface State of Semiconductor Electrodes

5.12.1 Surface states

The interface of semiconductor electrodes frequently contains more or less localized electron levels which are called either the *surface states* or the *interface states*. In this textbook we use the term of *surface states*.

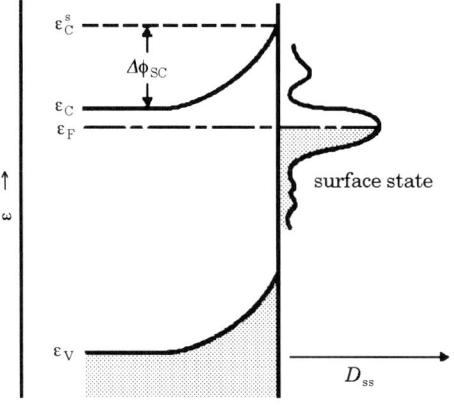

Fig. 5-57. Surface states in the band gap of semiconductors: D_{ss} = surface state density.

The surface states of interest here are those located within the band gap as illustrated in Fig. 5–57. The surface states distributed in the band gap, particularly those of high state density (high density of electron levels), influence the property of the interfacial electric double layer. When the Fermi level of electrode is pinned at a surface state level of high state density (*Fermi level pinning*), the surface state is charged or discharged to change the surface state charge σ_{ss}, thereby changing the potential difference $\Delta\phi_H$ of the compact layer as shown in Fig. 5–41 (B). When the Fermi level at the interface is located away from the surface state level of high state density (*band edge level pinning*), the surface state is either fully occupied or vacant with electrons and the charge in the surface state remains constant, thus causing no change in the potential difference $\Delta\phi_H$ across the compact layer at the interface.

In Eqns. 5–85 and 5–86, we have distinguished the charge σ_{ss} in the surface states from the charge σ_{ad} of adsorbed dehydrated ions. Since these two charges are on the same semiconductor surface, it is apparently difficult to distinguish one from the other. In general, the surface states at interfaces of semiconductor electrodes are classified into two groups: (a) surface states of short relaxation times ($\tau_{ss} \doteq 10^{-3}$ to 10^{-6} s), which belong to the intrinsic surface state such as the surface dangling levels; and (b) surface states of long relaxation times ($\tau_{ss} \doteq 1$ s), which belong to the extrinsic surface state due to adsorbed ions. For ordinary semiconductor electrodes, the surface state of long relaxation time usually predominates over the surface state of short relaxation time.

For covalent bonding semiconductors such as silicon and germanium, the surface dangling bond (Refer to Sec. 2.7.) combines with adsorbed particles such as interfacial hydroxyl groups to form both the bonding orbital (donor level) and the antibonding orbital (acceptor level) as shown in Fig. 5–58. Electrons missing the bonding orbital or occupying the antibonding orbital weaken the covalent bonding between the adsorbed particles and the substrate electrode, thus leading to the desorption of adsorbed particles. For ionic compounds such as metal oxides, the ionic surface states (Refer to Sec. 2.7.) provide the adsorption sites for interfacial hydroxyl groups to form the acid sites at the cationic surface state (metal ions) and the base sites at the anionic surface state (oxide ions).

The interface of the semiconductor electrode adsorbs both ionized basic hydroxyl groups $-OH_2^+(b)$ and associated acidic hydroxyl groups $-OH(a)$ in acidic solutions; these hydroxyl groups introduce the *two dimensional* redox electron levels as identified in Eqns. 5–88 and 5–89:

$$-OH_2^+(b) + e = -OH(a) + H_{ad} , \tag{5–88}$$

$$-OH(a) + e = OH^-_{ad} . \tag{5–89}$$

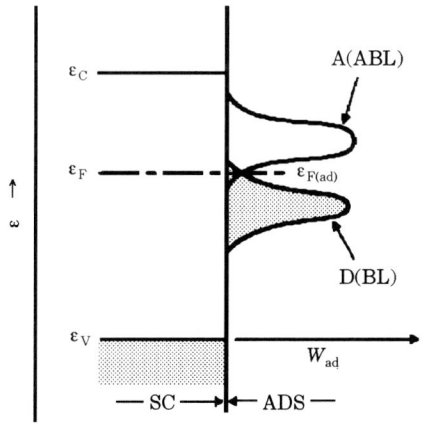

Fig. 5–58. Surface states created by covalently adsorbed particles on semiconductor electrodes: BL = bonding level in adsorption = electron donor level D; ABL = antibonding level in adsorption = electron acceptor level A; W_{ad} = probability density of the adsorption-induced surface state.

These levels of interfacial redox electrons are connected with the hydrogen and oxygen electrode reactions. As noted in Sec. 5.1.2, the electron level of adsorbate particles is broadened by contact adsorption and undergoes the Franck-Condon *level splitting* due to the difference in adsorption energy between the oxidized particle and the reduced particle on the interface of semiconductor electrodes as shown in Fig. 5–59.

The contact adsorption of particles other than hydroxyl groups also generates the energy levels of interfacial redox electrons in the same reason as occurs with the hydroxyl groups.

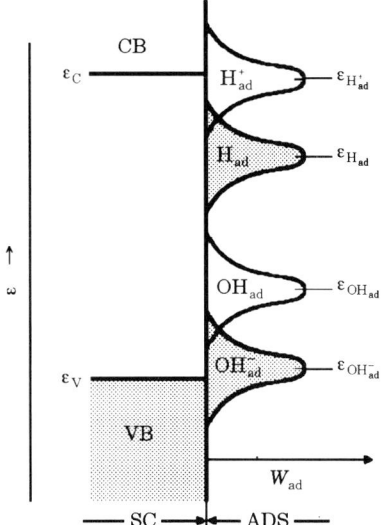

Fig. 5–59. Probability density for the energy level of interfacial redox electrons in the adsorbed redox particles of proton-hydrogen and hydroxyl-hydroxide on the electrode interface of semiconductors: ADS = adsorption; $\varepsilon_{H^+_{ad}}$ = most probable vacant electron level of adsorbed protons.

5.12.2 Differential electric capacity of surface states

As the Fermi level of the electrode approaches the surface state level of high state density, the surface state is charged or discharged as a capacitor. To simplify the treatment, we express the sum of σ_{ss} and σ_{ad} in Eqn. 5–86 as the surface state charge Q_{ss} and the capacity due to the surface state charge as the *surface state capacity* C_{ss}. Then, the interfacial capacity C is represented by the capacity of an equivalent circuit shown in Fig. 5–60.

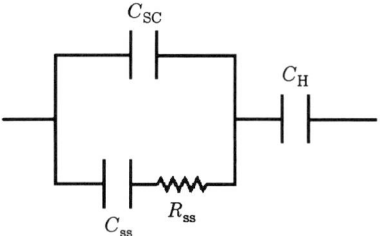

Fig. 5–60. Equivalent circuit for the interfacial electric double layer comprising a space charge layer, a surface state and a compact layer at semiconductor electrodes: C_{SC} = capacity of the space charge layer; C_{ss} = capacity of the surface state; C_H = capacity of the compact layer; R_{ss} = resistance of charging and discharging the surface state.

The surface state capacity, C_{ss}, is apparently zero in the range of potential where the Fermi level is located away from the surface state level (the state of band edge level pinning). As the Fermi level is pinned at the surface state, C_{ss} increases to its maximum which is equivalent to the capacity C_H of the compact layer, because the surface state charging is equivalent to the compact layer charging in the state of Fermi level pinning.

We consider, now, an electron-depleted space charge layer that is gradually polarized in the anodic direction. As long as the Fermi level is located away from the surface state, the interfacial capacity is determined by the capacity of the depletion layer that obeys the Mott-Schottky relation as shown in Fig. 5–61.

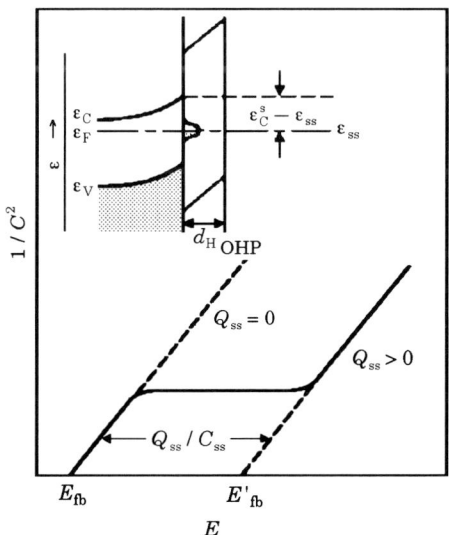

Fig. 5–61. Mott-Schottky plot of an n-type semiconductor electrode in the presence of surface states: E_{fb} = flat band potential with the surface state fully vacant of positive charge; E'_{fb} = flat band potential with the surface state fully occupied by positive charge; Q_{ss} = maximum charge of the surface state; ε_{ss} = surface state level, C_{ss} = capacity of the surface state ($\doteq C_H$).

As the Fermi level reaches the surface state level, the interfacial capacity is determined by the capacity of the compact layer (the maximum capacity of the surface state) and remains constant in the range of potential where the Fermi level is pinned. A further increase in the anodic polarization leads again to the capacity of the depletion layer in accordance with another Mott-Schottky plot parallel to the former plot as shown in Fig. 5–61. The flat band potential, which is obtained from the Mott-Schottky plot, shifts in the anodic direction as a result of anodic charging of the surface state. This shift of the flat band potential equals a change of potential of the compact layer, $\delta\Delta\phi_H$ ($Q_{ss}/C_{ss} = Q_{ss}/C_H$), due to the anodic charging of the surface state.

5.13 The Flat Band Potential of Semiconductor Electrodes

5.13.1 Flat band potential

The electrode potential at which the electron energy band is flat in semiconductor electrodes is called the *flat band potential*, E_{fb}. The flat band potential is used as a *characteristic* potential of individual semiconductor electrodes in the same way as the potential of zero charge is used for metal electrodes. At the flat band potential the space charge, σ_{SC}, is zero but the interfacial charge, $\sigma_{ss} + \sigma_H + \sigma_{ad}$, is not zero. The electrode interface is composed of only the compact layer at the flat band potential if no diffuse layer exists on the solution side.

The flat band potential can be estimated from the Mott-Schottky plot of the electrode capacity in the range of electrode potential where the depletion layer is formed as shown in Fig. 5–47 and in Fig. 5–49. The flat band potential can also be estimated by measuring the photopotential of semiconductor electrodes as shown in Fig. 5–62; the photopotential is zero at the flat band potential.

In the state of band edge level pinning, the *band edge levels* ε_C and ε_V are fixed at certain levels *characteristic* of individual semiconductor electrodes, irrespective of the type of impurity (n or p) in the electrode. On the other hand, the flat band potential, E_{fb}, (the Fermi level of the flat band electrode) depends on whether the semiconductor electrode is n-type or p-type, because the Fermi level is located close to the conduction band edge in the n-type but close to the valence band edge in the p-type semiconductor. Note that n-type E_{fb} is more negative than p-type E_{fb}. As is shown in Figs. 5–47 and 5–62, the flat band potential of semiconductor electrodes is more cathodic (negative) for the n-type than for the p-type.

As is described in Sec. 2.4, the Fermi level of semiconductors approaches the band edge level with increasing impurity concentration in the semiconductor. Therefore, the flat band potential is more cathodic in the n-type electrode, as the

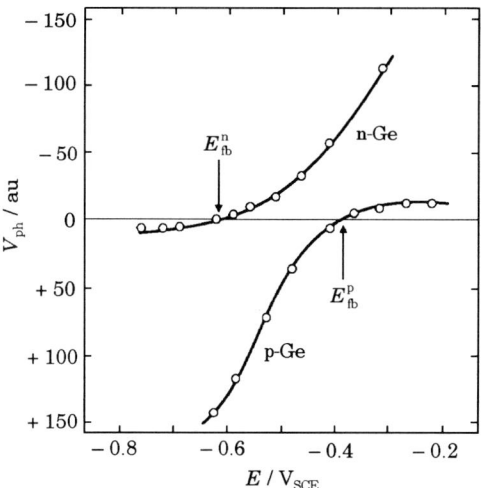

Fig. 5–62. Photopotential observed for n-type and p-type semiconductor electrodes of germanium as a function of the electrode potential: E_{fb}^n = flat band potential of n-type germanium; E_{fb}^p = flat band potential of p-type germanium. [From Pleskov, 1980.]

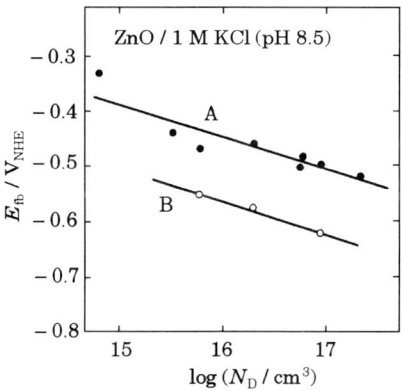

Fig. 5–63. The flat band potential of two n-type semiconductor electrodes of zinc oxide in 1 M KCl (pH 8.5) as a function of donor concentration: A = surface finished in 85 % H_3PO_4; B = surface finished in 2 M KOH; N_D = donor concentration. [From Dewald, 1960.]

concentration of donors increases; whereas, the flat band potential is more anodic in the p-type electrode, as the concentration of acceptors increases. Fig. 5–63 shows the flat band potential of two n-type semiconductor electrodes of zinc oxide observed as a function of the donor concentration in which a linear relationship occurs between the flat band potential and the logarithm of the donor concentration, in agreement with Eqn. 2–22.

The potential difference of the compact layer at the flat band potential depends on the concentration of potential-determining ions in the electrolyte solution. For most cases of the semiconductor electrodes in aqueous solutions, the potential difference across the compact layer is determined by the dissociation of surface hydroxyl groups; hence, the flat band potential is given as a linear function of pH. From Eqn. 5–87 we obtain Eqn. 5–90:

$$E_{fb} = \text{const.} - 2.3\, k\, T\, (\text{pH}) \,. \tag{5–90}$$

However, for chalcogenide compounds (metal sulfides, selenides and tellurides etc.) the potential difference across the compact layer is determined by the concentration of hydrated chalcogenide ions so that the flat band potential does not necessarily depend on pH.

Tables 5–5 and 5–6 show the flat band potentials observed for typical semiconductor electrodes in aqueous solutions.

TABLE 5–5. The flat band potential E_{fb} for typical semiconductor electrodes in aqueous solutions: SC = semiconductor electrode; ε_g = band gap. [From Morrison, 1980.]

SC	ε_g / eV	pH	E_{fb} / V_{NHE}
p-Si	1.1	1	– 0.6
n-SiC	2.9	1	– 1.35
p-SiC	2.9	1	1.7
		14	1.35
n-GaAs	1.43	2.1	– 0.95
		9.2	– 1.35
p-GaAs	1.43	2.1	0.4
		9.2	0.0
n-InP	1.25	2.1	– 0.35
		9.2	– 0.65
		14	– 0.8
p-InP	1.25	2.1	0.9
		9.2	0.7
		14	– 0.1
n-GaP	2.3	2.1	– 1.1
		4.8	– 1.2
		14	– 1.9
p-GaP	2.3	2.1	0.8
		4.8	0.7
		14	0.2

TABLE 5-6. The flat band potential E_{fb}, the iso-electric point pH_{iep}, the potential of the conduction band edge $E^s_{C(iep)}$ at pH_{iep} for metal oxide semiconductor electrodes in aqueous solutions: ε_g = band gap of metal oxides; pH= solution pH at which the flat band potential is measured. [From Morrison, 1980.]

SC	ε_g / eV	pH	E_{fb} / V_{NHE}	pH_{iep}	$E^s_{C(iep)}$ / V_{NHE}
n-Fe$_2$O$_3$	2	9	0.15	8.6	0.2
		13		− 0.1	
n-CdO	2.1	13	0.0	12	0.1
n-WO$_3$	2.6	1	0.1	0.4	0.8
		9	− 0.5		
		13	0.0		
n-TiO	3	2.1	− 0.05	5.8	− 0.3
		13	− 0.6		
n-Ta$_2$O$_5$	3.1(?)	13	− 1.2	2.9	− 0.6
n-ZnO	3.2	4.8	− 0.1	8.8	− 0.6
		13	− 0.85		
n-SnO$_2$	3.7	1	0.5	4.3	0.1
		13	− 0.45		
n-ZrO$_2$	5.0	13	− 1.8	6.7	− 1.4

5.13.2 Band edge potential

The flat band potential E_{fb} (the Fermi level ε_F at the flat band) depends on the impurity concentration; but the band edge potentials E^s_C and E^s_V (the band edge levels ε^s_C and ε^s_V) are characteristic of individual semiconductor electrodes, irrespective of the type of impurity (n-type or p-type). The relationship between the flat band potential and the band edge potentials is given in Eqn. 5–91:

$$E^s_C = E^n_{fb} - \frac{\Delta\varepsilon_{C/F}}{e}, \quad E^s_V = E^p_{fb} - \frac{\Delta\varepsilon_{V/F}}{e}, \tag{5-91}$$

where $\Delta\varepsilon_{C/F}$ ($= \varepsilon^s_C - \varepsilon^n_F$) is the difference between the conduction band edge ε^s_C and the Fermi level ε^n_F for the n-type; and $\Delta\varepsilon_{V/F}$ ($= \varepsilon^s_V - \varepsilon^p_F$) is the difference between the Fermi level ε^n_F and the valence band edge ε^s_C for the p-type semiconductors. In ordinary semiconductors, $\Delta\varepsilon_{C/F} \doteq 0.1$ eV and $\Delta\varepsilon_{V/F} \doteq - 0.1$ eV.

Referring to the iso-electric point pH_{iep}, we obtain Eqns. 5–92 and 5–93:

$$E^s_C = E^s_{C, iep} - 2.3 \, k \, T \, (pH - pH_{iep}), \quad E^s_{C, iep} = E^n_{fb, iep} - \frac{\Delta\varepsilon_{C/F}}{e}, \tag{5-92}$$

$$E^s_V = E^s_{V, iep} - 2.3 \, k \, T \, (pH - pH_{iep}), \quad E^s_{V, iep} = E^p_{fb, iep} - \frac{\Delta\varepsilon_{V/F}}{e}. \tag{5-93}$$

Table 5–6 shows for semiconductor metal oxides the flat band potential E_{fb} in aqueous solutions at various pH values and the conduction band edge potential E_C^s at the iso-electric point pH_{iep}.

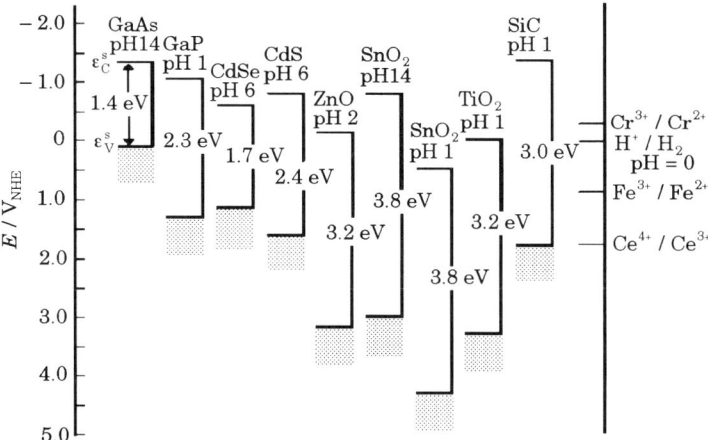

Fig. 5–64. Band edge levels of compound semiconductor electrodes in aqueous solutions at different pH values: hydrated redox particles and their standard redox potentials are on the right hand side. [From Gleria-Memming, 1975.]

Fig. 5–64 shows the band edge potential for compound semiconductor electrodes in aqueous solutions, in which the standard redox potentials (the Fermi levels) of some hydrated redox particles are also shown on the right hand side. In studying reaction kinetics of redox electron transfer at semiconductor electrodes, it is important to find the relationship between the band edge level (the band edge potential) and the Fermi level of redox electrons (the redox potential) as is described in Chap. 8.

References

[Amokrane-Badiali, 1992]: S. Amokrane and J. P. Badiali, *Modern Aspects of Electrochemistry*, **No. 22**, (Edited by J. O'M. Bockris et al.), p. 1, Plenum Press, New York, (1992).

References

[Anderson-Bockris, 1964]: T. N. Anderson and J. O'M. Bockris, *Electrochimica Acta*, **9**, 347(1964).

[Bange-Straehler-Sass-Parsons, 1987]: K. Bange, B. Straehler, J. K. Sass and R. J. Parsons, *J. Electroanal. Chem.*, **229**, 87(1987).

[Benard, 1983]: J. Benard, *Adsorption on Metal Surface, Studies in Surface Science and Catalysis* **13**, p. 150, Elsevier Sci. Pub. Co., Amsterdam, (1983).

[Blakely, 1973]: J. M. Blakely, *Introduction to the Properties of Crystal Surfaces*, p. 79-106, Pergamon Press, Oxford, (1973).

[Bockris-Devanathan-Muller, 1963]: J. O'M. Bockris, M. A. V. Devanathan and K. Muller, *Proc. Royal Soc., London Ser.*, **A274**, 55(1963).

[Bode, 1972]: D. D. Bode Jr., *J. Phys. Chem.*, **76**, 2915(1972).

[Boem, 1971]: H. P. Boem, *Disc. Faraday Soc.*, **52**, 264(1971).

[Bonzel-Pirug-Ritke, 1991]: H. P. Bonzel, G. Pirug and C. Ritke, *Langmuir*, **7**, 3006(1991).

[Borkowska-Stimming, 1993]: Z. Borkowska and U. Stimming, *Structure of Electrified Interfaces*, (Edited by J. Lipkowski and P. N. Ross), p. 277, VCH Publishers, Inc., New York, (1993).

[Chandrasekaran-Kainthla-Bockris, 1988]: K. Chandrasekaran, R. C. Kainthla and J. O'M. Bockris, *Electrochimica Acta*, **33**, 327(1988).

[Dewald, 1960]: J. F. Dewald, *Bell System Tech. J.*, **39**, 615(1960).

[Ertl, 1985]: G. Ertl, *Surface Science*, **152/153**, 328(1985).

[Frumkin, 1925]: A. N. Frumkin, *Z. Phys. Chem.*, **166**, 466(1925).

[Gerischer, 1961]: H. Gerischer, *Advances in Electrochemistry and Electrochemical Engineering.*, **Vol. 1**, (Edited by P. Delahay), p. 139, Interscience Publishers, New York, (1961).

[Gerischer, 1990]: H. Gerischer, *Electrochimica Acta*, **35**, 1677(1990).

[Gies-Gerhardt, 1985]: P. Gies and R. R. Gerhardt, *Phys. Review*, **B 31**, 6843(1995): **B 33**, 982(1996).

[Ginley-Butler, 1978]: D. S. Ginley and M. A. Butler, *J. Electrochem. Soc.*, **125**, 1968(1978).

[Gleria-Memming, 1975]: M. Gleria and R. Memming, *J. Electroanal. Chem.*, **65**, 167(1975).

[Gomer-Swanson, 1963]: R. Gomer and L. W. Swanson, *J. Chem. Phys.*, **38**, 1613(1963).

[Gouy, 1917]: A. Gouy, *Ann. Phys.*, **7**, 129(1917).

[Grahame, 1947]: D. C. Grahame, *Chem. Review*, **41**, 441(1947).

[Grahame, 1954]: D. C. Grahame, *J. Am. Chem. Soc.*, **76**, 4819(1954).

[Gurney, 1935]: R. W. Gurney, *Phys. Review*, **47**, 479(1935).

[Heinzinger, 1993]: K. Heinzinger, *Structure of Electrified Interfaces*, (Edited by J. Lipkowski and P. N. Ross), VCH Publishers, p. 239, Inc., New York, (1993).

[Helmholtz, 1879]: H. Helmholtz, *Wied. Ann.*, **7**, 337(1879).
[Hoffman-Gerischer, 1960]: M. Hoffman and H. Gerischer, *Z. Elektrochem.*, **65**, 771(1960).
[Jiang-Seo-Sato, 1990]: X. C. Jiang, M. Seo and N. Sato, *J. Electrochem. Soc.*, **137**, 3804(1990).
[Kiskinova-Pirug-Bonzel, 1985]: M. Kiskinova G. Pirug and H. P. Bonzel, *Surface Science*, **150**, 319(1985).
[Kolb, 1993]: D. M. Kolb, *Structure of Electrified Interfaces*, (Edited by J. Lipkowski and P. N. Ross), p. 65, VCH Publishers, Inc., New York, (1993).
[Kolb-Schneider, 1985]: D. M. Kolb and J. Schneider, *Surface Science*, **162**, 764(1985).
[Kolb-Schneider, 1988]: D. M. Kolb and J. Schneider, *Surface Science*, **193**, 579(1988).
[Kornyshev-Schmickler, 1985]: A. A. Kornyshev and W. Schmickler, *J. Electroanal. Chem.*, **185**, 253(1985).
[Kummer-Young, 1963]: J. T. Kummer and J. D. Young, *J. Phys. Chem.*, **67**, 107(1963).
[Lang-Kohn, 1973]: N. D. Lang and W. Kohn, *Phys. Rev.*, **B 8**, 6010(1973).
[Langmuir, 1918]: I. Langmuir, *J. Am. Chem. Soc.*, **40**, 1361(1918).
[Liu, 1983]: S. H. Liu, *J. Electroanal. Chem.*, **150**, 305(1983).
[Memming, 1969]: R. Memming, *J. Electrochem. Soc.*, **116**, 785(1969).
[Memming, 1983]: R. Memming, *Comprehensive Treatise of Electrochemistry*, **Vol. 7**, (Edited by B. E. Conway et al.), p. 59, Plenum Press, New York, (1983).
[Morrison, 1980]: S. R. Morrison, *Electrochemistry at Semiconductor and Oxidized Electrodes*, Plenum Press, New York, (1980).
[Myamlin-Pleskov, 1967]: V. A. Myamlin and Y. V. Pleskov, *Electrochemistry of Semiconductors*, Plenum Press, New York, (1967).
[Nii-Yoshihara, 1980]: K. Nii and K. Yoshihara, *Trans. Japan Inst. Metals*, **19**. 857(1980).
[Nyberg-Uvdal, 1986]: C. Nyberg P. Uvdal, *J. Chem. Phys.*, **84**, 4631(1986).
[Pleskov, 1980]: Yu. V. Pleskov, *Comprehensive Treatise of Electrochemistry*, **Vol. 1**, (Edited by J. O' M. Bockris et al.), p. 291, Plenum Press, New York, (1980).
[Reeves, 1974]: R. M. Reeves, *Modern Aspects of Electrochemistry*, **No. 9**, (Edited by B. E. Conway et al.), p. 239, Plenum Press, New York, (1974).
[Reeves, 1980]: R. M. Reeves, *Comprehensive Treatise of Electrochemistry*, **Vol. 1**, (Edited by J. O' M. Bockris et al.), p. 83, Plenum Press, New York, (1980).
[Russier-Rosinberg, 1988]: V. Russier and M. L. Rosinberg, *J. Physics*, **C 21**, L333(1988).
[Sato, 1989]: N. Sato, *Corrosion*, **45**, 354(1989).
[Sato, 1990]: N. Sato, *Iron Steel Inst. Japan (Tetsu-to-Hagane)*, **76**, 1423(1990).

[Sato, 1993]: N. Sato, *Electrode Chemistry*, **Vol. 1**, p. 253, p. 261, Japan Technical Information Service, Tokyo, (1993).

[Schindler-Stumm, 1987]: P. W. Schindler and W. Stumm, *Aquatic Surface Chemistry*, p. 83, John Wiley & Sons, New York, (1987).

[Schmickler, 1983]: W. Schmickler, *J. Electroanal. Chem.*, **157**, 1(1983).

[Schmickler, 1993]: W. Schmickler, *Structure of Electrified Interfaces*, (Edited by J. Lipkowski and P. N. Ross), p. 201, VCH Publishers, Inc., New York, (1993).

[Soriaga, 1993]: M. P. Soriaga, *Structure of Electrified Interfaces*, (Edited by J. Lipkowski and P. N. Ross), p. 103, VCH Publishers, Inc., New York, (1993).

[Stern, 1924]: O. Stern, *Z. Elektrochem.*, **30**, 508(1924).

[Stumm, 1992]: W. Stumm, *Chemistry of the Solid-Water Interface*, John Wiley & Sons, New York, (1992).

[Tamura-Furuichi, 1991]: H. Tamura and R. Furuichi, *Bunseki Kagaku*, **40**, 635(1991).

[Thiel-Madey, 1987]: P. A. Thiel and T. E. Madey, *Surface Science Rep.*, **7**, 211(1987).

[Tosi-Doyama, 1966]: M. P. Tosi and M. Doyama, *Phys. Rev.*, **151**, 642(1966).

[Trasatti, 1971]: S. Trasatti, *J. Electroanal. Chem.*, **33**, 351(1971).

[Trasatti, 1975]: S. Trasatti, *J. Electroanal. Chem.*, **64**, 128(1975).

[Trasatti, 1979]: S. Trasatti, *Modern Aspects of Electrochemistry*, No. **13**, (Edited by B. E. Conway et al.), p. 81, Plenum Press, New York, (1979).

[Van Hove, 1993]: M. A. Van Hove, *Structure of Electrified Interfaces*, (Edited by J. Lipkowski and P. N. Ross), p. 1, VCH Publishers, Inc., New York, (1993).

[Vetter-Schultze, 1972]: K. J. Vetter and J. W. Schultze, *Ber. Bunsenges. Phys. Chem.*, **76**, 920, 927(1972).

CHAPTER 6

ELECTROCHEMICAL CELLS

6.1 Electrochemical Cells

A pair of electrodes immersed in an electrolyte solution constitutes an *electrochemical cell*, which is occasionally called a *Galvanic cell* after L. Galvani (electrochemist, 1737–1798) as shown in Fig. 6–1. The term of *Voltaic cell*, which was originally assigned to copper-zinc piles in sulfuric acid (invented by A. Volta, electrochemist, 17457–1827), is occasionally used for electricity-producing cells. When the Fermi level of one electrode M_L on the left hand side is higher (the electrode potential is lower) than the Fermi level of the other electrode M_R on the right hand side as shown in Fig. 6–1, the positive electric charge flows from the electrode M_L through the electrolyte to the electrode M_R with a simultaneous flow of electrons from the electrode M_L through an external cell circuit (containing no electric power source) to the electrode M_R.

The electrode from which the positive electric charge flows into the electrolyte is termed the *anode*; and the electrode to which the positive electric charge flows from the electrolyte is termed the *cathode* as described in Sec. 4.1.

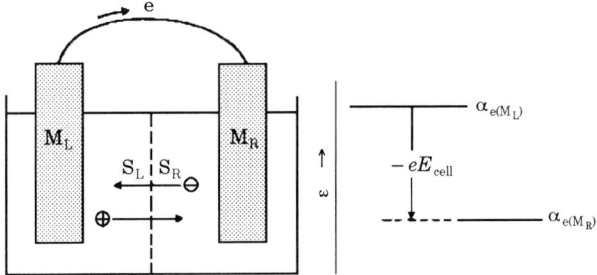

Fig. 6–1. Electrochemical cell, electric charge flow in the closed cell circuit, and the electron levels of two electrodes in an open cell circuit: M = electrode; S = electrolyte solution; α_e = real potential of electrons in the electrode, E_{cell} = electromotive force.

From a practical viewpoint, as shown in Fig. 6–2, the electrochemical cells can be classified into two groups: one is the *chemical cell* in which electricity is produced by consuming the chemical energy of substances; the other is the *electrolytic cell* in which chemical substances are produced by consuming the electrical energy. In practice, the chemical cell is connected to an external load and the electrolytic cell is connected to an external electric power source.

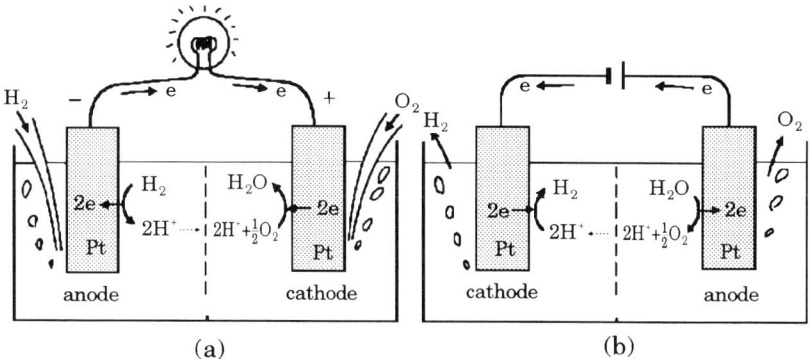

Fig. 6–2. Chemical cell and electrolytic cell: (a) hydrogen–oxygen fuel cell (chemical cell), (b) water decomposition cell (electrolytic cell).

The hydrogen–oxygen fuel cell is a typical example of the chemical cell for which the cell reactions are represented in Eqn. 6–1:

$$H_{2,\,gas} \rightarrow 2\,H^+_{aq} + 2\,e\,, \qquad \text{left electrode, anode,}$$
$$\tfrac{1}{2}\,O_{2,\,gas} + 2\,H^+_{aq} + 2\,e \rightarrow H_2O_{aq}\,, \qquad \text{right electrode, cathode,}$$

$$H_{2,\,gas} + \tfrac{1}{2}\,O_{2,\,gas} \rightarrow H_2O_{aq}\,, \qquad A_{H_2/O_2} = -\Delta G_{H_2/O_2}\,, \qquad (6\text{–}1)$$

where A_{H_2/O_2} is the affinity of the reaction and $\Delta G_{H_2/O_2}$ is the free enthalpy of the reaction. This cell may also regarded as a cell of water synthesis which produces the electric energy in the course of synthesizing water.

On the other hand, a typical electrolytic cell is the cell of water decomposition for which the cell reactions are given in Eqn. 6–2:

$2H^+_{aq} + 2e \rightarrow H_{2,gas}$,   left electrode, cathode,
$H_2O_{aq} \rightarrow \frac{1}{2}O_{2,gas} + 2H^+_{aq}$,   right electrode, anode,

$$H_2O_{aq} \rightarrow H_{2,gas} + \frac{1}{2}O_{2,gas} . \tag{6-2}$$

This reaction of water electrolysis is the same as, but reverse in its direction to, the reaction of hydrogen–oxygen fuel cell in Eqn. 6–1. Note that the anode and the cathode are reversed in the chemical and the electrolytic cells.

The "whole cell reaction" that occurs when the cell circuit is closed consists of two electrochemical reactions: one at the anode and the other at the cathode. As an example, the following anodic and cathodic ion transfer reactions,

$Zn \rightarrow Zn^{2+}_{aq} + 2e$,   anodic reaction,

$Cu^{2+}_{aq} + 2e \rightarrow Cu$,   cathodic reaction,

can be coupled to produce the whole cell reaction given by Eqn. 6–3:

$$Zn + Cu^{2+}_{aq} \rightarrow Zn^{2+}_{aq} + Cu . \tag{6-3}$$

According to the convention in IUPAC (International Union of Pure and Applied Chemistry) [Whiffen, 1979], the cell of Eqn. 6–3 is illustrated by the cell diagram shown in Eqn. 6–4:

$$Zn \mid Zn^{2+}_{aq} : Cu^{2+}_{aq} \mid Cu , \tag{6-4}$$

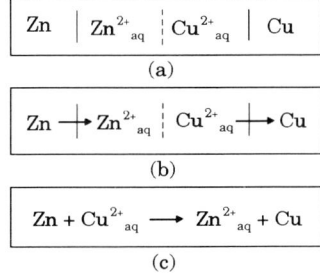

Fig. 6–3. Expression of cell diagrams and cell reactions (IUPAC convention): (a) cell diagram, (b) flow of electric charge, (c) cell reaction.

where the vertical solid line denotes the interface between electrode and electrolyte, and the vertical dotted line denotes the junction of two miscible electrolytes.

As shown in Fig. 6–3, it is also in the same IUPAC convention that the positive electric charge flows from the left hand electrode through the electrolyte to the right hand electrode, as the cell reaction proceeds in the direction as written in Eqn. 6–3. This defines the sign of the electromotive force of electrochemical cells.

The charge number, n, which represents the number of electrons involved in a unit advancement of the cell reaction, is also important; in the cell reactions of Eqns. 6–1 and 6–3 $n = 2$.

6.2 Electromotive Force of Electrochemical Cells

The *electromotive force* of an electrochemical cell is the difference of electrode potential between the two electrodes in the cell. According to the IUPAC convention, the electromotive force is the potential of the right hand electrode referred to the potential of the left hand electrode of the cell. We consider, for example, a hydrogen–oxygen cell shown in Fig. 6–4; the cell reaction is given by Eqn. 6–1 and the cell diagram is given by Eqn. 6–5:

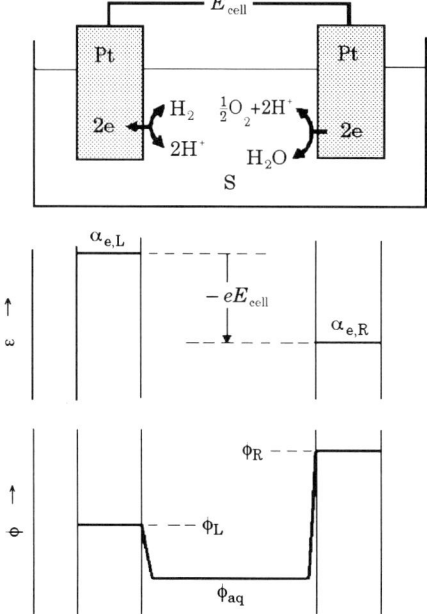

Fig. 6–4. Electromotive force E_{cell}, electron level α_e, and electrostatic potential profile for an electrochemical cell: ϕ = inner potential.

$$Pt_L \mid H_{2,\,gas} \mid H^+_{aq},\, H_2O \mid O_{2,\,gas} \mid Pt_R \quad . \tag{6-5}$$

From the reaction equilibria at the left hand electrode (anode) and at the right hand electrode (cathode), we obtain the real potential, α_e, of electrons in the two electrodes as shown in Eqns. 6–6 and 6–7:

$$\alpha_{e,L} = \frac{\mu_{H_2,\,gas} - 2\,\alpha_{H^+_{aq}}}{2} \quad , \tag{6-6}$$

$$\alpha_{e,R} = \frac{\mu_{H_2O} - 2\,\alpha_{H^+_{aq}} - \frac{1}{2}\,\mu_{O_2,\,gas}}{2} \quad . \tag{6-7}$$

The electromotive force, E_{H_2/O_2}, of the cell in Fig. 6–4 is given by the difference in the Fermi levels between the two electrodes, $(\varepsilon_{F,R} - \varepsilon_{F,L}) = -e\,E_{H_2/O_2}$. As described in Sec. 4.5, the difference in the Fermi levels, $(\varepsilon_{F,R} - \varepsilon_{F,L})$, is equivalent to the difference in the real potentials of electrons between the two electrodes, $(\alpha_{e,R} - \alpha_{e,L})$; hence, $(\alpha_{e,R} - \alpha_{e,L}) = -e\,E_{H_2/O_2}$. The electromotive force is then expressed in Eqn. 6–8:

$$E_{H_2/O_2} = \frac{-(\alpha_{e,R} - \alpha_{e,L})}{e} = \frac{\mu_{H_2O,\,aq} - \mu_{H_2,\,gas} - \frac{1}{2}\mu_{O_2,\,gas}}{2\,e} = \frac{-\Delta G_{H_2/O_2}}{2\,e}$$

$$= E^0_{H_2/O_2} + \frac{k\,T}{2\,e}\ln p_{H_2} + \frac{k\,T}{4\,e}\ln p_{O_2} \quad , \tag{6-8}$$

where $E^0_{H_2/O_2}$ is the standard electromotive force (1.23 V at 25°C) and p is the fugacity of the hydrogen and oxygen gases.

In general, the cell reaction may be written in Eqn. 6–9:

$$\Sigma\, \nu_i\, i = 0 \quad , \qquad \Delta G \quad , \tag{6-9}$$

where i denotes a particle in the reaction, ν_i denotes the stoichiometric coefficient, and ΔG denotes the free enthalpy of the reaction. The electromotive force E_{cell} is then given by Eqn. 6–10:

$$E_{cell} = \frac{-\Delta G}{n\,e} = \frac{-\Sigma\,\nu_i\,\mu_i}{n\,e} = \frac{-\Sigma\,\nu_i\,\mu_i^0}{n\,e} - \frac{k\,T}{n\,e}\Sigma\,\nu_i\ln a_i$$

$$= E^0_{cell} - \frac{k\,T}{n\,e}\Sigma\,\nu_i\ln a_i \quad , \tag{6-10}$$

where a_i is the activity of particle i, n is the charge number of the cell reaction, and E^0_{cell} is the *standard electromotive force*.

The effect of temperature on the electromotive force can be derived from Eqn. 6–10 to obtain Eqn. 6–11:

$$\left(\frac{\partial E_{cell}}{\partial T}\right)_{a_i} = -\frac{1}{ne}\left(\frac{\partial \Delta G}{\partial T}\right)_{a_i} = \frac{\Delta S}{ne}, \tag{6–11}$$

where $\Delta S\, (= \delta Q/T)$ is the entropy change of the cell reaction, and Q is the heat absorbed in the cell. Therefore, the cell reaction is endothermic if the temperature dependence of E_{cell} is positive ($\Delta S > 0$); whereas, it is exothermic if the temperature dependence of E_{cell} is negative ($\Delta S < 0$). The reaction of the hydrogen–oxygen fuel cell, which is described in Figs. 6–2 and 6–4 as well as Eqns 6–1 and 6–2, is an exothermic reaction. Such heat generation or absorption in cell reactions is called the "Peltier effect" after C. A. Peltier (physicist, 1785–1845).

6.3 Equilibrium Potential of Electrode Reactions

6.3.1 Equilibrium potential of electron transfer reactions

The electrode potential of an electrode reaction at equilibrium can be measured as the electromotive force of an electrochemical cell composed of both the reaction electrode and the normal hydrogen electrode. The potential of the reaction electrode thus measured is taken as the equilibrium potential of the electrode reaction relative to the normal hydrogen electrode.

To illustrate this relationship to the normal hydrogen electrode, we consider an electrode reaction of redox electron transfer as shown in Eqn. 6–12:

$$OX + n\,e = RED. \tag{6–12}$$

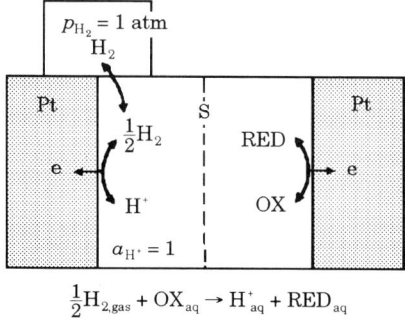

Fig. 6–5. Electrochemical cell composed of an electrode of redox electron transfer on the right hand side and the normal hydrogen electrode on the left hand side.

Fig. 6–5 shows the electrochemical cell of the redox reaction involving electron transfer coupled with the normal hydrogen electrode reaction. The cell diagram and cell reaction can be written, respectively, in Eqns. 6–13 and 6–14:

$$\text{Pt}_L \mid \text{H}_{2,\text{gas}} \mid \text{H}^+_{\text{aq}} : \text{RED}_{\text{aq}}, \text{OX}_{\text{aq}} \mid \text{Pt}_R \ , \qquad (6\text{–}13)$$

$$\frac{n}{2}\text{H}_{2,\text{gas}} + \text{OX}_{\text{aq}} = n\,\text{H}^+_{\text{aq}} + \text{RED}_{\text{aq}} \ , \qquad \Delta G_{\text{REDOX}} \qquad (6\text{–}14)$$

where $\Delta G_{\text{REDOX}} = (n\,\mu_{\text{H}^+} + \mu_{\text{RED}} - 0.5\,n\,\mu_{\text{H}_2} - \mu_{\text{OX}})$ is the free enthalpy of the reaction. In the normal hydrogen electrode, it is conventional to assign a value of zero to the energy levels of both the hydrated proton and the gaseous hydrogen molecule, respectively; $\mu_{\text{H}^+} = 0$ and $\mu_{\text{H}_2} = 0$. The electromotive force of this cell gives the equilibrium potential E_{REDOX} of the redox reaction as shown in Eqn. 6–15:

$$E_{\text{REDOX}} = \frac{-\Delta G_{\text{REDOX}}}{n\,e} = \frac{-\Delta G^0_{\text{REDOX}}}{n\,e} + \frac{kT}{n\,e}\ln\frac{a_{\text{OX}}}{a_{\text{RED}}}$$

$$= E^0_{\text{REDOX}} + \frac{kT}{n\,e}\ln\frac{a_{\text{OX}}}{a_{\text{RED}}} \ , \qquad (6\text{–}15)$$

where E^0_{REDOX} is the *standard redox potential* at which the activities of the reductant and oxidant particles are equal to unity.

TABLE 6–1. The standard equilibrium potentials for redox electrode reactions of hydrated redox particles at 25°C: E_{NHE} = relative electrode potential referred to the normal hydrogen electrode. [Handbooks of electrochemistry.]

Electrode	Reaction	n	E_0 / V_{NHE}
F_2/F^-	$F_2 + 2\,e \rightarrow 2\,F^-$	2	2.87
Co^{3+}/Co^{2+}	$Co^{3+} + e \rightarrow Co^{2+}$	1	1.82
Ce^{4+}/Ce^{3+}	$Ce^{4+} + e \rightarrow Ce^{3+}$	1	1.61
Mn^{3+}/Mn^{2+}	$Mn^{3+} + e \rightarrow Mn^{2+}$	1	1.51
Cl_2/Cl^-	$Cl_2 + 2\,e \rightarrow 2\,Cl^-$	2	1.359
MnO_2/Mn^{2+}	$MnO_2 + 4\,H^+ + 4\,e \rightarrow Mn^{2+} + 2\,H_2O$	2	1.23
O_2/H_2O	$O_2 + 4\,H^+ + 4\,e \rightarrow 2\,H_2O$	4	1.229
Fe^{3+}/Fe^{2+}	$Fe^{3+} + e \rightarrow Fe^{2+}$	1	0.771
O_2/H_2O_2	$O_2 + 2\,H^+ + 2\,e \rightarrow H_2O_2$	2	0.682
Cu^{2+}/Cu^+	$Cu^{2+} + e \rightarrow Cu^+$	1	0.153
Sn^{4+}/Sn^{3+}	$Sn^{4+} + 2\,e \rightarrow Sn^{2+}$	2	0.15
H^+/H_2	$2\,H^+ + 2\,e \rightarrow H_2$	2	0.00
Ti^{3+}/Ti^{2+}	$Ti^{3+} + e \rightarrow Ti^{2+}$	1	−0.37
Cr^{3+}/Cr^{2+}	$Cr^{3+} + e \rightarrow Cr^{2+}$	1	−0.41

Table 6–1 shows the standard equilibrium potentials of several redox reactions of hydrated redox particles.

6.3.2 Equilibrium potential of ion transfer reactions

Next, we consider a transfer reaction of metal ions across the interface of metal electrodes as in Eqn. 6–16:

$$M^{2+}_M \rightleftarrows M^{2+}_{aq} \ ,\tag{6-16}$$

where M^{2+}_M is the metal ion in the state of metallic bonding in metal electrodes, and M^{2+}_{aq} is the hydrated metal ion in aqueous solution. Fig. 6–6 shows the cell in which the transfer reaction of metal ions is coupled with the normal hydrogen electrode reaction; the cell reaction is written in Eqn. 6–17 as:

$$H_{2,gas} + M^{2+}_{aq} \to 2 H^+_{aq} + M_M \ , \qquad \Delta G_{M^{2+}/M} \ . \tag{6-17}$$

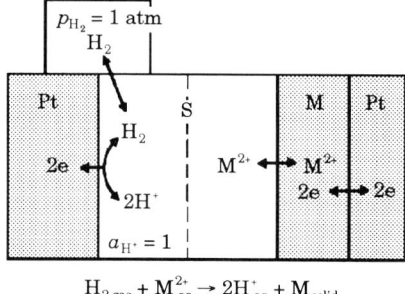

Fig. 6–6. Electrochemical cell composed of an electrode of metal ion transfer and the normal hydrogen electrode: M^{2+} = metal ion in the state of metallic bonding and in hydrated state.

The electromotive force of this cell gives the equilibrium potential $E_{M^{2+}/M}$ for the transfer reaction of metal ions as shown in Eqn. 6–18:

$$E_{M^{2+}/M} = \frac{-\Delta G_{M^{2+}/M}}{2e} = \frac{-\Delta G^0_{M^{2+}/M}}{2e} + \frac{kT}{2e} \ln a_{M^{2+}} = E^0_{M^{2+}/M} + \frac{kT}{2e} \ln a_{M^{2+}} \ , \tag{6-18}$$

where $E^0_{M^{2+}/M}$ is the *standard equilibrium potential* for the metal ion transfer and $a_{M^{2+}}$ is the activity of hydrated metal ions. Eqn. 6–18 indicates that the equilibrium potential depends on the activity of hydrated metal ions. Table 6–2 shows the standard equilibrium potentials of the transfer reactions of metal ions at metal electrodes in aqueous solutions.

TABLE 6-2. The standard equilibrium potentials for the transfer reactions of metal ions at metal electrodes at 25°C. [Handbooks of electrochemistry.]

Electrode	E_0 / V_{NHE}	Electrode	E_0 / V_{NHE}
Au^{3+}/Au	1.50	Cd^{2+}/Cd	-0.403
Pd^{2+}/Pd	0.987	Fe^{2+}/Fe	-0.44
Ag^+/Ag	0.799	Ga^{3+}/Ga	-0.53
Cu^+/Cu	0.521	Cr^{3+}/Cr	-0.74
Cu^{2+}/Cu	0.337	Zn^{2+}/Zn	-0.763
H^+/H_2	0.0	Mn^{2+}/Mn	-1.18
Fe^{3+}/Fe	-0.036	Zr^{4+}/Zr	-1.53
Pb^{2+}/Pb	-0.126	Ti^{2+}/Ti	-1.63
Sn^{2+}/Sn	-0.136	Al^{3+}/Al	-1.66
Ni^{2+}/Ni	-0.250	Mg^{2+}/Mg	-2.37
Co^{2+}/Co	-0.277	Na^+/Na	-2.714
In^{3+}/In	-0.342	Li^+/Li	-3.045

Further, we consider an ion transfer reaction at a multiphase silver–silver chloride electrode in an aqueous chloride solution as shown in Fig. 6–7. The ion transfer reactions at the $AgCl/Cl^-_{aq}$ interface and at the $Ag/AgCl$ interface are, respectively, given in Eqn. 6–19:

$$Cl^-_{AgCl} \rightleftarrows Cl^-_{aq} \, , \quad Ag^+_{AgCl} \rightleftarrows Ag^+_{Ag} \, . \tag{6-19}$$

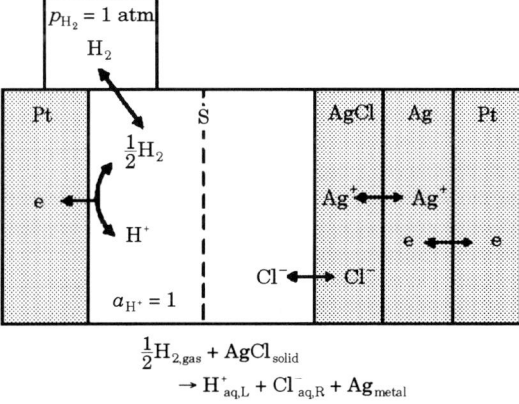

Fig. 6-7. Electrochemical cell composed of a silver-silver chloride multiphase electrode of ion transfer and the normal hydrogen electrode: Ag = silver metal; AgCl = silver chloride.

The overall cell reaction is given in Eqn. 6–20:

$$\tfrac{1}{2} H_{2,\,gas} + AgCl_{AgCl} = H^+_{aq} + Ag_{Ag} + Cl^-_{aq}\,, \qquad \Delta G_{AgCl/Cl^-}\,. \qquad (6\text{--}20)$$

Then, the equilibrium potential E_{AgCl/Cl^-} is obtained as shown in Eqn. 6–21:

$$E_{AgCl/Cl^-} = \frac{-\Delta G_{AgCl/Cl^-}}{2\,e} = \frac{-\Delta G^0_{AgCl/Cl^-}}{2\,e} + \frac{kT}{2\,e}\ln a_{Cl^-}$$

$$= E^0_{AgCl/Cl^-} + \frac{kT}{2\,e}\ln a_{Cl^-}\,, \qquad (6\text{--}21)$$

where the activities of solid Ag and AgCl are assigned a value of unity. The standard equilibrium potential calculated from Eqn. 6–21 is $E^0_{AgCl/Cl^-} = +\,0.224$ V at 25°C.

6.4 Electrochemical Reference Level for Hydrated Ions

In electrochemistry, the chemical potential of hydrated ions has been determined from the equilibrium potential of ion transfer reactions referred to the normal hydrogen electrode. For the reaction of metal ion transfer (metal dissolution-deposition reaction) of Eqns. 6–16 and 6–17, the standard equilibrium potential $E^0_{M^{2+}/M}$ is give in terms of the standard chemical potential, μ^0, by Eqn. 6–22:

$$E^0_{M^{2+}/M} = \frac{\mu^0_{H_2} + \mu^0_{M^{2+}} - 2\,\mu^0_{H^+} - \mu^0_{M}}{2\,e} = \frac{\mu^0_{M^{2+}} - 2\,\mu^0_{H^+}}{2\,e}\,, \qquad (6\text{--}22)$$

where the chemical potentials of gaseous hydrogen molecules and of solid metal in the standard state have been assigned a value of zero following the convention in Sec. 6.3.1 ($\mu^0_{H_2} = 0$, $\mu^0_M = 0$).

It follows from Eqn. 6–22 that the standard chemical potential of hydrated ions determined from the standard equilibrium potential of the ion transfer reaction is a relative value that is to the standard chemical potential of hydrated protons at unit activity, which, by convention in aqueous electrochemistry, is assigned a vale of zero on the electrochemical scale of ion levels.

The reaction equilibrium of the normal hydrogen electrode, $H_{2,\,gas} = 2\,H^+_{aq} + 2\,e_{H^+/H_2}$, where e_{H^+/H_2} is the equilibrium redox electron, can be obtained by the equilibration of the free enthalpy of the reactants and that of the products as follows: $2\,\alpha^0_{H^+} + 2\,\alpha^0_{e,H^+/H_2} = 2\,\mu^0_{H^+} + 2\,e\,\chi_S + 2\,\mu^0_{e,H^+/H_2} - 2\,e\,\chi_S = \mu^0_{H_2} = 0$, where α^0 is the standard real potential and χ_S is the surface potential difference of the aqueous solution. Hence, we obtain Eqn. 6–23:

$$2\,\alpha^0_{H^+} + 2\,\alpha^0_{e,H^+/H_2} = \mu^0_{H_2} = 0 \quad . \tag{6-23}$$

As has been estimated in Sec. 2.11, the real potential of an equilibrium redox electron in the normal hydrogen electrode reaction is $\alpha^0_{e,H^+/H_2} = -4.5$ eV. The standard chemical potential of a hydrated proton, $\mu^0_{H^+}$, is given by Eqn. 6-24:

$$\mu^0_{H^+} = \alpha^0_{H^+} - e\,\chi_S = -\alpha^0_{e,H^+/H_2} - e\,\chi_S \quad . \tag{6-24}$$

From $\chi_S = 0.13$ V, we obtain $\mu^0_{H^+} = -4.37$ eV, which is the standard ion level of a hydrated proton on the scale of chemical thermodynamics and which is assigned a value of zero on the electrochemical scale of ion levels in aqueous electrochemistry.

References

[Whiffen, 1979]: D. H. Whiffen, *Manual of Symbols and Terminology for Physicochemical Quantities and Units*, p. 28, Pergamon Press, Oxford, (1979).

CHAPTER 7

ELECTRODE REACTIONS

7.1 Electrode Reactions

7.1.1 Electron transfer and ion transfer reactions

Electrode reactions can be classified into two groups: one in which an *electron transfer* takes place across the electrode interface, such as ferric–ferrous redox reaction ($Fe^{3+}_{aq} + e = Fe^{2+}_{aq}$); and the other in which an *ion transfer* takes place across the electrode interface, such as the iron dissolution-deposition reaction ($Fe^{2+}_{M} = Fe^{2+}_{aq}$). Since electrons are Fermi particles in contrast to ions that obey the Boltzmann statistics as described Chap. 1, the reaction kinetics of the two groups differ in their electrode reactions.

In either group of electrode reactions, the energy level of reacting particles (electrons or ions) in the electrode depends linearly on the electrode potential. Hence, the reaction affinity ($A = -\Delta G$) can be varied over a wide range by simply controlling the electrode potential. This is one of the characteristics of electrode reactions in contrast with ordinary chemical reactions of which the affinity can be varied in a relatively narrow range by controlling the temperature and the concentration of reaction particles.

7.1.2 Cathodic and anodic reactions

As shown in Fig. 7–1, the electrode reaction in which a particle of negative charge (electron or anion) transfers from the electrode to the electrolyte (aqueous solution) is called the *cathodic reaction*; and the electrode reaction in which a particle of positive charge (hole or cation) transfers from the electrode to the electrolyte is called the *anodic reaction*. Further, the electrode at which the cathodic reaction takes place is called the *cathode;* and the electrode at which then anodic reaction takes place is called the *anode*.

For redox reactions of electron transfer, the transfer of electrons from the electrode to an oxidant particle to form a reductant particle is the cathodic reaction (the electron-accepting reduction of oxidants); and the transfer of electrons from a reductant particle to the electrode to form an oxidant particle is the anodic reaction (the electron-donating oxidation of reductants). These cathodic and anodic reactions are compared in Eqn. 7–1 and 7–2:

$$OX + e \rightarrow RED, \quad \text{cathodic reaction (reduction)} \tag{7–1}$$

$$OX \rightarrow RED + e. \quad \text{anodic reaction (oxidation)} \tag{7–2}$$

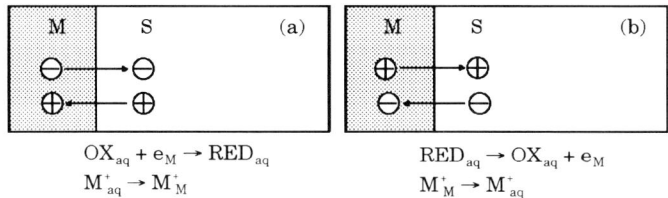

Fig. 7–1. (a) Cathodic reaction and (b) anodic reaction: M = metal electrode; S = aqueous solution (electrolyte); M^+_M = metal ion in the metallic bonding state; M^+_{aq} = metal ion in the hydrated state; e_M = electron in metals.

Similarly, for reactions of metal ion transfer the cathodic and anodic reactions are expressed by Eqn. 7–3 and Eqn. 7–4, respectively:

$$M^+_{aq} \rightarrow M^+_M, \quad \text{cathodic reaction (reduction)} \tag{7–3}$$

$$M^+_M \rightarrow M^+_{aq}. \quad \text{anodic reaction (oxidation)} \tag{7–4}$$

From the necessity of electrical neutrality of the electrode, the cathodic transfer of metal ions requires the electrode to accept electrons from an external cell circuit, and the anodic transfer of metal ions requires the electrode to donate electrons to an external cell circuit. However, no electron transfer takes place across the electrode interface; this is the reason why no electrons are involves in the metal ion transfer reactions in Eqns. 7–3 and 7–4.

7.1.3 Electron transfer of hydrated particles and adsorbed particles

The reaction of electron transfer at electrodes in aqueous electrolytes proceeds either with hydrated redox particles at the plane of closest approach of hydrated ions to the electrode interface (OHP, the outer Helmholtz plane) or with dehydrated and adsorbed redox particles at the plan of contact adsorption on the electrode interface (IHP, the inner Helmholtz plane) as shown in Fig. 7–2.

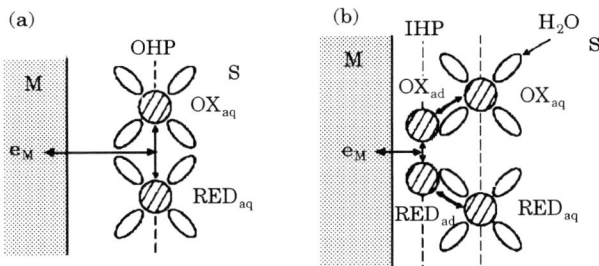

Fig. 7–2. Electron transfer of hydrated redox particles and of dehydrated adsorbed redox particles on the electrode interface: (a) electron transfer of hydrated redox particles, (b) electron transfer of dehydrated and adsorbed redox particles on electrodes. (RED_{aq}, OX_{aq}) = hydrated redox particles; (RED_{ad}, OX_{ad}) = dehydrated and adsorbed redox particles on electrodes; OHP = outer Helmholtz plane, IHP = inner Helmholtz plane.

A typical reaction of electron transfer of hydrated redox particles is the ferric–ferrous redox reaction in Eqn. 7–5:

$$Fe^{3+}_{aq} + e = Fe^{2+}_{aq} \; , \tag{7-5}$$

in which the electron transfer occurs across the compact double layer between the electrode interface and the outer Helmholtz plane at which hydrated ferric–ferrous redox particles exist. It is known that six water molecules are coordinated with an iron ion in the hydrated state.

As the adsorption affinity of redox particles on electrodes increases, the hydrated redox particles is adsorbed in the dehydrated state (chemical adsorption, contact adsorption) rather than in the hydrated state (physical adsorption) as shown in Fig. 7–2 (b). Typical reactions of redox electron transfer of dehydrated and adsorbed redox particles on electrodes are the hydrogen and the oxygen electrode reactions in Eqns. 7–6 and 7–7:

$$H_{2, gas} + 2e = 2H^+_{aq} \; , \tag{7-6}$$

$$O_{2, gas} + 4H^+_{aq} + 4e = 2H_2O_{aq} \; , \tag{7-7}$$

in which the electron transfer occurs between the electrode and the adsorbed reaction intermediates at the plane of adsorption (the inner Helmholtz plane).

The electron transfer of hydrated redox particles at the outer Helmholtz plane is occasionally called the *outer-sphere electron transfer*; while the electron transfer of dehydrated and adsorbed redox particles on electrodes is called the *inner-sphere electron transfer*.

For the electron transfer of hydrated redox particles (the outer-sphere electron transfer), the electrode acts merely as a source or sink of electrons transferring across the compact double layer so that the nature of the electrode exerts little or no influence on the reaction kinetics; this lack of influence by the substrate has been observed for the ferric–ferrous redox reaction. On the other hand, the electron transfer of adsorbed redox particles (the inner-sphere electron transfer) is affected by the state of adsorption so that the nature of the electrode exerts a definite influence on the reaction kinetics, as has been observed with the hydrogen electrode reaction where the reaction rate depends on the property of the electrode.

7.2 Reaction Rate

7.2.1 Forward and backward reaction affinities

We consider a simple reaction from reactant R to product P in Eqn. 7–8; these R and P are the particles of Boltzmann statistics:

$$\nu_R R \rightarrow n_P P , \qquad (7\text{–}8)$$

where ν is the stoichiometric coefficient of the reaction. The reaction affinity, A, can be expressed by the electrochemical potential, $\bar{\mu}$, of the reactant and product particles as shown in Eqn. 7–9:

$$A = \nu_R \bar{\mu}_R - \nu_P \bar{\mu}_P = \nu_R (\bar{\mu}_R^0 + k T \ln a_R) - \nu_P (\bar{\mu}_P^0 + k T \ln a_P) = A^+ - A^- , \qquad (7\text{–}9)$$

where $\bar{\mu}^0$ is the standard electrochemical potential and a is the activity of particles.

In Eqn. 7–9 A^+ and A^- are defined as the *forward* and the *backward* reaction affinities, respectively; A^+ corresponds to the energy (the free enthalpy) of the reactant particles and A^- corresponds to the energy (the free enthalpy) of the product particles as shown in Eqn. 7–10:

$$A^+ = A^{0+} + \nu_R k T \ln a_R , \quad A^- = A^{0-} + \nu_P k T \ln a_P , \qquad (7\text{–}10)$$

where A^{0+} and A^{0-} are the *standard* forward and backward reaction affinities, respectively.

At the reaction equilibrium ($A = 0$), we obtain Eqns. 7–11 and 7–12:

$$A^{0+} - A^{0-} = - \{ k T \ln (a_{R,eq})^{\nu_R} - k T \ln (a_{P,eq})^{\nu_P} \} , \qquad (7\text{–}11)$$

$$A^{0+} = -kT \ln\left(\frac{(a_{R,eq})^{\nu_R}}{a^0}\right), \quad A^{0-} = -kT \ln\left(\frac{(a_{P,eq})^{\nu_P}}{a^0}\right), \quad (7\text{--}12)$$

where a_{eq} is the activity of particles in the reaction equilibrium and a^0 is a constant common to the forward and backward affinities.

In general, the forward and backward reaction affinities A^+ and A^- are directly related to the forward and backward reaction rates, respectively.

7.2.2 Reaction rate

We consider a simple reaction composed of only a single elemental step of reacting particles that obey the Boltzmann distribution function. Then, the reaction rate, v, is given in Eqn. 7–13 [Rysselberghe, 1963]:

$$v = \frac{1}{\lambda}\left\{\exp\left(\frac{A^+}{kT}\right) - \exp\left(\frac{A^-}{kT}\right)\right\}, \quad (7\text{--}13)$$

where λ is a parameter depending on temperature and pressure, but not on the activities of the reactant and product particles. From Eqns. 7–10, 7–12 and 7–13, the forward and backward reaction rates v^+ and v^- can be obtained, respectively, as in Eqns. 7–14 and 7–15:

$$v^+ = \frac{1}{\lambda}\exp\left(\frac{A^+}{kT}\right) = \frac{1}{\lambda}(a_R)^{\nu_R}\exp\left(\frac{A^{0+}}{kT}\right) = \frac{1}{\lambda a^0}\left(\frac{a_R}{a_{R,eq}}\right)^{\nu_R}, \quad (7\text{--}14)$$

$$v^- = \frac{1}{\lambda}\exp\left(\frac{A^-}{kT}\right) = \frac{1}{\lambda}(a_P)^{\nu_P}\exp\left(\frac{A^{0-}}{kT}\right) = \frac{1}{\lambda a^0}\left(\frac{a_P}{a_{P,eq}}\right)^{\nu_P}. \quad (7\text{--}15)$$

The ratio of the forward rate to the backward rate is given by Eqn. 7–16:

$$\frac{v^+}{v^-} = \exp\left(\frac{A^+ - A^-}{kT}\right) = \exp\left(\frac{A}{kT}\right) = \left(\frac{a_R}{a_{R,eq}}\right)^{\nu_R}\left(\frac{a_{P,eq}}{a_P}\right)^{\nu_P}, \quad (7\text{--}16)$$

indicating that the ratio of v^+/v^- is an exponential function of the reaction affinity A. Eqn. 7–16 is important in that it relates the ratio of forward/backward reaction rates to the affinity of a single step reaction in the nonequilibrium state ($A > 0$). Obviously, the forward reaction rate equals the backward reaction rate at equilibrium ($A = 0$).

Introducing the exchange rate v_{eq} in the reaction equilibrium, we obtain Eqn. 7–17 from Eqn. 7–13:

$$v = v_{eq}\left\{\exp\left(\frac{A^+ - A^+_{eq}}{kT}\right) - \exp\left(\frac{A^- - A^-_{eq}}{kT}\right)\right\}, \quad (7\text{--}17)$$

where A^+_{eq} and A^-_{eq} are the forward and backward reaction affinities in reaction equilibrium ($A^+_{eq} = A^-_{eq}$), respectively. In nonequilibrium states away from equilibrium ($A^+ > A^+_{eq}$ and $A^- < A^-_{eq}$), the backward reaction rate can be disregarded so that the reaction rate is determined merely by the forward rate.

In the affinity regime near reaction equilibrium, both $A^+ - A^+_{eq}$ and $A^- - A^-_{eq}$ are close to zero so that Eqn. 7–17 can be expressed, to a first approximation, by Eqn. 7–18:

$$v = v_{eq} \frac{A}{kT} \ . \tag{7–18}$$

This equation indicates that the reaction rate is proportional to the reaction affinity A near the equilibrium ($A < kT$) as shown in Fig. 7–3 (a). Eqn. 7–18 is a *linear kinetics* similar to Ohm's law in the electric current versus voltage relationship.

On the other hand, in the affinity regime away from the equilibrium ($A \gg kT$), a nonlinear *exponential kinetics* applies as shown in Fig. 7–3 (b).

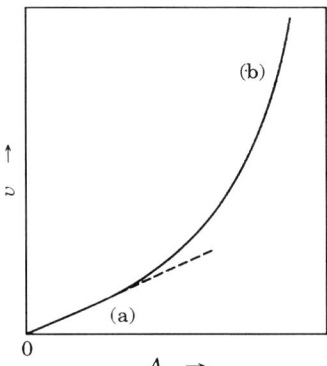

Fig. 7–3. Reaction rate as a function of reaction affinity: curve (a) = regime of linear kinetics near the reaction equilibrium; curve (b) = regime of nonlinear exponential kinetics away from the reaction equilibrium; v = reaction rate; A = affinity.

7.2.3 Polarization curve of electrode reactions

In electrochemical kinetics, the plot of the reaction current (the reaction rate) as a function of the electrode potential is conventionally called the *polarization curve*. Fig. 7–4 shows schematic polarization curves of cathodic and anodic electrode reactions. The term of *polarization* means shifting the electrode potential from a certain specified potential, e.g. the equilibrium potential of an electrode reaction, to more negative (cathodic) or more positive (anodic) potentials. The term of polarization also occasionally applies to the *magnitude* of potential shift from the specified potential.

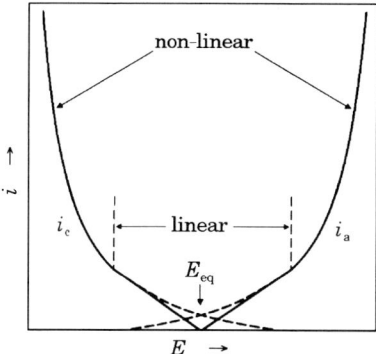

Fig. 7–4. Anodic and cathodic polarization curves of an electrode reaction: E = electrode potential; E_{eq} = equilibrium potential; i = reaction current; i_a = anodic current.; i_c = cathodic current.

Polarization in the cathodic direction accelerates the cathodic reaction and is called *cathodic* polarization; polarization in the anodic direction accelerates the anodic reaction and is called *anodic* polarization. In Fig. 7–4 the polarization curve is cathodic at potentials more negative and is anodic at potentials more positive than the equilibrium potential E_{eq}. In the kinetics of electrode reactions the magnitude of polarization (the potential change in polarization) is called the *overvoltage* or *overpotential* and conventionally expressed by symbol η, which is negative in cathodic polarization and positive in anodic polarization.

For ordinary electrode reactions, as is described in Sec. 7.2.2, the kinetics are linear (a linear affinity versus reaction rate relationship and a linear potential versus current relationship) in the vicinity of the equilibrium potential E_{eq}; and

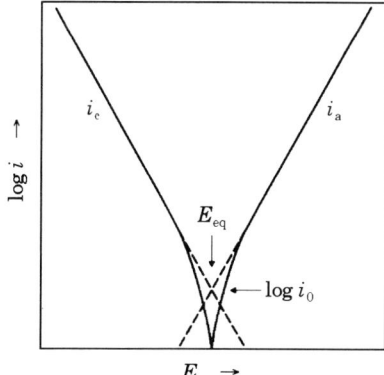

Fig. 7–5. Tafel plot of reaction current: i_0 = exchange reaction current at equilibrium.

the kinetics gradually become nonlinear and exponential with increasing polarization as shown in Fig. 7–4 and in Fig. 7–5.

In the range of electrode potential in which the kinetics are exponential, the overvoltage η is a linear function of the logarithm of the reaction current, $\log i$, as shown in Eqn. 7–19 and in Fig. 7–5:

$$\eta = a + b \log i \,, \qquad (7\text{–}19)$$

where a and b are kinetic parameters. Eqn. 7–19 is called the *Tafel equation* and the η–$\log i$ plot is called the *Tafel plot* after J. Tafel who observed it about a century ago [Tafel, 1905]. Parameter b represents the slope of the Tafel line and is called the *Tafel slope*.

7.3 Reaction Mechanism

7.3.1 The stoichiometric number of reactions

A reaction consisting of a single elementary step *alone* is uncommon, and most reactions involve a number of elementary steps with reaction-intermediates (multistep reactions). For a reaction consisting of a series connection of several elementary steps, the number of repetitions that an elementary step proceeds in a unit extent (advancement) of the overall reaction is defined as the *stoichiometric number*, ν^*, of the step [Horiuti-Ikushima, 1939]. For example, if the cathodic reaction of the hydrogen electrode consists of the following two steps,

$H^+_{aq} + e = H_{ad}$	discharge step
$2\,H_{ad} \rightarrow H_{2,\,gas}$	combination step
$2\,H^+_{aq} + 2\,e \rightarrow H_{2,\,gas}$	reaction ,

the stoichiometric number is $\nu^* = 2$ for the discharge step and $\nu^* = 1$ for the combination step.

The overall reaction rate v_T is related to the rate v_r of elementary step r as expressed in Eqn. 7–20:

$$v_T = \frac{v_r}{\nu^*_r} \,. \qquad (7\text{–}20)$$

Further, the overall reaction affinity, $-\Delta G$, is represented by the sum total of the step affinities, $-\Delta g_r$, multiplied by the respective stoichiometric number, ν^*_r, of elementary steps r as expressed in Eqn. 7–21:

$$\Delta G = \Sigma \, v_r^* \, \Delta g_r \; . \tag{7-21}$$

If there is an elementary step r which determines the overall reaction rate because its rate is much smaller than the rates of the other elementary steps, the overall reaction affinity $-\Delta G$ would be located at the rate-determining step r as expressed in Eqn. 7-22:

$$\Delta G = v_r^* \, \Delta g_r \; . \tag{7-22}$$

Such a single rate-determining step scarcely occurs in ordinary reactions; usually, the overall reaction affinity is distributed in multiple rate-determining steps rather than localized in a single step as is described in Sec. 7.4.

7.3.2 The activation energy

The reaction path from the initial state to the final state of an elementary step is represented by the potential energy curves of the initial and final states of a reacting particle as shown in Fig. 7–6, where the reaction coordinate x denotes the position of a reaction particle moving across the compact double layer on the electrode interface.

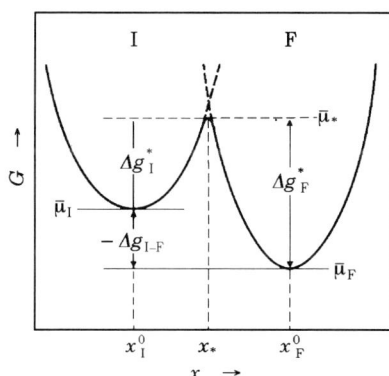

Fig. 7–6. Potential energy curves for an elementary step of reaction: G = particle energy; x = reaction coordinate; $\bar{\mu}$ = electrochemical potential of particles; $\bar{\mu}_*$ = electrochemical potential of the activated particle; $-\Delta g_{I-F}$ = step affinity; Δg^* = activation energy.

The intersection of the potential energy curves of reacting particles in the initial state (x_I, $\bar{\mu}_I$) and in the final state (x_F, $\bar{\mu}_F$) of an elementary step is the activated state (x_*, $\bar{\mu}_*$) of the step; $\bar{\mu}_*$ is the electrochemical potential of the activated particle. From Fig. 7–6 the activation energies Δg_I^* and Δg_F^* in the forward and backward directions are given, respectively, as follows by Eqn. 7–23:

$$\Delta g_I^* = \bar{\mu}_* - \bar{\mu}_I , \quad \Delta g_F^* = \bar{\mu}_* - \bar{\mu}_F . \tag{7-23}$$

The difference between activation energies of the forward and the backward steps equal the affinity, $-\Delta g_{I-F}$, of the step as expressed in Eqn. 7–24:

$$-\Delta g_{I-F} = -\Delta g_I^* + \Delta g_F^* . \tag{7-24}$$

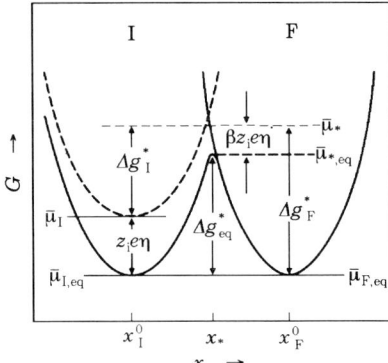

Fig. 7–7. Potential energy curves for an elementary step of reaction in the equilibrium state (solid curve) and in the nonequilibrium state (dotted curve): Δg_{eq}^* = activation energy in equilibrium; Δg_I^* = forward activation energy in nonequilibrium; $\bar{\mu}_{*,eq}$ = electrochemical potential of the activated particle in equilibrium; β = symmetry factor; z_i = charge number of the reacting particle.

In the case of electrode reactions, the activation energy depends on the electrode potential. We now consider an elementary step in which a charged particle (charge number, z_i) transfers across the compact double layer on the electrode interface as shown in Fig. 7–7. In the reaction equilibrium, where the electrochemical potentials of reacting particles are equilibrated between the initial state and the final states ($\bar{\mu}_{I(eq)} = \bar{\mu}_{F(eq)}$), the forward activation energy equals the backward activation energy ($\bar{\mu}_{*,eq} - \bar{\mu}_I = \bar{\mu}_{*,eq} - \bar{\mu}_F$); $\bar{\mu}_{*,eq}$ is the electrochemical potential of the reacting particle in the activated state in equilibrium.

When the energy level of the initial state is raised from $\bar{\mu}_{I(eq)}$ to $\bar{\mu}_I$ by changing the electrode potential, the affinity $-\Delta g_{I-F}$ of an elementary step is also changed as given by Eqn. 7–25:

$$-\Delta g_{I-F} = -\bar{\mu}_I - \bar{\mu}_{I(eq)} , \tag{7-25}$$

which corresponds to the overvoltage η in Eqn. 7–26:

$$z_i e \, \eta = -\Delta g_{\text{I-F}} = \bar{\mu}_\text{I} - \bar{\mu}_{\text{I(eq)}} , \qquad (7\text{--}26)$$

where z_i is the charge number of the activated particle. Simultaneously, the energy level of the reacting particle in the activated state is raised from $\bar{\mu}_{*,\text{eq}}$ to $\bar{\mu}_*$ as shown in Fig. 7–7.

From the potential energy curves shown in Fig. 7–7, it appears that the reduction in the activation energy, $\Delta g_\text{I}^* - \Delta g_{\text{eq}}^*$, caused by the electrode polarization is not equal to but smaller than the reaction affinity, $z_i e \, \eta$, imposed by the polarization. Hence, we obtain Eqn. 7–27:

$$\Delta g_\text{I}^* - \Delta g_{\text{eq}}^* = -(1 - \beta) z_i e \, \eta , \qquad (7\text{--}27)$$

where β ($0 < \beta < 1$) is called the *symmetry factor* [Horiuti-Polanyi, 1935]. If the intermediate at the activated state is located midway in the reaction coordinate between the initial and final states, the symmetry factor β would be close to 0.5.

7.3.3 Quantum tunneling and activated flow of particles

The transfer of a particle across an interfacial energy barrier proceeds either in the quantum dynamic *tunneling* or in the statistically *activated flow* as shown in Fig. 7–8. The probability w_t of the tunneling transfer across an energy barrier rectangular in shape Δg^* high (activation energy) and x wide is given, to a first approximation, by Eqn. 7–28 [Gerischer, 1960; Vetter, 1967]:

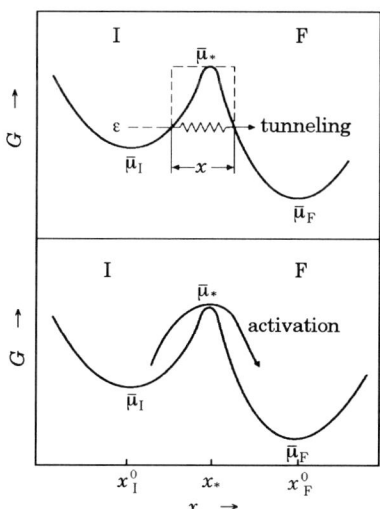

Fig. 7–8. Quantum dynamic tunneling and statistically activated flow of particles across an energy barrier: ε = energy of a tunneling particle

$$w_t \doteq \exp\left\{-\frac{2x}{\hbar}\sqrt{2m(\Delta g^* - \varepsilon)}\right\}, \tag{7-28}$$

where m is the mass of transferring particles, ε is the energy of the transferring particles, $\hbar = h/2\pi$, and h is the Planck constant.

In contrast, the probability w_a of the activated flow is given by the Boltzmann distribution function in Eqn. 7–29:

$$w_a = \exp\left(\frac{-\Delta g^*}{kT}\right). \tag{7-29}$$

For the transfer of electrons, of which the mass is extremely small ($m_e = 9.1 \times 10^{-12}$ g), simple calculations give the tunneling probability w_t to be much greater than the activation probability w_a ($w_t \gg w_a$) for an energy barrier of 1 to 4 eV with the transfer distance of $x = 0.5$ nm. For electrons, therefore, the tunneling transfer predominates greatly over the activated flow, and the activated flow can be disregarded. On the other hand, for the transfer of ions, of which the mass is 10^5 times greater than the electron mass, the activation probability w_a is much greater than the tunneling probability w_t ($w_t \ll w_a$); hence, the activated flow predominates greatly over the tunneling transfer.

In the tunneling electron transfer, the transfer rate v_t at the energy level ε can be given by the product of the state density occupied by electrons $D_{\text{I,occ}}$ in the initial state, the state density vacant for electrons $D_{\text{F,vac}}$ in the final state of the step, the tunneling probability w_t, and the transmission coefficient κ_t as expressed in Eqn. 7–30:

$$v_t = \kappa_t w_t D_{\text{I,occ}} D_{\text{F,vac}}, \tag{7-30}$$

where κ_t, w_t, $D_{\text{I,occ}}$, and $D_{\text{F,vac}}$ are all functions of the energy level ε at which the electron transfer takes place. Accordingly, the whole electron transfer current, i, is expressed by the integration of Eqn. 7–30 to obtain Eqn. 7–31:

$$i = e\int_{-\infty}^{+\infty} v_t(\varepsilon)\, d\varepsilon = e\int_{-\infty}^{+\infty} \kappa_t w_t D_{\text{I,occ}} D_{\text{I,vac}}\, d\varepsilon, \tag{7-31}$$

where e is the elemental charge of electron. A similar equation to Eqn. 7–31 can also be obtained for the electron transfer in the revere direction.

As is described in Chap. 8, the current of electron transfer (the rate of electron transfer) can be given, to a first approximation, as an exponential function of the overvoltage η as shown in Eqn. 7–32:

$$i = i_0 \exp\left(\frac{\alpha e \eta}{kT}\right), \tag{7-32}$$

where i_0 is the exchange current (the exchange reaction rate), and α is a kinetics parameter called the *transfer coefficient* or the *Tafel constant*.

For the activated transfer of ions, the transfer current can be derived from the theory of absolute reaction rates as shown in Eqn. 7–32 [Horiuti-Nakamura, 1967]:

$$i = z e \kappa_a \frac{kT}{h} w_a = z e \kappa_a \frac{kT}{h} \exp\left\{\frac{-(\bar{\mu}^* - \bar{\mu}_I)}{kT}\right\} = i_0 \exp\left\{\frac{(1-\beta) z e \eta}{kT}\right\}, \quad (7\text{–}33)$$

where z is the ionic valence (the charge number of ion), κ_a is the transmission coefficient at the activated state, and h is the Planck constant.

These reaction currents given by Eqns. 7–32 and 7–33 are formally in agreement with the Tafel equation of Eqn. 7–19 obtained by experimental observations. Note that the rate equations in Eqns. 7–32 and 7–33 apply to the forward reaction *only* and disregard the backward reaction rate.

7.3.4 The reaction order

The reaction order is a kinetic parameter representing the effect of the activity of a reaction particle on the reaction rate and is used to elucidate the mechanism of the reaction. The reaction order, ζ_k, with respect to particle k is defined in Eqn. 7–34:

$$\zeta_k \equiv kT \left(\frac{\partial \ln i}{\partial \bar{\mu}_k}\right) = \left(\frac{\partial \ln i}{\partial \ln \lambda_k}\right), \quad (7\text{–}34)$$

where $\bar{\mu}_k$ is the electrochemical potential of particle k, and λ_k is the absolute activity of k.

We now consider a transfer reaction of charged particles across the interface of electrodes. For a hydrated particle in aqueous solution, the electrochemical potential of the particle is independent of the electrode potential; but it depend on the activity of the particle (the concentration of the particle), regardless of whether particle is charged (ion) or noncharged (neutral particle). In contrast, for a charged particle (electron or ion) in the electrode, the electrochemical potential of the particle depends on both the electrode potential E and the absolute activity λ_k; $\bar{\mu}_k = kT \ln \lambda_k + z e E$. Accordingly, from Eqn. 7–34 we obtain Eqn. 7–35 for the reaction order, ζ_k, with respect to a charged particle k in the electrode if the activity of k is constant:

$$\zeta_k \equiv kT \left(\frac{\partial \ln i}{\partial \bar{\mu}_k}\right) = \frac{kT}{ze}\left(\frac{\partial \ln i}{\partial E}\right) = \frac{kT}{ze}\left(\frac{\partial \ln i}{\partial \eta}\right) = \frac{\alpha}{z}. \quad (7\text{–}35)$$

Obviously, this reaction order corresponds to the transfer coefficient α (the

Tafel constant) obtained from the slope b of the Tafel line (Eqns. 7–19 and 7–32) and is defined in Eqn. 7–36:

$$z\,\zeta_k = \alpha = \frac{kT}{e}\left(\frac{\partial \ln i}{\partial \eta}\right) = \frac{2.3\,kT}{b\,e} \quad . \tag{7-36}$$

It follows that the Tafel slope of an electrode reaction is connected in its physical meaning to the *reaction order* of transferring electrons or ions from the electrode (metal electrode).

7.4 Rate-Determining Steps of Reactions

7.4.1 Reaction of elementary steps in series

We consider a reaction of R → P consisting of a series of connected elementary steps, 1, 2, 3, $\cdots$, i, $\cdots$, r, $\cdots$, as shown in Fig. 7–9. The overall reaction affinity, $-\Delta G$, is the sum total of the affinities, $-\Delta g_i$, each multiplied by the stoichiometric number, v_i^*, corresponding to the elementary steps as expressed in Eqn. 7–37:

$$-\Delta G = -\Sigma\, v_i^*\, \Delta g_i \quad . \tag{7-37}$$

$$R \circledcirc \xrightarrow{\text{step 1}} \bullet \xrightarrow{\text{step 2}} \bullet - - - \to \bullet \xrightarrow{\text{step S}} \bullet - - - - \to \circledcirc P$$
$$\quad\quad -\Delta g_1 \quad -\Delta g_2 \quad\quad\quad -\Delta g_S$$
$$\quad\quad\quad v_1^* \quad\quad v_2^* \quad\quad\quad\quad v_S^*$$

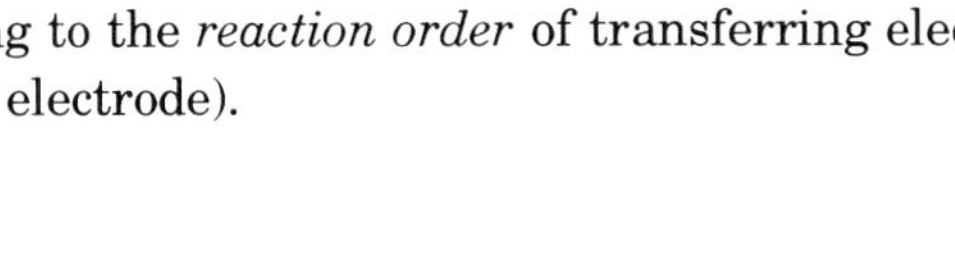

Fig. 7–9. Reaction path consisting of a series of connected elementary steps: R = particles in the initial state of reaction; P = particles produced in the final state; $-\Delta g_i$ = affinity of step i, v_i^* = stoichiometric number of step i.

The overall rates of the forward and backward reactions, v_T^+ and v_T^-, are related with the rates of the respective forward and backward elementary steps, v_i^+ and v_i^-, as shown in Eqns. 7–38 and 7–39:

$$v_T^+ = \frac{v_1^+}{v_1^*} = \frac{v_2^+}{v_2^*} = \cdots = \frac{v_i^+}{v_i^*} = \cdots \quad , \tag{7-38}$$

$$v_T^- = \frac{v_1^-}{v_1^*} = \frac{v_2^-}{v_2^*} = \cdots = \frac{v_i^-}{v_i^*} = \cdots , \quad (7\text{–}39)$$

where v_i^* is the stoichiometric number of the respective elementary steps.

Further, Eqn. 7–16 yields the ratio of the rates of forward step to backward step as expressed in Eqn. 7–40:

$$\frac{v_i^+}{v_i^-} = \exp\left(\frac{-\Delta g_i}{kT}\right). \quad (7\text{–}40)$$

We consider next a simple reaction composed of two elemental steps shown in Fig. 7–10. The rates of the overall reaction in the forward and backward directions can be obtained as a function of the rates of the two elemental steps as given in Eqn. 7–14:

$$v_T^+ = \frac{v_1^+ v_2^+}{v_2^+ + v_1^-} , \quad v_T^- = \frac{v_1^- v_2^-}{v_2^+ + v_1^-} . \quad (7\text{–}41)$$

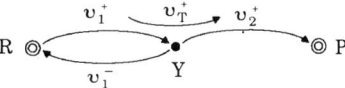

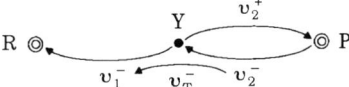

Fig. 7–10. Rate of a two-step reaction: v_T^+ = forward rate of the overall reaction, v_1^+ = forward rate of step 1.

Consequently, the ratio of forward to backward reaction rates is represented in Eqn. 7–42 by the product of the ratios for the two elementary steps:

$$\frac{v_T^+}{v_T^-} = \frac{v_1^+}{v_1^-} \frac{v_2^+}{v_2^-} . \quad (7\text{–}42)$$

For multistep reactions, in general, the following relationship has been obtained [Matsuda-Horiuti, 1962]:

$$\frac{v_T^+}{v_T^-} = \frac{v_1^+}{v_1^-} \frac{v_2^+}{v_2^-} \cdots = \prod_i \left(\frac{v_i^+}{v_i^-}\right) . \quad (7\text{–}43)$$

7.4.2 Reaction rate determined by a single step

When the overall rate of a multistep reaction is determined solely by a single elementary step of which the rate is extremely small compared with the rates of the other elementary steps, the multistep reaction is called the reaction of a single rate-determining step. In such a multistep reaction, as shown in Fig. 7–11 (a), all the elementary steps except for the rate-determining step are considered to be in quasi-equilibrium. Note that the multistep reaction of a single rate-determining step is rather uncommon in practice.

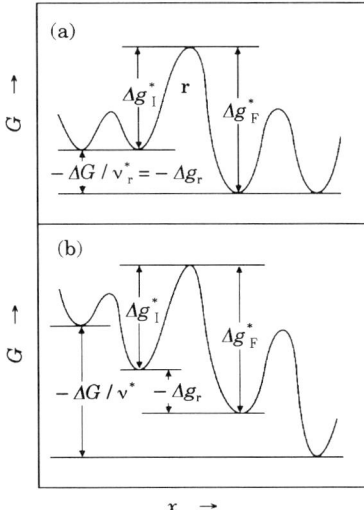

Fig. 7–11. Potential energy curves for a multistep reaction of (a) single rate-determining step and (b) multiple rate-determining steps: v_r^* = stoichiometric number of the single rate-determining step; v^* = mean stoichiometric number of the multiple rate-determining steps.

In the reaction with a single rate-determining step, the affinity of elementary steps other than the rate-determining step is negligible, and the overall reaction affinity $-\Delta G$ approximately equals the affinity $-\Delta g_r$ multiplied by the stoichiometric number v_r^* of the rate-determining step in Eqn. 7–44 as has been shown in Eqn. 7–22:

$$-\Delta G = -v_r^* \Delta g_r \ . \tag{7-44}$$

Since the rate of overall reaction is given by $v_T = v_T^+ - v_T^- = (v_r^+ - v_r^-)/v_r$ and the rate ratio is given by $v_T/v_T^- = v_r^+/v_r^- = \exp(-\Delta g_r / kT)$, the stoichiometric number v_r^* of the rate-step can be derived to obtain Eqn. 7–45:

$$v_r^* = \frac{-\Delta G}{kT \ln\left(\frac{v_T^+}{v_T^-}\right)} \quad . \tag{7-45}$$

Consequently, we obtain from Eqn. 7–45 the rate of overall reaction $v_T = v_T^+ - v_T^-$ as expressed in Eqn. 7–46:

$$v_T = v_T^+ \left\{ 1 - \exp\left(\frac{-\Delta G}{v_r^* kT}\right) \right\} . \tag{7-46}$$

In the regime of affinity close to the equilibrium ($-\Delta G \rightarrow 0$), Eqn. 7–46 gives the approximate equation in Eqn. 7–47:

$$v_T \doteq v_{T,eq}^+ \frac{-\Delta G}{v_r^* kT} \quad , \tag{7-47}$$

where $v_{T,eq}^+$ is the exchange reaction rate at equilibrium. This is the rate equation in the regime of linear kinetics already described in Sec. 7.2. From Eqn. 7–47 we obtain the stoichiometric number v_r^* of the rate-determining step as the product of the exchange reaction rate $v_{T,eq}^+$ and the linear reaction resistance $[\partial(-\Delta G)/\partial v_T]_{eq}$ in Eqn. 7–48:

$$v_r^* = \frac{v_{T,eq}}{kT} \left(\frac{\partial(-\Delta G)}{\partial v_T}\right) . \tag{7-48}$$

In the case of electrode reactions, Eqn. 7–48 is expressed by Eqn. 7–49:

$$v_r^* = \frac{i_0 n e}{kT} \left(\frac{\partial \eta}{\partial i}\right)_{eq} , \tag{7-49}$$

where i_0 is the exchange current, η is the overvoltage, and n is the charge number of the reaction. Eqn. 7–49 has been used to estimate the stoichiometric number of the rate-determining step from the linear reaction resistance $(\partial \eta / \partial i)_{eq}$ in the vicinity of reaction equilibrium

7.4.3 Reaction rate determined by multiple steps

For reactions in which the rate is determined not by a single elemental step but by *multiple elementary steps*, the overall affinity of the reaction is distributed among the multiple steps, which determine the reaction rate as shown in Fig. 7–11 (b). The ratio of forward to backward reaction rates is then derived from Eqns. 7–40 and 7–43 to give Eqn. 7–50:

$$\frac{v_T^+}{v_T^-} = \prod_i \left(\frac{v_i^+}{v_i^-} \right) = \exp\left(\frac{-\Sigma \Delta g_i}{kT} \right). \tag{7-50}$$

In the same way as Eqn. 7–45 was developed, the formal stoichiometric number, v^*, of the rate-determining steps can be expressed by Eqn. 7–51:

$$v^* = \frac{-\Delta G}{kT \ln\left(\frac{v_T^+}{v_T^-} \right)} = \frac{-\Delta G}{-\Sigma \Delta g_i} = \frac{-\Sigma v_i^* \Delta g_i}{-\Sigma \Delta g_i}, \tag{7-51}$$

This number v^* is the *mean stoichiometric number* of the rate-determining multiple steps. The mean stoichiometric number of the rate-determining steps is thus represented by the energy average (affinity average) of the stoichiometric numbers v_i^* weighed with the step affinity Δg_i in the respective rate-determining steps.

The reaction affinity $-\Delta G$ and the ratio v_T^+/v_T^- of forward to backward rates can be estimated, regardless of whether the reaction rate is determined by a single step or multiple steps. Therefore, Eqn. 7–51 can be used to determine the mean stoichiometric number of multistep reactions.

In the reaction kinetics, the mean stoichiometric number is important to elucidate the reaction mechanism.

7.4.4 Affinity distributed to elementary steps

We consider a simple reaction composed of two elementary steps in series with the *affinity distribution ratio* $m = \Delta g_1 / \Delta g_2$ and the stoichiometric numbers v_1^* and v_2^*. The mean stoichiometric number v^* is then given by Eqn. 7–52:

$$v^* = \frac{v_1^* \Delta g_1 + v_2^* \Delta g_2}{\Delta g_1 + \Delta g_2} = \frac{v_1^* m + v_2^*}{m + 1}, \tag{7-52}$$

which relates the mean stoichiometric number v^* with the affinity distribution ratio m.

Since the rates of the two steps are related to each other in that $(v_1^+ - v_1^-)/v_1^* = (v_2^+ - v_2^-)/v_2^*$, we obtain Eqn. 7–53 from Eqn. 7–40:

$$\frac{v_1^+}{v_1^*} \left\{ 1 - \exp\left(\frac{-\Delta g_1}{kT} \right) \right\} = \frac{v_2^+}{v_2^*} \left\{ 1 - \exp\left(\frac{-\Delta g_2}{kT} \right) \right\}. \tag{7-53}$$

In the regime of linear kinetics near the equilibrium, Eqn. 7–53 reduces to the approximate equation in Eqn. 7.54:

$$\frac{v_{1,\,eq}^+}{v_1^*} \left(\frac{-\Delta g_1}{kT} \right)_{eq} = \frac{v_{2,\,eq}^+}{v_2^*} \left(\frac{-\Delta g_2}{kT} \right)_{eq}, \tag{7-54}$$

where $v_{1,\mathrm{eq}}$ and $v_{2,\mathrm{eq}}$ are the exchange rates of the two steps in the reaction equilibrium.

The ratio of the exchange rates ($v_{1,\mathrm{eq}}/v_{2,\mathrm{eq}}$) and the ratio of the exchange currents ($i_{1,0}/i_{2,0}$) are related to each other in that $(v_{1,\mathrm{eq}}/v_1^*)/(v_{2,\mathrm{eq}}/v_2^*)$ = ($i_{1,0}/i_{2,0}$). Hence, we obtain Eqn. 7–55:

$$i_{1,0}\left(\frac{-\Delta g_1}{kT}\right)_{\mathrm{eq}} = i_{2,0}\left(\frac{-\Delta g_2}{kT}\right)_{\mathrm{eq}}, \quad m_0 = \left(\frac{\Delta g_1}{\Delta g_2}\right) = \frac{i_{1,0}}{i_{2,0}}, \qquad (7\text{–}55)$$

indicating that the ratio of affinity distribution m_0 in the reaction equilibrium equals the ratio of exchange current of the two elementary steps.

The relation between the overall affinity of the reaction $-\Delta G$ and the ratio of affinity distribution m can be derived from the rate equations of the elementary steps. According to the result of calculations carried out for a two-step reaction [Enyo, 1983], the reaction in which the rate is determined solely by *one* of the two steps (a single rate-determining step, $m > 1$) near the equilibrium turns to be the reaction in which the rate is determined not by one but *two* steps (multiple rate-determining steps, $m \doteq 1$) with increasing affinity of the reaction as illustrated in Fig. 7–12. In general, the rate of multistep reactions is determined not by a single rate-step but multiple rate determining steps in the state of high reaction affinity, even if the rate is determined by a single rate-step near the equilibrium.

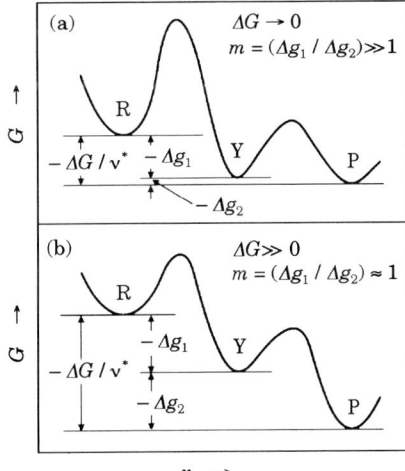

Fig. 7–12. Potential energy curves for a two-step reaction (a) near the equilibrium indicating the reaction affinity distributed to step 1 and (b) away from the equilibrium indicating the reaction affinity distributed to step 1 and step 2: $-\Delta G$ = affinity of the overall reaction; $m = (\Delta g_1 / \Delta g_2)$.

7.4.5 Rate of multistep reactions

For a two-step reaction illustrated in Fig. 7–10, the rate of overall reaction is obtained from Eqn. 7–41 as given by Eqn. 7–56:

$$v_T = v_T^+ - v_T^- = \frac{v_1^+ \omega}{1 + \omega}\left(1 - \frac{v_1^- v_2^-}{v_1^+ v_2^+}\right), \qquad \omega \equiv \frac{v_2^+}{v_1^-}. \tag{7-56}$$

As the reaction affinity increases in the forward direction, the forward rates v_1^+ and v_2^+ of the two steps increase, while the backward rates v_1^- and v_2^- decrease, giving rise to an increase of $\omega \gg 1$ and a decrease of $(v_1^- v_2^-)/(v_1^+ v_2^+) \ll 1$. Therefore, for the range of affinity in which the forward affinity is high, Eqn. 7–56 yields Eqn. 7–57:

$$v_T \doteq v_1^+ \tag{7-57}$$

On the other hand, an increase of the reaction affinity in the backward direction leads to a decrease of $\omega \ll 1$ and an increase of $(v_1^- v_2^-)/(v_1^+ v_2^+) \gg 1$. Therefore, for the range of affinity in which the backward affinity is high, we obtain Eqn. 7–58:

$$v_T \doteq v_2^- \tag{7-58}$$

Consequently, it follows that, in the regime of high reaction affinity, the rate of the forward or backward reactions is determined by the *first leading elementary step* in a series connection of the steps constituting the forward or backward reactions. This means that the rate-determining step is not necessarily the same in the forward and backward reactions. Such an apparent alternation of the rate-determining step depending on the direction (forward or backward) of reaction takes place without any change in the exchange reaction rate at equilibrium.

For multistep reactions in general, it may be concluded that (1) the mechanism of *single rate determining step* is valid *only* in the affinity regime near the equilibrium; and that (2) the mechanism of *multiple rate determining steps* predominates in the high affinity regime in which the first leading step in the forward or backward reactions controls the overall reaction rate in the forward or backward directions. The first leading step controlling the overall reaction rate is not what is called the rate-determining step, because the overall reaction affinity is not located at the first leading step *alone* but distributed among *all* elementary steps of the reaction.

References

[Enyo, 1983]: M. Enyo, *Comprehensive Treatise of Electrochemistry*, **Vol. 7** (Edited by B. E. Conway et al.), p. 241, Plenum Press, New York, (1983).
[Gerischer, 1960]: H. Gerischer, *Z. Physik. Chem. N. F.*, **26**, 223, 325(1960).
[Horiuti-Ikushima, 1939]: J. Horiuti and M. Ikushima, *Proc. Acad. Tokyo*, **15**, 39(1939).
[Horiuti-Polanyi, 1935]: J. Horiuti and M. Polanyi, *Acta Physicochim.*, **2**, 505(1935).
[Horiuti-Nakamura, 1967]: J. Horiuti and T. Nakamura, *Advance in Catalysis*, **Vol. 17**, (Edited by D. D. Eley), p. 2, Academic Press, New York, (1967).
[Matsuda-Horiuti, 1962]: A. Matsuda and J. Horiuti, *J. Res. Inst. Hokkaido University.*, **10**, 14(1962).
[Rysselberghe, 1963]: P. Van Rysselberghe, *Thermodynamics of Irreversible Process*, p. 63, Hermann, Paris; Blaisdell Publ. Co., London, (1963).
[Tafel, 1905]: J. Tafel, *Z. Physik. Chem.*, **50**, 641(1905).
[Vetter, 1967]: K. J. Vetter, *Electrochemical Kinetics — Theoretical and Experimental Aspects —*, p. 121, Academic Press, New York, (1967).

CHAPTER 8

ELECTRODE REACTIONS IN ELECTRON TRANSFER

8.1 Electron Transfer at Metal Electrodes

8.1.1 Kinetics of electron transfer

We consider a simple electron transfer reaction (an outer-sphere electron transfer) between hydrated redox particles OX_{aq}/RED_{aq} and a metal electrode M as shown in Eqn. 8–1:

$$OX_{aq} + e_M \rightleftarrows RED_{aq} , \tag{8-1}$$

where e_M is the electron in the metal electrode.

Fig. 8–1 shows the potential energy barrier for the transfer reaction of redox electrons across the interface of metal electrodes. On the side of metal electrode, the allowed electron energy band is occupied by electrons up to the Fermi level and vacant for electrons above the Fermi level. On the side of hydrated redox particles, the reductant particle RED_{aq} is occupied by electrons in its highest occupied molecular orbital (HOMO); and the oxidant particle OX_{aq} is vacant for electrons in its lowest unoccupied molecular orbital (LUMO). As is described in Sec. 2.10, the highest occupied electron level (HOMO) of reductants and the lowest unoccupied electron level (LUMO) of oxidants are formed by the Franck-Condon level splitting of the same frontier orbital of the redox particles

The plane of closest approach of hydrated ions, the outer Helmholtz plane (OHP), is located 0.3 to 0.5 nm away from the electrode interface; hence, the thickness d_H of the interfacial compact layer across which electrons transfer is in the range of 0.3 to 0.5 nm. Electron transfer across the interfacial energy barrier occurs by the quantum tunneling mechanism at the identical energy level between the metal electrode and the hydrated redox particles as shown in Fig. 8–1.

The rate of cathodic electron transfer by tunneling, at an electron level ε, from the electrode to the oxidant particles is proportional to the product of the state density occupied by electrons $D_{M(e)}(\varepsilon)$ in the electrode and the state density vacant for electrons $D_{OX}(\varepsilon)$ in the oxidant particles as has been shown in Eqn.

7–30. Similarly, the rate of anodic electron transfer by tunneling from the reductant particles to the electrode is proportional to the product of the state density occupied by electrons $D_{RED}(\varepsilon)$ in the reductant particles and the state density vacant for electrons $D_{M(h)}(\varepsilon)$ in the electrode.

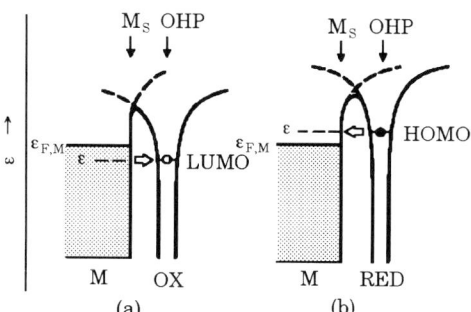

Fig. 8–1. Potential energy barrier for the tunneling transfer of electrons across the interface of metal electrodes: (a) the cathodic electron transfer from the occupied level of the metal electrode to the vacant level of hydrated oxidant particles, (b) the anodic electron transfer from the occupied level of hydrated reductant particles to the vacant level of the metal electrode. M_s = electrode surface; OHP = outer Helmholtz plane; $\varepsilon_{F(M)}$ = Fermi level of electrons in metal electrodes. [From Gerischer, 1960.]

The state density of electrons $D(\varepsilon)$ in the reductant and oxidant particles is given in Eqn. 8–2 by the product of the probability density $W(\varepsilon)$ and the particle concentration c as has been shown in Eqns. 2–48 and 2–49:

$$D_{RED}(\varepsilon) = c_{RED}\, W_{RED}(\varepsilon)\,, \qquad D_{OX}(\varepsilon) = c_{OX}\, W_{OX}(\varepsilon)\,. \tag{8–2}$$

The total state density of redox electrons is then expressed in Eqn. 8–3:

$$D_{REDOX}(\varepsilon) = D_{RED}(\varepsilon) + D_{OX}(\varepsilon)\,. \tag{8–3}$$

Similarly, the total state density in the metal electrode is expressed by the sum of the state densities of electrons and vacant electrons as shown in Eqn. 8–4:

$$D_M(\varepsilon) = D_{M(e)}(\varepsilon) + D_{M(h)}(\varepsilon)\,. \tag{8–4}$$

Fig. 8–2 illustrates the distribution of the state density of electrons in the metal electrode and in the redox particles on both sides of the interface of the electrode in equilibrium with the redox electron transfer.

The cathodic and anodic currents of redox electron transfer, $i^-(\varepsilon)$ and $i^+(\varepsilon)$, are given, respectively, in Eqn. 8–5 and Eqn. 8–6 as follows:

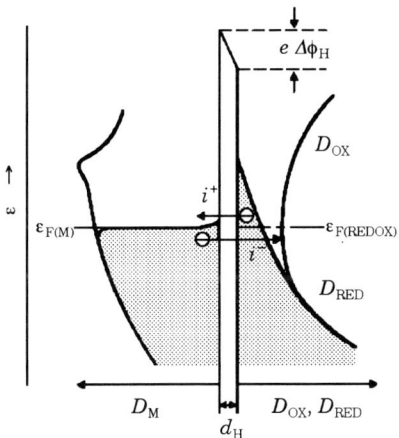

Fig. 8–2. Electron state density in metal electrodes and in hydrated redox particles on both sides of the electrode interface in equilibrium with redox electron transfer: D_M = state density of electrons in the metal electrode; D_{OX} (D_{RED}) = state density of redox electrons in the hydrated oxidant (reductant) particle; $\varepsilon_{F(REDOX)}$ = Fermi level of redox electrons; i^+ (i^-) = anodic (cathodic) electron transfer current; $\Delta\phi_H$ = potential difference across the interfacial compact layer; d_H = thickness of the interfacial compact layer across which electrons transfer.

$$i^-(\varepsilon) = e\, k_t^-(\varepsilon)\, D_{M(e)}(\varepsilon)\, D_{OX}(\varepsilon)\ , \tag{8–5}$$

$$i^+(\varepsilon) = e\, k_t^+(\varepsilon)\, D_{M(h)}(\varepsilon)\, D_{RED}(\varepsilon)\ , \tag{8–6}$$

where $k_t(\varepsilon)$ is the rate constant for electron tunneling. Introducing the Fermi function $f(\varepsilon - \varepsilon_F)$ into Eqns. 8–5 and 8–6, we obtain Eqns. 8–7 and 8–8:

$$i^-(\varepsilon) = e\, k_t^-(\varepsilon)\, D_M(\varepsilon)\, f(\varepsilon - \varepsilon_{F(REDOX)})\, D_{REDOX}(\varepsilon)\, \{\, 1 - f(\varepsilon - \varepsilon_{F(REDOX)})\,\}\ , \tag{8–7}$$

$$i^+(\varepsilon) = e\, k_t^-(\varepsilon)\, D_M(\varepsilon)\, \{\, 1 - f(\varepsilon - \varepsilon_{F(REDOX)})\,\}\, D_{REDOX}(\varepsilon)\, f(\varepsilon - \varepsilon_{F(REDOX)})\ , \tag{8–8}$$

where $\varepsilon_{F(M)}$ and $\varepsilon_{F(REDOX)}$ are the Fermi levels of metal electrodes and redox particles, respectively.

The overall current of electron transfer is then obtained by integrating Eqns. 8–7 and 8–8 with respect to the energy level ε of electrons as shown in Eqn. 8–9:

$$i^- = \int_{-\infty}^{+\infty} i^-(\varepsilon)\, d\varepsilon\ ,\qquad i^+ = \int_{-\infty}^{+\infty} i^+(\varepsilon)\, d\varepsilon\ . \tag{8–9}$$

The net current of electron transfer i is the difference between the cathodic current and the anodic current: $i = i^+ - i^-$.

From Eqns. 8–7, 8–8, and 8–9 we can derive the anodic and cathodic transfer currents i^+ and i^- of electrons, provided that k_t, $D_M(\varepsilon)$ and $D_{REDOX}(\varepsilon)$ are known as functions of the electron energy level, ε. The state density $D_{REDOX}(\varepsilon)$ of redox electrons in hydrated redox particles has already been described in Sec. 2.10; k_t and $D_M(\varepsilon)$ have been assumed nearly constant in a relatively narrow range of electron energy [Gerischer, 1960]. In Sec. 8.1.2, we discuss the state density $D_{REDOX}(\varepsilon)$ of redox electrons in the energy range near the Fermi level of redox electrons, i.e. near the equilibrium electrode potential of redox reactions.

8.1.2 The state density of redox electrons

As is shown in Eqns. 2–48 and 2–49, the *probability density* $W(\varepsilon)$ of electron energy states in the reductant or oxidant particles is represented as a normal distribution function (Gaussian distribution) centered at the *most probable* electron level in the reductant or oxidant particles (See Fig. 2–39.) as expressed in Eqns. 8–10 and 8–11:

$$W_{RED}(\varepsilon) = \frac{1}{\sqrt{4 \lambda_{RED} k T}} \exp \frac{-(\varepsilon - \varepsilon_{RED})^2}{4 \lambda_{RED} k T}, \tag{8-10}$$

$$W_{OX}(\varepsilon) = \frac{1}{\sqrt{4 \lambda_{OX} k T}} \exp \frac{-(\varepsilon - \varepsilon_{OX})^2}{4 \lambda_{OX} k T}, \tag{8-11}$$

where ε_{RED} and ε_{OX} are the most probable electron levels, and λ_{RED} and λ_{OX} are the reorganization energies of coordinated water molecules in the reductant and oxidant particles, respectively. The electron level at which the two probability densities $W_{RED}(\varepsilon)$ and $W_{OX}(\varepsilon)$ equal each other corresponds to the *standard Fermi level of redox electrons*, $\varepsilon^0_{F(REDOX)}$. The standard Fermi level is related with the *most probable levels* of the reductant and oxidant particles as follows: $\varepsilon_{RED} = \varepsilon^0_{F(REDOX)} - \lambda_{RED}$ and $\varepsilon_{OX} = \varepsilon^0_{F(REDOX)} + \lambda_{OX}$.

Eqn. 2–51 gives the Fermi level of redox electrons as a function of the concentration of redox particles as shown in Eqn. 8–12:

$$\varepsilon_{F(REDOX)} = \varepsilon^0_{F(REDOX)} + k T \ln \left(\frac{c_{RED}}{c_{OX}} \right). \tag{8-12}$$

Introducing $\varepsilon_{RED} = \varepsilon^0_{F(REDOX)} - \lambda_{RED}$, $\varepsilon_{OX} = \varepsilon^0_{F(REDOX)} + \lambda_{OX}$, and Eqn. 8–12 into Eqns. 8–10 and 8–11, and taking into account $\lambda_{OX} \gg \varepsilon^0_{F(REDOX)} - \varepsilon$ and $\lambda_{OX} \gg k T \ln(c_{RED} / c_{OX})$ in a narrow range of energy near $\varepsilon^0_{F(REDOX)}$, we obtain the probability densities $W_{RED}(\varepsilon)$ and $W_{OX}(\varepsilon)$ of electron energy states in the reductant and oxidant particles, respectively, as expressed in Eqns. 8–13 and 8–14 [Gerischer, 1960]:

$$W_{\text{RED}}(\varepsilon) = W_{\text{RED}}(\varepsilon_{\text{RED}}) \sqrt{\frac{c_{\text{OX}}}{c_{\text{RED}}}} \exp\left(\frac{-\lambda_{\text{RED}}}{4kT}\right) \exp\left(\frac{-(1-\beta)(\varepsilon - \varepsilon_{\text{F(REDOX)}})}{kT}\right). \quad (8\text{–}13)$$

$$W_{\text{OX}}(\varepsilon) = W_{\text{OX}}(\varepsilon_{\text{OX}}) \sqrt{\frac{c_{\text{RED}}}{c_{\text{OX}}}} \exp\left(\frac{-\lambda_{\text{OX}}}{4kT}\right) \exp\left(\frac{\beta(\varepsilon - \varepsilon_{\text{F(REDOX)}})}{kT}\right), \quad (8\text{–}14)$$

where β is the *symmetry factor* associated with λ_{RED} and λ_{OX}; β is close to β ≑ 0.5, provided that $\lambda_{\text{RED}} \doteqdot \lambda_{\text{OX}}$. As is shown in Eqn. 8–2, the electron state density $D(\varepsilon)$ is represented by the product of the probability density $W(\varepsilon)$ of electron energy states and the concentration of redox particles, c.

At the Fermi level of redox electrons the whole state density $D_{\text{REDOX}}(\varepsilon_{\text{F(REDOX)}})$ of redox particles is half occupied by electrons that provide the occupied state density $D_{\text{RED}}(\varepsilon_{\text{F(REDOX)}})$ of reductant particles, and the remaining half is vacant for electrons in the unoccupied state density $D_{\text{OX}}(\varepsilon_{\text{F(REDOX)}})$ of oxidant particles. Consequently, to a first approximation, the *state densities* $D_{\text{RED}}(\varepsilon)$ and $D_{\text{OX}}(\varepsilon)$ at energy level ε near the Fermi level of redox electrons can be derived, respectively, to produce Eqns. 8–15 and 8–16 [Gerischer, 1960]:

$$D_{\text{RED}}(\varepsilon) = D_{\text{REDOX}}(\varepsilon) f(\varepsilon - \varepsilon_{\text{F(REDOX)}})$$

$$\doteqdot D_{\text{RED}}(\varepsilon_{\text{F(REDOX)}}) \exp\left(\frac{-(1-\beta)(\varepsilon - \varepsilon_{\text{F(REDOX)}})}{kT}\right)$$

$$= \frac{1}{2} D_{\text{REDOX}}(\varepsilon_{\text{F(REDOX)}}) \exp\left(\frac{-(1-\beta)(\varepsilon - \varepsilon_{\text{F(REDOX)}})}{kT}\right), \quad (8\text{–}15)$$

$$D_{\text{OX}}(\varepsilon) = D_{\text{REDOX}}(\varepsilon) \{1 - f(\varepsilon - \varepsilon_{\text{F(REDOX)}})\}$$

$$\doteqdot D_{\text{OX}}(\varepsilon_{\text{F(REDOX)}}) \exp\left(\frac{\beta(\varepsilon - \varepsilon_{\text{F(REDOX)}})}{kT}\right)$$

$$= \frac{1}{2} D_{\text{REDOX}}(\varepsilon_{\text{F(REDOX)}}) \exp\left(\frac{\beta(\varepsilon - \varepsilon_{\text{F(REDOX)}})}{kT}\right). \quad (8\text{–}16)$$

Furthermore, the total state density of redox electrons $D_{\text{REDOX}}(\varepsilon)$ in hydrated redox particles near the Fermi level is expressed in Eqn. 8–17 by the sum of $D_{\text{RED}}(\varepsilon)$ and $D_{\text{OX}}(\varepsilon)$:

$$D_{\text{REDOX}}(\varepsilon) = D_{\text{RED}}(\varepsilon) + D_{\text{OX}}(\varepsilon) \doteqdot \frac{1}{2} D_{\text{REDOX}}(\varepsilon_{\text{F(REDOX)}})$$

$$\times \left\{ \exp\left(\frac{\beta(\varepsilon - \varepsilon_{\text{F(REDOX)}})}{kT}\right) + \exp\left(\frac{-(1-\beta)(\varepsilon - \varepsilon_{\text{F(REDOX)}})}{kT}\right) \right\}. \quad (8\text{–}17)$$

Introducing Eqn. 8–17 into Eqns. 8–7 and 8–8, we can derive from integration of Eqn. 8–9 the cathodic and anodic electron transfer currents, i^- and i^+.

Fig. 8–3 shows an energy diagram of the total electron state density $D_{REDOX}(\varepsilon)$ of redox particles, the occupied electron state density $D_{RED}(\varepsilon)$ of reductant particles, and the unoccupied electron state density $D_{OX}(\varepsilon)$ of oxidant particles as functions of the electron energy ε.

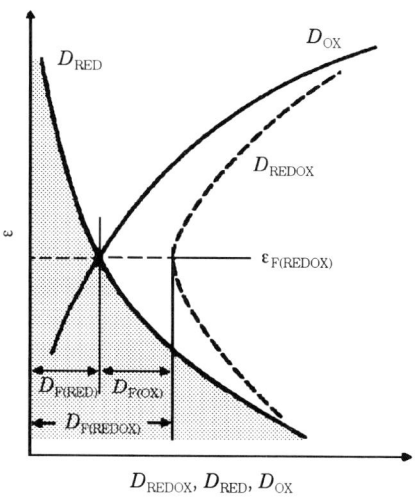

Fig. 8–3. Electron state density in hydrated reductant and oxidant particles near the Fermi level of redox electrons: $D_{REDOX} = D_{RED} + D_{OX}$ = electron state density in redox particles; $D_{F(REDOX)} = D_{F(RED)} + D_{F(OX)}$ = electron state density in redox particles at the Fermi level of redox electrons.

8.1.3 Exchange reaction current at the equilibrium potential

For the equilibrium of redox reactions, the Fermi level of the electrode equals the Fermi level of the redox particles ($\varepsilon_{F(M)} = \varepsilon_{F(REDOX)}$); and the anodic reaction current equals, but in the opposite direction, the cathodic reaction current ($i_0 = i_0^+ = i_0^-$). It follows from the principle of *micro-reversibility* that the forward and backward reaction currents equal each other not only as a whole current but also as a differential current at constant energy level ε; $i_0(\varepsilon) = i_0^+(\varepsilon) = i_0^-(\varepsilon)$. Therefore, referring to Eqns. 8–7 and 8–8, we obtain the exchange reaction current as shown in Eqn. 8–18:

$$i_0^-(\varepsilon) = e\, k_t^-(\varepsilon)\, D_M(\varepsilon)\, f(\varepsilon - \varepsilon_{F(M)})\, D_{REDOX}(\varepsilon)\, \{\, 1 - f(\varepsilon - \varepsilon_{F(REDOX)})\,\}$$

$$= e\, k_t^+(\varepsilon)\, D_M(\varepsilon)\, \{\, 1 - f(\varepsilon - \varepsilon_{F(M)})\,\}\, D_{REDOX}(\varepsilon)\, f(\varepsilon - \varepsilon_{F(REDOX)}) = i_0^+(\varepsilon)\,. \quad (8\text{–}18)$$

At equilibrium the Fermi level is equilibrated between the electrode and the redox particles ($\varepsilon_{F(M)} = \varepsilon_{F(REDOX)}$); hence, Eqn. 8–18 gives Eqn. 8–19:

$$k_t^- = k_t^+ = k_t\,. \quad (8\text{–}19)$$

Eqn. 8–19 indicates that the rate constant of electron transfer by tunneling is the same in the forward and the backward directions at constant electron energy. The same rate constant of electron transfer by tunneling in the forward and backward directions is valid not only in the equilibrium state but also in the nonequlibrium state of electron transfer [Gerischer, 1960].

Integration of Eqn. 8–18 with respect to electron energy ε to produce Eqn. 8–20 yields the *exchange reaction current* i_0:

$$i_0 = e \int_{-\infty}^{+\infty} k_t(\varepsilon)\, D_M(\varepsilon)\, f(\varepsilon - \varepsilon_{F(M)})\, D_{REDOX}(\varepsilon)\, \{\, 1 - f(\varepsilon - \varepsilon_{F(REDOX)})\,\}\, d\varepsilon\,, \quad (8\text{–}20)$$

where $k_t(\varepsilon)$ and $D_M(\varepsilon)$ may be assumed nearly constant in a limited range of electron energy around the Fermi level ($D_M(\varepsilon_{F(M)}) = \varepsilon_{F(REDOX)}$)). In Eqn. 8–20 $f(\varepsilon - \varepsilon_F)$ is the Fermi function given by $f(\varepsilon - \varepsilon_F) = 1/[1 + \exp\{(\varepsilon - \varepsilon_F)/(k\,T)\}]$.

Introducing Eqn. 8–17 into Eqn. 8–20, we obtain Eqn. 8–21 as an approximate equation for the exchange reaction current i_0:

$$i_0 = e\, k_t(\varepsilon_{F(M)})\, D_M(\varepsilon_{F(M)})\, \frac{1}{2}\, D_{REDOX}(\varepsilon_{F(REDOX)})\, B(\beta)\,, \quad (8\text{–}21)$$

where

$$B(\beta) = \int_{-\infty}^{+\infty} \exp\left(\frac{\beta\,(\varepsilon - \varepsilon_{F(REDOX)})}{k\,T}\right) f(\varepsilon - \varepsilon_{F(M)})\, d\varepsilon\,. \quad (8\text{–}22)$$

The major part (90 %) of the integral $B(\beta)$ is in a narrow range of energy ($\varepsilon_F \pm 0.25$ eV) around the Fermi level ε_F (= $\varepsilon_{F(M)} = \varepsilon_{F(REDOX)}$), and is approximated by $B(\beta) \doteq \pi\, k\, T\, /\sin\{(1 - \beta)\,\pi\}$ [Gerischer, 1960]. Then, to a first approximation, we obtain from Eqn. 8–21 the exchange reaction current i_0 as expressed in Eqn. 8–23:

$$i_0 \doteq \frac{e\, \pi\, k\, T}{2\, \sin\{(1 - \beta)\,\pi\}}\, k_t(\varepsilon_{F(M)})\, D_M(\varepsilon_{F(M)})\, D_{REDOX}(\varepsilon_{F(REDOX)}) \quad (8\text{–}23)$$

Fig. 8–4 is an energy diagram that includes as functions of electron energy ε: the electron state density $D_M(\varepsilon)$ in the metal electrode; the electron state density $D_{REDOX}(\varepsilon)$ in the redox particles; the rate constant $k_t(\varepsilon)$ of tunneling electron transfer; and the exchange reaction current $i_0(\varepsilon)$. The exchange reaction current is distributed *only* in a narrow range of electron energy with its maximum centered at the Fermi level.

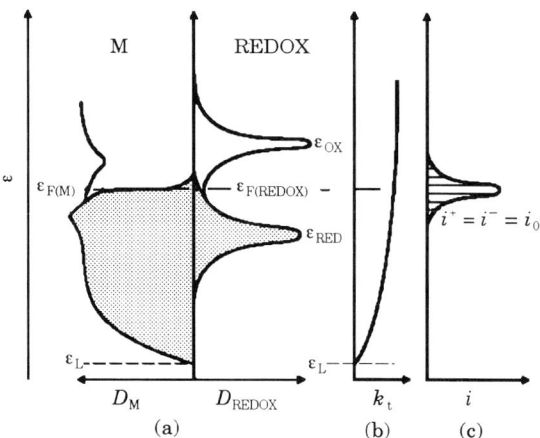

Fig. 8–4. (a) Electron state density D in metal electrodes and in hydrated redox particles, (b) rate constant for electron tunneling k_t, and (c) exchange reaction current i_0 of electron transfer in equilibrium with the redox reaction: ε_L = lower edge of the allowed band of metal electrons. [From Gerischer, 1960.]

8.1.4 Reaction current under polarization

As the electrode potential is polarized from the equilibrium potential of the redox reaction, the Fermi level $\varepsilon_{F(M)}$ of electrons in the metal electrode is shifted from the Fermi level $\varepsilon_{F(REDOX)}$ of redox electrons in the redox particles by an energy equivalent to the overvoltage η as expressed in Eqn. 8–24:

$$\varepsilon_{F(M)} = \varepsilon_{F(REDOX)} - e\eta \ , \tag{8–24}$$

where η is positive in anodic polarization and negative in cathodic polarization. The overall reaction current, $i(\eta)$, at an overvoltage η is given by the algebraic sum of the anodic current, $i^+(\eta)$, and the cathodic current, $i^-(\eta)$, as in Eqn. 8–25:

$$i(\eta) = i^+(\eta) - i^-(\eta) \ . \tag{8–25}$$

From the discussion of this section, the anodic and cathodic reaction currents can be derived, respectively, to produce Eqns. 8–26 and 8–27:

$$i^-(\eta) = e \int_{-\infty}^{+\infty} k_t(\varepsilon, \eta)\, D_M(\varepsilon)\, f(\varepsilon - \varepsilon_{F(REDOX)} + e\,\eta)$$

$$\times D_{REDOX}(\varepsilon)\, \{1 - f(\varepsilon - \varepsilon_{F(REDOX)})\}\, d\varepsilon \quad , \tag{8-26}$$

$$i^+(\eta) = e \int_{-\infty}^{+\infty} k_t(\varepsilon, \eta)\, D_M(\varepsilon)\, \{1 - f(\varepsilon - \varepsilon_{F(REDOX)} + e\,\eta)\}$$

$$\times D_{REDOX}(\varepsilon)\, f(\varepsilon - \varepsilon_{F(REDOX)})\, d\varepsilon \quad . \tag{8-27}$$

Introducing Eqns. 8–15 and 8–16 into Eqns. 8–26 and 8–27, we obtain Eqns. 8–28 and 8–29:

$$i^-(\eta) = e\, k_t(\varepsilon_{F(M)})\, D_M(\varepsilon_{F(M)})\, \frac{1}{2} D_{REDOX}(\varepsilon_{F(REDOX)})$$

$$\times \int_{-\infty}^{+\infty} \left\{ \frac{\exp\left(\dfrac{\beta(\varepsilon - \varepsilon_{F(REDOX)})}{kT}\right)}{1 - \exp\left(\dfrac{\varepsilon - \varepsilon_{F(REDOX)} + e\,\eta}{kT}\right)} \right\} d\varepsilon$$

$$\doteq i_0\, \frac{k_t(\varepsilon_{F(M)})}{k_t(\varepsilon_{F(REDOX)})}\, \exp\left(\frac{-\beta e\,\eta}{kT}\right), \tag{8-28}$$

$$i^+(\eta) = e\, k_t(\varepsilon_{F(M)})\, D_M(\varepsilon_{F(M)})\, \frac{1}{2} D_{REDOX}(\varepsilon_{F(REDOX)})$$

$$\times \int_{+\infty}^{-\infty} \left\{ \frac{\exp\left(\dfrac{-(1-\beta)(\varepsilon - \varepsilon_{F(REDOX)})}{kT}\right)}{1 - \exp\left(\dfrac{\varepsilon_{F(REDOX)} - \varepsilon - e\,\eta}{kT}\right)} \right\} d\varepsilon$$

$$\doteq i_0\, \frac{k_t(\varepsilon_{F(M)})}{k_t(\varepsilon_{F(REDOX)})}\, \exp\left(\frac{(1-\beta)e\,\eta}{kT}\right). \tag{8-29}$$

Figs. 8–5 and 8–6 are energy diagrams, as functions of the electron energy ε under the conditions of anodic and cathodic polarization, respectively, for: the electron state density $D_M(\varepsilon)$ in the metal electrode; the electron state density $D_{REDOX}(\varepsilon)$ in the redox particles; and the differential reaction current $i(\varepsilon)$. From these Figures it is revealed that *most* of the reaction current of redox electron transfer occurs in a narrow range of energy centered at the *Fermi level* of metal electrodes *even* in the state of polarization. Further, polarization of the electrode potential causes the ratio to change between the occupied state density

$D_{\text{RED}}(\varepsilon_{F(M)})$ and the unoccupied state density $D_{\text{OX}}(\varepsilon_{F(M)})$ in the redox particles at the Fermi level of the metal electrode; hence, either the cathodic or the anodic reaction currents can predominate.

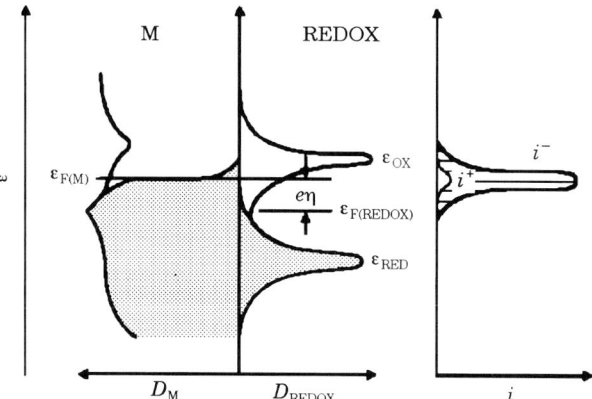

Fig. 8–5. Electron state density in metal electrodes and in hydrated redox particles, and the anodic and cathodic currents of redox electron transfer under cathodic polarization: η = cathodic overvoltage (negative); i^+ = anodic current; i^- = cathodic current. [From Gerischer, 1960.]

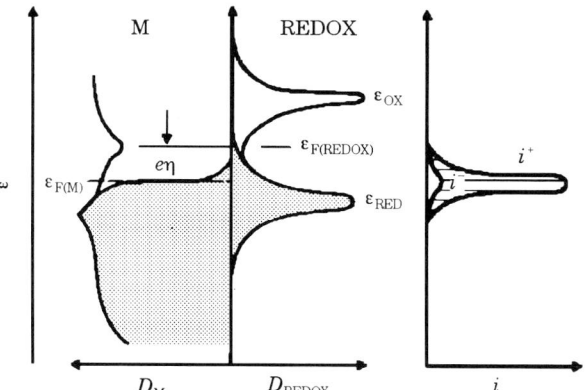

Fig. 8–6. Electron state density in metal electrodes and in hydrated redox particles, and the anodic and cathodic currents of redox electron transfer under anodic polarization: η = anodic overvoltage (positive). [From Gerischer, 1960.]

It is characteristic of metal electrodes that the reaction current of redox electron transfer, under the anodic and cathodic polarization conditions, occurs mostly at the Fermi level of metal electrodes rather than at the Fermi level of redox particles. In contrast to metal electrodes, as is discussed in Sec. 8.2, semiconductor electrodes exhibits no electron transfer current at the Fermi level of the electrodes.

Since the rate constant $k_t(\varepsilon)$ of electron tunneling may be assumed constant, the reaction current of electron transfer, $i^\pm$, in Eqns. 8–28 and 8–29 can be expressed in the form of Eqn. 8–30:

$$\eta = a + b \log i^\pm , \qquad (8\text{–}30)$$

which corresponds to the *Tafel equation* for the current of electrode reactions.

Fig. 8–7 shows the anodic and cathodic polarization curves observed for a redox couple of hydrated titanium ions Ti^{4+}/Ti^{3+} on an electrode of mercury in a sulfuric acid solution; the Tafel relationship is evident in both anodic and cathodic reactions. From the slope of the Tafel plot, we obtain the symmetry factor β nearly equal to 0.5 ($\beta \doteqdot 0.5$).

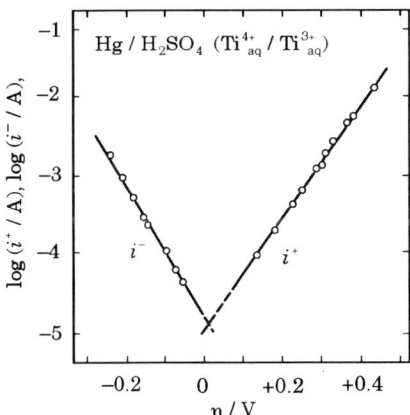

Fig. 8–7. Cathodic and anodic polarization curves observed for a transfer reaction of redox electrons of hydrated Ti^{4+}/Ti^{3+} particles at a mercury electrode in 1 M H_2SO_4 solution containing 0.17 M Ti^{4+}_{aq} and 0.03 M Ti^{3+}_{aq} at 25 °C: electrode surface area = 0.15 cm^{-2}. [From Vetter, 1967.]

8.1.5 Diffusion and reaction rate

The transfer reaction of redox electrons at the electrode interface accompanies a mass transport of hydrated redox particles through the interfacial diffusion layer to and from the electrode interface as shown in Fig. 8–8. If the rate of mass transport of hydrated redox particles is large, the reaction current is determined

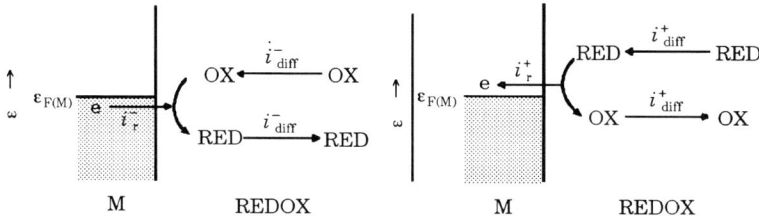

Fig. 8–8. Diffusion of redox particles before and after electron transfer at metal electrodes: i_r^- (i_r^+) = cathodic (anodic) electron transfer current, i_{diff}^- (i_{diff}^+) = cathodic (anodic) diffusion current.

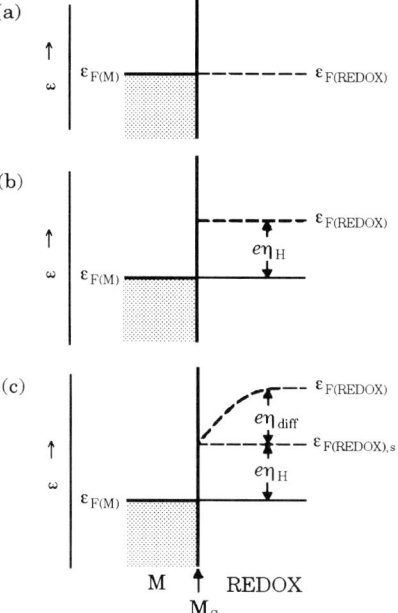

Fig. 8–9. Energy levels for redox electron transfer reaction at metal electrodes: (a) in equilibrium, (b) in anodic polarization with the reaction rate determined by interfacial electron transfer, (c) anodic polarization with the reaction rate determined by both interfacial electron transfer and diffusion of hydrated particles. $\varepsilon_{\text{F(REDOX), s}}$ = Fermi level of redox electrons at the interface.

by the electron transfer at the interface; and the total overvoltage η occurs across the interfacial compact layer, $\eta = \eta_H$, as shown in Fig. 8–9(b).

As the resistance of mass transport increases, the diffusion overvoltage becomes significant; and the total overvoltage η is distributed to both interfacial overvoltage η_H and diffusion overvoltage η_{diff} as expressed in Eqn. 8–31:

$$\eta = \eta_H + \eta_{\text{diff}} \tag{8-31}$$

Then, a concentration gradient of hydrated redox particles arises in the interfacial diffusion layer; and the Fermi level $\varepsilon_{F(REDOX)s}$ of redox particles at the interface becomes different from the Fermi level $\varepsilon_{F(REDOX)}$ of redox particles outside the diffusion layer as shown in Fig. 8–9(c). The partial overvoltages η_H and η_{diff} are then given by Eqn. 8–32:

$$\eta_H = -\frac{\varepsilon_{F(M)} - \varepsilon_{F(REDOX)s}}{e} \quad , \quad \eta_{diff} = -\frac{\varepsilon_{F(REDOX)s} - \varepsilon_{F(REDOX)}}{e} \quad . \tag{8–32}$$

We consider a transfer reaction of redox electrons in which the interfacial transfer of electrons is in quasi-equilibrium ($\eta_H = 0$) and the diffusion of redox particles determines the overall reaction rate. The anodic diffusion current, i_{diff}^+, and the anodic limiting current of diffusion, i_{lim}^+, in the stationary state of the electrode reaction are given, respectively, in Eqns. 8–33 and 8–34:

$$i_{diff}^+ = e\, m_{RED}\, (c_{RED} - c_{REDs}) = e\, m_{OX}\, (c_{OXs} - c_{OX}) \quad , \tag{8–33}$$

$$i_{lim}^+ = e\, m_{RED}\, c_{RED} \quad , \tag{8–34}$$

where m is a *transport coefficient* (the diffusion constant divided by the thickness of diffusion layer); c and c_s are the concentrations of hydrated particles in the solution away from the electrode and at the electrode interface, respectively. Eqns. 8–33 and 8–34 yield Eqn. 8–35 if the concentration of hydrated oxidant particles is negligibly small ($c_{OX} \to 0$ and $i_{lim}^+ = e\, m_{RED}\, c_{RED} = e\, m_{OX}\, c_{OXs}$):

$$\frac{c_{REDs}}{c_{RED}} = \frac{i_{lim}^+ - i_{diff}^+}{i_{lim}^+} \quad , \quad c_{REDs} = \frac{i_{lim}^+ - i_{diff}^+}{e\, m_{RED}} \quad . \tag{8–35}$$

In the transfer equilibrium of redox electrons, the Fermi level of the electrode $\varepsilon_{F(M)}$ equals the Fermi level of the redox particles $\varepsilon_{F(REDOX)s}$ at the electrode interface. Hence, incorporating the standard Fermi level, $\varepsilon^0_{F(REDOX)}$, of redox electrons, we obtain the Fermi level $\varepsilon_{F(M)}$ as expressed in Eqn. 8–36:

$$\varepsilon_{F(M)} = \varepsilon_{F(REDOX)s} = \varepsilon^0_{F(REDOX)} + k\,T\,\ln\left(\frac{c_{REDs}}{c_{OXs}}\right) . \tag{8–36}$$

For a negligibly small concentration of oxidant particles ($c_{OX} \to 0$), therefore, we obtain Eqn. 8–37:

$$\varepsilon_{F(M)} = \varepsilon_{F(1/2)} + k\,T\,\ln\left(\frac{i_{lim}^+ - i_{diff}^+}{i_{diff}^+}\right) \quad , \quad \varepsilon_{F(1/2)} = \varepsilon^0_{F(REDOX)} - k\,T\,\ln\left(\frac{m_{RED}}{m_{OX}}\right) , \tag{8–37}$$

where $\varepsilon_{F(1/2)}$ is the electron level (the electrode potential) at which the reaction current is half the limiting current of diffusion as shown in Fig. 8–10.

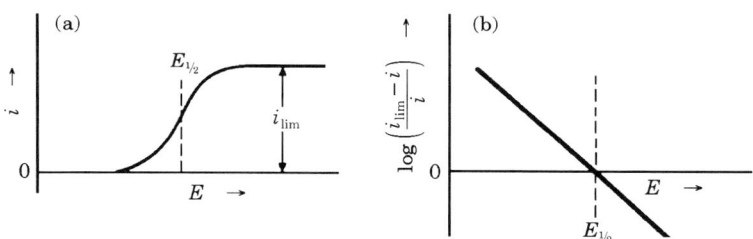

Fig. 8–10. Anodic reaction current vs. potential curve for a redox electron transfer of hydrated redox particles at metal electrodes: i_{lim} = limiting current of the diffusion of redox particles; $E_{1/2}$ = potential at which the reaction current is half the limiting diffusion current ($i = 0.5\, i_{\text{lim}}$). [From Bard-Faulkner, 1980.]

This electron level $\varepsilon_{F(1/2)}$ is independent of the concentration of the redox particle and, hence, is characteristic of individual redox reactions [Bard-Faulkner, 1980].

When the rate of electron transfer is determined by both transfer of interfacial electrons and by diffusion of redox particles, the resistance of overall reaction, $1/i$, may be represented by the sum of the resistance of electron transfer, $1/i_r$, and the resistance of diffusion of redox particles, $1/i_{\text{diff}} \doteq 1/i_{\text{lim}}$, as shown in Eqn. 8–38 [Reineke-Memming, 1992]:

$$\frac{1}{i} = \frac{1}{i_r} + \frac{1}{i_{\text{diff}}} \doteq \frac{1}{i_r} + \frac{1}{i_{\text{lim}}}, \tag{8-38}$$

where i_r is the reaction current of electron transfer at the interface.

Disregarding the backward reaction current, we obtain in Eqn. 8–39 the anodic reaction current, i_r^+, of electron transfer:

$$i_r \doteq i_r^+ = e\, k_0\, c_{\text{RED s}} \exp\left(\frac{-(1-\beta)(\varepsilon_{F(M)} - \varepsilon^0_{F(\text{REDOX})})}{kT} \right). \tag{8-39}$$

Furthermore, Eqns. 8–35 and 8–38 and 8–39 yield Eqn. 8–40:

$$kT \ln\left(\frac{i_{\text{lim}}^+ - i}{i} \right) = (1-\beta)(\varepsilon_{F(M)} - \varepsilon'_{F(1/2)}), \quad \varepsilon'_{F(1/2)} = \varepsilon^0_{F(\text{REDOX})} - kT \ln\left(\frac{m_{\text{RED}}}{k_0} \right). \tag{8-40}$$

Eqn. 8–40 indicates a linear relationship between the electrode potential and the logarithm of $(i_{\text{lim}}^+ - i)/i$ as shown in Fig. 8–10.

8.2 Electron Transfer at Semiconductor Electrodes

8.2.1 Semiconductor electrodes compared with metal electrodes

We consider a simple redox electron transfer of hydrated redox particles (an outer-sphere electron transfer) of Eqn. 8–1 at semiconductor electrodes. The kinetics of electron transfer reactions is the same in principal at both metal and semiconductor electrodes; but the rate of electron transfer at semiconductor electrodes differs considerably from that at metal electrodes because the electron occupation in the electron energy bands differs distinctly with metals and semiconductors.

Fig. 8–11 shows as a function of electron energy ε; the electron state density $D_{SC}(\varepsilon)$ in semiconductor electrodes, and the electron state density $D_M(\varepsilon)$ in metal electrodes. Both $D_{SC}(\varepsilon)$ and $D_M(\varepsilon)$ are in the state of electron transfer equilibrium with the state density $D_{REDOX}(\varepsilon)$ of hydrated redox particles; the Fermi level is equilibrated between the redox particles and the electrode. For metal electrodes the electron state density $D_M(\varepsilon)$ is high at the Fermi level, and most of the electron transfer current occurs at the Fermi level $\varepsilon_{F(M)}$. For semiconductor electrodes the Fermi level $\varepsilon_{F(SC)}$ is located in the band gap where no electron level is available for the electron transfer ($D_{SC}(\varepsilon_{F(SC)}) = 0$) and, hence, no electron transfer current can occur at the Fermi level $\varepsilon_{F(SC)}$. The electron transfer is allowed to occur *only* within the conduction and valence bands where the state density of electrons is high ($D_{SC}(\varepsilon) > 0$).

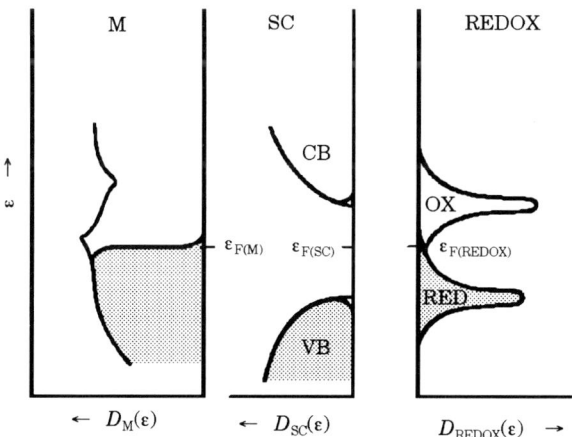

Fig. 8–11. Electron state density in metal electrodes, semiconductor electrodes, and redox particles in equilibrium with the redox electron transfer reaction. [From Gerischer, 1961.]

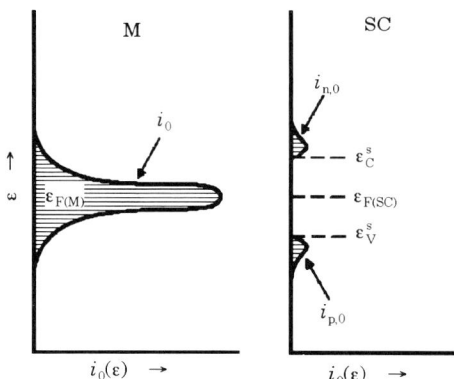

Fig. 8–12. The exchange reaction current of redox electron transfer at metal electrodes and at semiconductor electrodes in equilibrium with the redox electron transfer reaction shown in Fig. 8–11: i_0 = exchange current at the metal electrode; $i_{n,0}$ ($i_{p,0}$) = exchange current of the electron (hole) transfer via the conduction (the valence) band of semiconductor; ε_C^s (ε_V^s) = lower edge level of the conduction band (upper edge level of the valence band) at the electrode interface.

The distribution of the exchange transfer current of redox electrons $i_0(\varepsilon)$, which corresponds to the state density curves shown in Fig. 8–11, is illustrated for both the metal and semiconductor electrodes in Fig. 8–12 (See also Fig. 8–4.). Since the state density of electrons in the electrode available for electron transfer exists only in the conduction and valence bands fairly away from the Fermi level $\varepsilon_{F(M)}$ and since the state density of redox electrons in the redox particles decreases remarkably with increasing deviation of the electron level (with increasing polarization) from the Fermi level $\varepsilon_{F(REDOX)}$ of the redox electrons, the exchange transfer current of redox electrons is fairly small at semiconductor electrodes compared with that at metal electrodes as shown in Fig. 8–12.

8.2.2 The conduction band and the valence band mechanisms

As shown in Fig. 8–12 and in Fig. 8–13, the transfer current of redox electrons can be divided into the transfer current, $i_n(\varepsilon)$, carried by electrons e_{CB} in the conduction band and the transfer current, $i_p(\varepsilon)$, carried by holes h_{VB} in the valence band as shown in Eqn. 8–41:

$$OX_{aq} + e_{CB} \rightleftarrows RED_{aq} , \qquad OX_{aq} \rightleftarrows RED_{aq} + h_{VB} . \qquad (8-41)$$

The transfer reaction current of redox electrons, $i(\varepsilon)$, is expressed in Eqn. 8–42 by the sum of the electron and hole transfer currents:

$$i^+(\varepsilon) = i_n^+(\varepsilon) + i_p^+(\varepsilon) , \qquad i^-(\varepsilon) = i_n^-(\varepsilon) + i_p^-(\varepsilon) . \qquad (8-42)$$

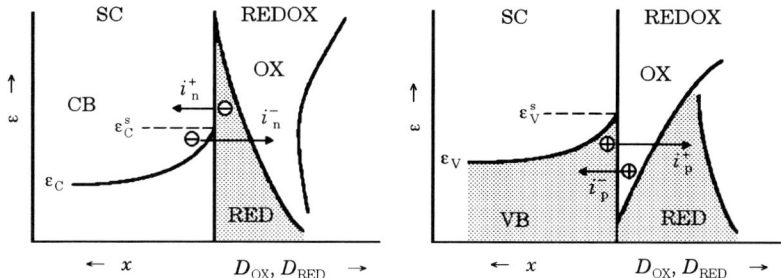

Fig. 8-13. Electron transfer via the conduction band and hole transfer via the valence band: D_{OX} (D_{RED}) = state density of oxidant (reductant) particles; x = distance from the interface; i^+ (i^-) = anodic (cathodic) current; i_n (i_p) = electron (hole) current.

The anodic currents, $i_n^+(\varepsilon)$ and $i_p^+(\varepsilon)$, and the cathodic currents, $i_n^-(\varepsilon)$ and $i_n^-(\varepsilon)$, via the conduction and valence bands are, respectively, given in Eqns. 8–43 through 8–46:

$$i_n^+(\varepsilon) = e\, k_{t,n}\, D_{SC(n)}(\varepsilon)\, \{\, 1 - f(\varepsilon - \varepsilon_{F(SC)})\, \}\, D_{RED}(\varepsilon)\,, \tag{8-43}$$

$$i_p^+(\varepsilon) = e\, k_{t,p}\, D_{SC(p)}(\varepsilon)\, \{\, 1 - f(\varepsilon - \varepsilon_{F(SC)})\, \}\, D_{RED}(\varepsilon)\,, \tag{8-44}$$

$$i_n^-(\varepsilon) = e\, k_{t,n}\, D_{SC(n)}(\varepsilon)\, f(\varepsilon - \varepsilon_{F(SC)})\, D_{OX}(\varepsilon)\,, \tag{8-45}$$

$$i_p^-(\varepsilon) = e\, k_{t,p}\, D_{SC(p)}(\varepsilon)\, f(\varepsilon - \varepsilon_{F(SC)})\, D_{OX}(\varepsilon)\,. \tag{8-46}$$

The overall transfer currents of redox electrons can be obtained by integrating these equations with respect to electron energy ε from the lower edge ε_{CB}^s of the conduction band to infinity for the *conduction band mechanism* and from minus infinity to the upper edge ε_{VB}^s of the valence band for the *valence band mechanism*.

In general, redox electron transfer occurs in a narrow range of energy of a few kT near the band edges as is shown in Fig. 8–12. The electron state density $D_{SC}(\varepsilon)$ in such a narrow range of energy near the band edges can be approximated by the *effective state densities* N_C and N_V (Refer to Sec. 2.4.). Further, the rate constant $k_t(\varepsilon)$ of tunneling transfer of electrons and holes at semiconductor electrodes may be assumed constant in a narrow range of energy, resembling the case of metal electrodes.

8.2.3 Electron state density in redox electrode reactions

In contrast to metal electrodes in which the electrostatic potential is constant within the electrode, in semiconductor electrodes a space charge layer exists that creates an electrostatic potential gradient. Therefore, the band edge levels ε_{CB}^{b} and ε_{VB}^{b} in the interior of semiconductor electrodes differ from the analogous ε_{CB}^{s} and ε_{VB}^{s} at the interface of electrodes; hence, the difference between the band edge level and the Fermi level in the interior of semiconductors is not the same as that at the interface of electrodes as shown in Fig. 8–14 and expressed in Eqn. 8–47:

$$\varepsilon_{C}^{b} - \varepsilon_{F(SC)} \neq \varepsilon_{C}^{s} - \varepsilon_{F(SC)} \ , \qquad \varepsilon_{F(SC)} - \varepsilon_{V}^{b} \neq \varepsilon_{F(SC)} - \varepsilon_{V}^{s} \ . \tag{8–47}$$

Furthermore, the band edge level, relative to the Fermi level at the interface of electrodes, remains unchanged in the state of *Fermi level pinning* (the interface degeneracy), but the same relative band edge level varies with the electrode potential in the state of *band edge level pinning* (the interface nondegeneracy) as

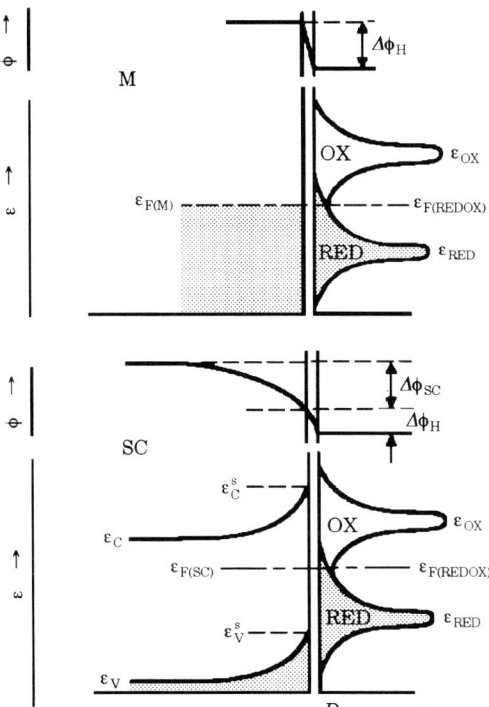

Fig. 8–14. Electron state density for a redox electron transfer reaction and profile of the electrostatic inner potential, ϕ, across the interface of metal and semiconductor electrodes: $\Delta\phi_H$ = potential drop across the interfacial compact layer; $\Delta\phi_{SC}$ = potential difference of the space charge layer.

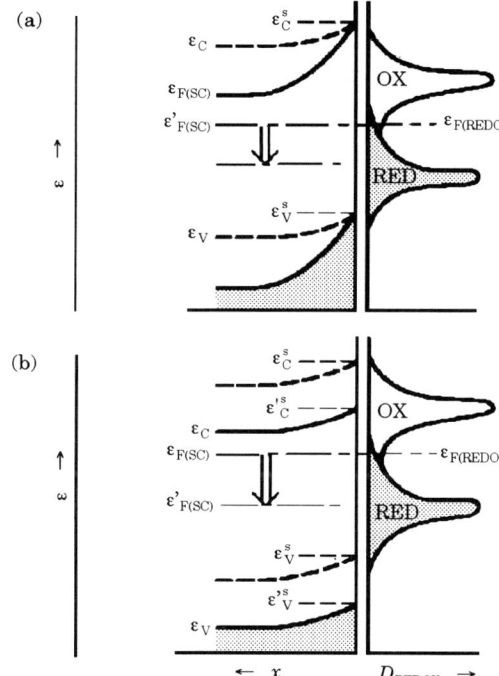

Fig. 8–15. Electron state density for a redox electron transfer reaction of hydrated redox particles at semiconductor electrodes (a) in the state of band edge level pinning and (b) in the state of Fermi level pinning: dotted curve = band edge levels in the reaction equilibrium; solid curve = band edge levels in the anodic polarization; $\varepsilon'_{F(SC)}$ = Fermi level of electrode in the anodic polarization; ε'^{s}_{V} and ε'^{s}_{C} = band edge levels in the anodic polarization.

shown in Fig. 8–15 (Refer to Sec. 5.9.). In the state of Fermi level pinning, any change of electrode potential overall occurs across the interfacial compact layer and produces no change in the concentration of interfacial electrons or holes in the electrode. In the state of band edge level pinning, any change of electrode potential occurs only across the space charge layer in the electrode producing a *potential-dependent* concentration of interfacial electrons or holes in the semiconductor electrode.

In the state of band edge level pinning, the electron level of the redox particles with the state density of $D_{REDOX}(\varepsilon)$ relative to the electron level of the semiconductor with the state density of $D_{SC}(\varepsilon)$ remains unchanged at the electrode interface irrespective of the electrode potential. However, in the state of Fermi level pinning, the electron level of the redox particles relative to the electron level of the semiconductor electrode depends on the electrode potential in the same way as occurs with metal electrodes (quasi-metallization of semiconductor electrodes).

Since semiconductor electrodes are usually in the state of band edge level pinning, the concentration of interfacial electrons or holes in the electrode changes with anodic or cathodic polarization, while the relative electron level between the redox particles, $D_{\text{REDOX}}(\varepsilon)$, and the electrode, $D_{\text{SC}}(\varepsilon)$, remains unchanged.

8.2.4 Exchange reaction current at the equilibrium potential

In the equilibrium state of electron transfer, the Fermi level of an electrode equals the Fermi level of the redox particles ($\varepsilon_{\text{F(SC)}} = \varepsilon_{\text{F(REDOX)}}$); the forward reaction current equals the backward reaction current which both equal the exchange reaction current i_0. Further, the exchange current is the sum of the conduction band current, $i_{\text{n},0}$, and the valence band current, $i_{\text{p},0}$ ($i_0 = i_{\text{n},0} + i_{\text{p},0}$). The exchange currents $i_{\text{n},0}$ and $i_{\text{p},0}$ are given, respectively, by Eqns. 8–48 and 8–49:

$$i_{\text{n},0} = e \int_{\varepsilon_{\text{C}}^{\text{s}}}^{\infty} k_{\text{t(n)}}(\varepsilon) \, D_{\text{SC(n)}}(\varepsilon) \, f(\varepsilon - \varepsilon_{\text{F(SC)}}) \, D_{\text{OX}}(\varepsilon) \, d\varepsilon$$

$$= e \int_{\varepsilon_{\text{C}}^{\text{s}}}^{\infty} k_{\text{t(n)}}(\varepsilon) \, D_{\text{SC(n)}}(\varepsilon) \, \{ 1 - f(\varepsilon - \varepsilon_{\text{F(SC)}}) \} \, D_{\text{RED}}(\varepsilon) \, d\varepsilon \quad , \quad (8\text{–}48)$$

$$i_{\text{p},0} = e \int_{-\infty}^{\varepsilon_{\text{V}}^{\text{s}}} k_{\text{t(p)}}(\varepsilon) \, D_{\text{SC(p)}}(\varepsilon) \, \{ 1 - f(\varepsilon - \varepsilon_{\text{F(SC)}}) \} \, D_{\text{RED}}(\varepsilon) \, d\varepsilon$$

$$= e \int_{-\infty}^{\varepsilon_{\text{V}}^{\text{s}}} k_{\text{t(p)}}(\varepsilon) \, D_{\text{SC(p)}}(\varepsilon) \, f(\varepsilon - \varepsilon_{\text{F(SC)}}) \, D_{\text{OX}}(\varepsilon) \, d\varepsilon \quad , \quad (8\text{–}49)$$

where $D_{\text{SC(n)}}(\varepsilon)$ and $D_{\text{SC(p)}}(\varepsilon)$ are the state densities of electrons in the conduction and valence bands, respectively.

Accounting the result of calculations showing that more than 90 % of the exchange reaction current flows in a narrow range of energy ($\varepsilon_{\text{C}}^{\text{s}}$ to $\varepsilon_{\text{C}}^{\text{s}} + 4\,k\,T$, and $\varepsilon_{\text{V}}^{\text{s}}$ to $\varepsilon_{\text{V}}^{\text{s}} - 4\,k\,T$) near the band edge level and that the maximum exchange current occurs about one $k\,T$ away from the band edge level ($\varepsilon_{\text{C}}^{\text{s}} + k\,T$, and $\varepsilon_{\text{V}}^{\text{s}} - k\,T$), we choose the state densities of $D_{\text{OX}}(\varepsilon_{\text{C}}^{\text{s}} + k\,T)$ and $D_{\text{RED}}(\varepsilon_{\text{V}}^{\text{s}} - k\,T)$ to represent the conduction band mechanism and the valence band mechanism, respectively. Then, the following approximate equations, Eqns. 8–50 and 8–51, can be obtained [Gerischer, 1961]:

$$i_{\text{n},0} = e\,k_{\text{n}}\,D_{\text{OX}}(\varepsilon = \varepsilon_{\text{C}}^{\text{s}} + k\,T)\,N_{\text{C}}\,\exp\!\left(\frac{-(\varepsilon_{\text{C}}^{\text{s}} - \varepsilon_{\text{F(SC)}})}{k\,T} \right) \quad , \quad (8\text{–}50)$$

$$i_{\text{p},0} = e\,k_{\text{p}}\,D_{\text{RED}}(\varepsilon = \varepsilon_{\text{V}}^{\text{s}} - k\,T)\,N_{\text{V}}\,\exp\!\left(\frac{-(\varepsilon_{\text{F(SC)}} - \varepsilon_{\text{V}}^{\text{s}})}{k\,T} \right) \quad , \quad (8\text{–}51)$$

where K is the rate constant, and N_{C} and N_{V} are the effective state densities of electrons and holes in the conduction and valence bands, respectively.

We further obtain the concentration, n_{s}, of interfacial electrons in the conduction band and the concentration, p_s, of interfacial holes in the valence band as shown in Eqn. 8–52:

$$n_s = N_C \exp\left(\frac{-(\varepsilon_C^s - \varepsilon_{F(SC)})}{kT}\right), \quad p_s = N_C \exp\left(\frac{-(\varepsilon_{F(SC)} - \varepsilon_V^s)}{kT}\right). \quad (8\text{–}52)$$

Accounting further that $D_{OX} = W_{OX} c_{OX}$, $D_{RED} = W_{RED} c_{RED}$, $\varepsilon_{F(REDOX)} = \varepsilon_{F(REDOX)}^0 + kT \ln(c_{RED}/c_{OX})$, and $\varepsilon_{F(SC)}^i \equiv (1/2)(\varepsilon_V^s + \varepsilon_C^s) + (1/2)kT \ln(N_V/N_C)$, we obtain from Eqns. 8–50 and 8–51 the ratio of the exchange current $i_{n,0}$ for the conduction band to the exchange current $i_{p,0}$ for the valence band as shown in Eqn. 8–53:

$$\frac{i_{n,0}}{i_{p,0}} = \frac{k_n}{k_p} \frac{W_{OX}(\varepsilon = \varepsilon_C^s + kT)}{W_{RED}(\varepsilon = \varepsilon_V^s - kT)} \times \frac{c_{RED}}{c_{OX}} \exp\left(\frac{2(\varepsilon_{F(REDOX)}^0 - \varepsilon_{F(SC)}^i)}{kT}\right), \quad (8\text{–}53)$$

where W is the probability density of electron energy states, c is the concentration of redox particles, $\varepsilon_{F(REDOX)}^0$ is the standard Fermi level of redox electrons, and $\varepsilon_{F(SC)}^i$ is the Fermi level of intrinsic semiconductors at the flat band potential (the middle level in the band gap).

It follows from Eqn. 8–53 that the ratio of participation of the conduction band to the valence band in the exchange reaction current depends on the standard Fermi level of the redox electrons relative to the middle level in the band gap at the interface of semiconductor electrodes.

Figs. 8–16 and 8–17 show the state density $D(\varepsilon)$ and the exchange reaction current $i_0(\varepsilon)$ as functions of the energy level ε of electrons in two different cases of the transfer reaction of redox electrons in equilibrium. In one case in which the Fermi level of redox electrons $\varepsilon_{F(REDOX)}$ is close to the conduction band edge (Fig. 8–16), the conduction band mechanism predominates over the valence band mechanism in the reaction equilibrium that occurs; because the Fermi level of electrode $\varepsilon_{F(SC)}$ (= $\varepsilon_{F(REDOX)}$) at the interface that is also close to the conduction band edge generates a higher concentration of interfacial electrons in the conduction band than interfacial holes in the valence band. In the other case in which the Fermi level of redox electrons is close to the valence band edge (Fig. 8–17), the valence band mechanism predominates over the conduction band mechanism; because the valence band holes are much more concentrated than the conduction band electrons at the electrode interface.

From these illustrations it follows, in general, that the transfer reaction of redox electrons at semiconductor electrodes occurs via the conduction band mechanism if its equilibrium potential is relatively low (high in the Fermi level of redox electrons); whereas, the transfer reaction of redox electrons proceeds via the valence band mechanism if the equilibrium redox potential is high (low in the Fermi level of redox electrons).

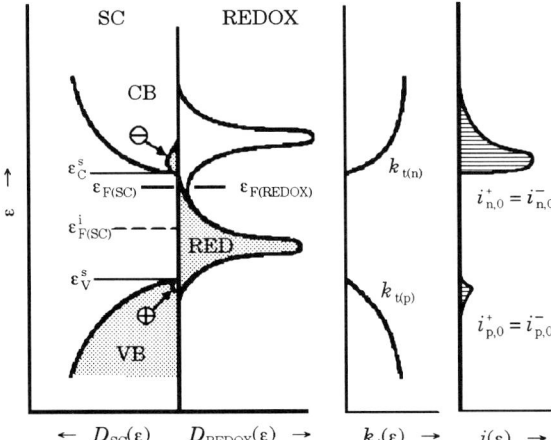

Fig. 8–16. Electron state density in semiconductor electrodes and in hydrated redox particles, rate constant of the electron tunneling, and exchange redox current in the equilibrium with a redox electron transfer reaction for which the Fermi level is close to the conduction band edge: $\varepsilon^i_{F(SC)}$ = Fermi level of intrinsic semiconductors at the flat band potential; $i_{n,0}$ ($i_{p,0}$) = exchange reaction current of electrons (holes); $k_{t(n)}$ ($k_{t(p)}$) = tunneling rate constant of electrons (holes).

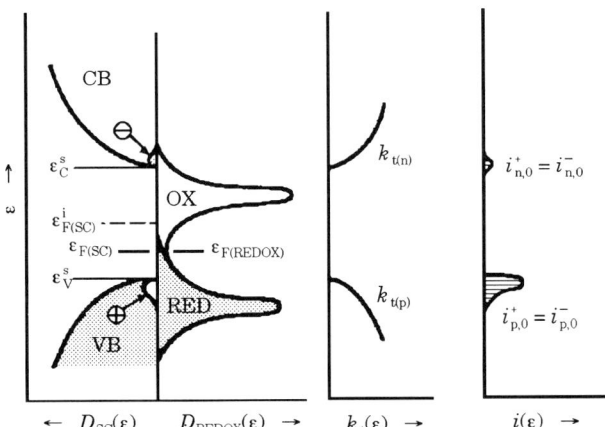

Fig. 8–17. Electron state density in semiconductor electrodes and in hydrated redox particles, rate constant of the electron tunneling, and exchange redox current in the equilibrium with a redox electron transfer reaction for which the Fermi level is close to the valence band edge.

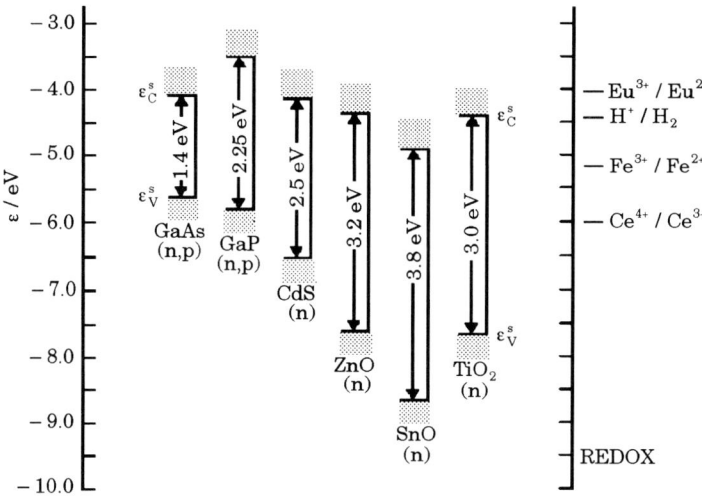

Fig. 8–18. Band edge levels of semiconductor electrodes in a solution of pH 1: (n) = n-type; (n, p) = n- and p-types; REDOX = redox reactions and the standard Fermi level. [From Gleria-Memming, 1975.]

TABLE 8-1. Preference for the conduction band mechanism (CB) and the valence band mechanism (VB) in outer-sphere electron transfer reactions of hydrated redox particles at semiconductor electrodes (SC): E_0 = standard redox potential referred to NHE; ε_g = band gap of semiconductors. [From Memming, 1983.]

SC	Redox (E_0/V_{NHE}, ε_g/eV)	Ce^{4+}/Ce^{3+} (1.6 V)	Fe^{3+}/Fe^{2+} (0.77 V)	V^{3+}/V^{2+} (−0.26 V)	Ti^{3+}/Ti^{2+} (−0.37 V)
Ge	0.88	VB	VB	CB	CB
Si	1.1	VB	VB	?	?
GaAs	1.4	VB	VB	?	?
GaP	2.25	VB	VB	?	?
CdS	2.5	VB	VB	CB	CB
ZnO	3.25	CB	CB	CB	CB
SnO	3.8	CB	CB	CB	CB

In the state of band edge level pinning, the band edge levels, ε_C^s and ε_V^s, at the interface of electrodes remain unchanged and are independent of the electrode potential. Fig. 8–18 shows both the band edge level of several semiconductor electrodes in an aqueous solution of pH 1 and the standard Fermi level of simple redox reactions. Table 8–1 identifies the prevailing mechanism of redox electron transfer (via the conduction band or the valence band) for simple redox reactions at several semiconductor electrodes.

In discussion of this section we have dealt only with the transfer of band electrons and holes, but there are cases, as shown in Fig. 8–19, in which electrons are transferred via the surface state level instead of the band level. For semiconductor electrodes where the band gap is large electron transfer at the surface states often predominates over the transfer of band electrons and holes.

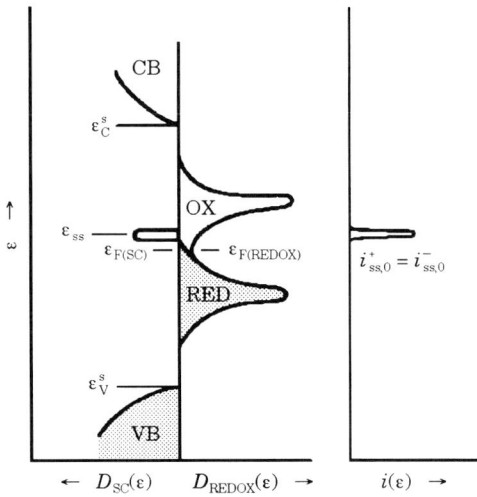

Fig. 8–19. Surface states participating in the reaction equilibrium of redox electron transfer at semiconductor electrodes with a wide band gap: ε_{ss} = surface state level; $i_{ss,0}$ = exchange current via the surface state.

8.3 Reaction Current at Semiconductor Electrodes

8.3.1 Reaction current under polarization

We consider the transfer of redox electrons at semiconductor electrodes which are polarized with an *overvoltage* η relative to the equilibrium redox potential (the Fermi level $\varepsilon_{F(REDOX)}$). The transfer current of redox electrons is given in Eqn. 8–54 by the algebraic sum of the electron current via the conduction band, $i_n^+(\eta) - i_n^-(\eta)$, and the hole current via the valence band, $i_p^+(\eta) - i_p^-(\eta)$:

$$i(\eta) = i_n^+(\eta) - i_n^-(\eta) + i_p^+(\eta) - i_p^-(\eta) \quad . \tag{8-54}$$

Polarization of electrodes shifts the Fermi level of the electrode $\varepsilon_{F(SC)}$ from the Fermi level of the redox electron $\varepsilon_{F(REDOX)}$ by an energy equivalent to the overvoltage η as described in Eqn. 8–55:

$$\varepsilon_{F(SC)} = \varepsilon_{F(REDOX)} - e\,\eta \quad . \tag{8-55}$$

The overvoltage η is distributed to the overvoltage of space charge layer η_{SC} and the overvoltage of compact layer η_H: $\eta = \eta_{SC} + \eta_H$. In the state of band edge level pinning, the whole overvoltage is located in the space charge layer, $\eta = \eta_{SC}$; and the electron level at the electrode interface relative to the redox electron level remains constant, irrespective of any change of the electrode potential. On the other hand, in the state of Fermi level pinning, the overvoltage is located entirely in the compact layer, $\eta = \eta_H$; hence, the electron level at the electrode interface relative to the redox electron level changes with the electrode potential.

The partial reaction currents, $i_n^+(\eta)$ and $i_n^-(\eta)$, carried by electrons in the conduction band may be derived, respectively, by integrating the differential currents given in Eqns. 8–43 and 8–45 to obtain Eqns. 8–56 and 8–57:

$$i_n^+(\eta) = e \int_{\varepsilon_C^s}^{\infty} k_{t,n}\, D_{SC(n)}(\varepsilon) \left\{ 1 - f(\varepsilon - \varepsilon_{F(SC)}) \right\} D_{RED}(\varepsilon)\, d\varepsilon \,, \tag{8-56}$$

$$i_n^-(\eta) = e \int_{\varepsilon_C^s}^{\infty} k_{t,n}\, D_{SC(n)}(\varepsilon)\, f(\varepsilon - \varepsilon_{F(SC)})\, D_{OX}(\varepsilon)\, d\varepsilon \quad . \tag{8-57}$$

Similarly, the partial reaction currents, $i_p^+(\eta)$ and $i_p^-(\eta)$, carried by the valence band holes in Eqn. 8–54 may also be obtained.

Next, we consider the anodic reaction current of redox electron transfer via the conduction band, where the exchange reaction current has been shown in Fig. 8–16. Application of a slight anodic polarization to the electrode reduces the Fermi level of electrodes from the equilibrium level ($\varepsilon_{F(SC)}(\eta = 0) = \varepsilon_{F(REDOX)}$) to the polarized level ($\varepsilon_{F(SC)}(\eta) = \varepsilon_{F(REDOX)} - e\,\eta$) without changing the electron level at the electrode interface relative to the redox electron level (the band edge level pinning) as shown in Fig. 8–20. As a result of anodic polarization, therefore, the concentration of interfacial electrons, n_s, in the conduction band decreases, and the concentration of interfacial holes, p_s, in the valence band increases with the following results; the cathodic transfer current of redox electrons, i_n^-, via the conduction band decreases (with the anodic electron injection current, i_n^+, being constant); the anodic transfer current of redox holes, i_p^+, via the valence band increases (with the cathodic hole injection current, i_p^-, being constant). When the anodic polarization is small, however, the conduction band mechanism prevails over the valence band mechanism as shown in Fig. 8–20.

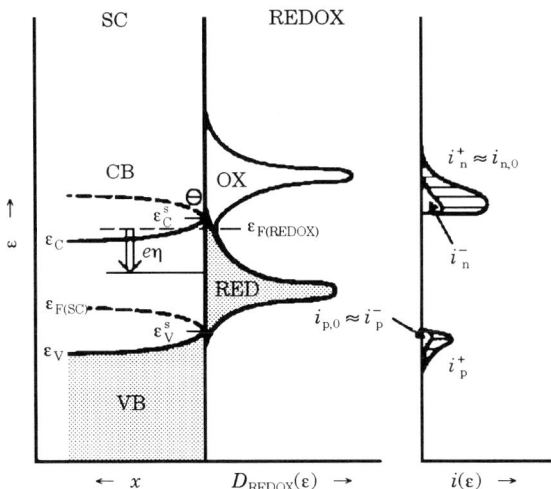

Fig. 8–20. State density of redox electrons and the reaction current for a redox electron transfer at semiconductor electrodes slightly polarized in the anodic direction: $\eta = \eta_{SC}$; dotted curve = band edge levels in the equilibrium (Fig. 8–16); solid curve = band edge levels in the anodic polarization.

A further increase in anodic polarization of the electrode reduces still further the Fermi level $\varepsilon_{F(SC)}(\eta)$ which gradually approaches the valence band edge ε_V^s at the electrode interface as shown in Fig. 8–21. Therefore, as the anodic polarization increases the concentration of interfacial holes in the valence band increases causing the anodic electron transfer to change from the conduction band mechanism to the valence band mechanism.

A further increase in anodic polarization brings the Fermi level $\varepsilon_{F(SC)}(\eta)$ of the electrode into the valence band at the electrode interface, causing the electrode interface to turn from band edge level pinning to *Fermi level pinning* at the valence band edge (the surface degeneracy) as shown in Fig. 8–22. Therefore, the excess anodic overvoltage, beyond a certain critical overvoltage corresponding to the onset of Fermi level pinning, is located across the interfacial compact layer;

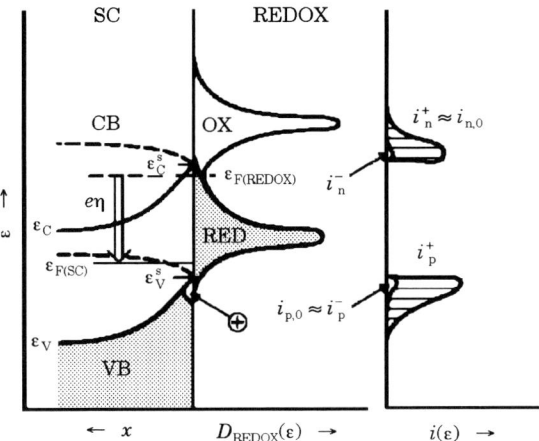

Fig. 8–21. State density of redox electrons and the reaction current for a redox electron transfer at semiconductor electrodes further polarized beyond that in Fig. 8–20 in the anodic direction: $\eta = \eta_{SC}$.

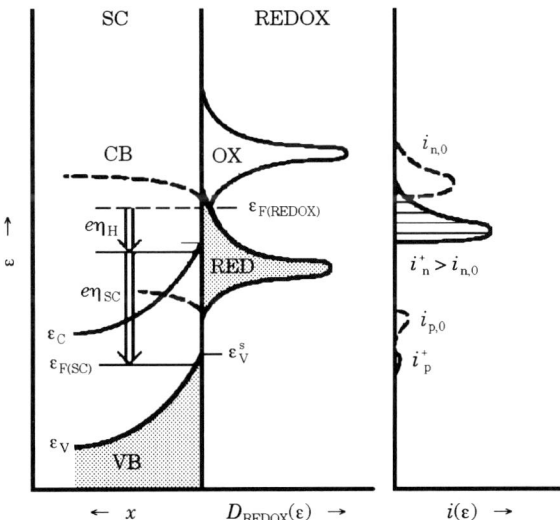

Fig. 8–22. State density of redox electrons and the reaction current for a redox electron transfer at semiconductor electrodes further polarized beyond that in Fig. 8–21 in the anodic direction: $\eta = \eta_H + \eta_{SC}$.

this raises the redox electron level relative to the electron level of the electrode at the electrode interface, thereby causing the electron current in the conduction band again to participate and predominate in the anodic reaction current as shown in Fig. 8–22.

The same discussion may be also applied to the cathodic electron transfer current of redox reactions in which the reaction changes from the conduction band mechanism, which applies at relatively low overvoltages in the state of band edge level pinning, to the valence band mechanism at high overvoltages in the state of Fermi level pinning.

8.3.2 Reaction current versus potential curve

We again consider the transfer of redox electrons via the conduction band mechanism as shown in Fig. 8–23. The anodic and cathodic transfer currents of redox electrons have been given in Eqns. 8–56 and 8–57, respectively. In these equations, the state density occupied by electrons in the conduction band is approximated by the concentration of the conduction band electrons at the electrode interface, $n_s = \int D_{SC}(\varepsilon) f(\varepsilon - \varepsilon_{F(SC)}) d\varepsilon$; and the state density vacant for electrons

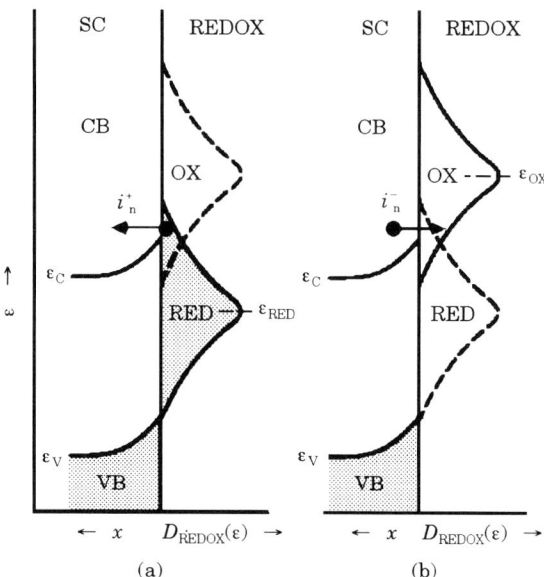

Fig. 8–23. Energy diagram for a redox electron transfer via the conduction band of semiconductor electrodes: (a) anodic redox electron transfer, (b) cathodic redox electron transfer.

in the conduction band is approximated by the effective state density of the conduction band, $N_C \doteq N_C - n_s = \int D_{SC}(\varepsilon)\{1 - f(\varepsilon - \varepsilon_{F(SC)})\}\,d\varepsilon$. Further, the state density of the redox electrons is distributed following the Gaussian distribution function centered at the most probable occupied level ε_{RED} of redox electrons in the reductant particle and at the most probable vacant level ε_{OX} of redox electrons in the oxidant particle as has been shown in Eqns. 2–48 and 2–49. Hence, Eqns. 8–56 and 8–57 yield Eqns. 8–58 and 8–59:

$$i_n^+(\eta) = \frac{e\, c_{RED}\, N_C}{\sqrt{4\pi\lambda k T}} \int_{\varepsilon_C^s}^{\infty} k_{t,n} \exp\left(\frac{-(\varepsilon - \varepsilon_{RED})^2}{4\lambda k T} \right) d\varepsilon \quad, \tag{8-58}$$

$$i_n^-(\eta) = \frac{e\, c_{RED}\, n_s}{\sqrt{4\pi\lambda k T}} \int_{\varepsilon_C^s}^{\infty} k_{t,n} \exp\left(\frac{-(\varepsilon - \varepsilon_{OX})^2}{4\lambda k T} \right) d\varepsilon \quad. \tag{8-59}$$

Since most of the electron transfer currents occur in a narrow range of energy near the band edge at the interface of the electrode, the integration of Eqns. 8–58 and 8–59 is limited to a narrow range of energy from ε_C^s to $\varepsilon_C^s + 4kT$; and the state density of the redox electrons may be represented by the density at the level equivalent to the interfacial band edge level of the electrode ε_C^s: $D_{RED}(\varepsilon_C^s) = c_{RED}\, W_{RED}(\varepsilon_C^s)$ and $D_{OX}(\varepsilon_C^s) = c_{OX}\, W_{OX}(\varepsilon_C^s)$. Therefore, to a first approximation, the anodic and cathodic reaction currents, $i_n^+(\eta)$ and $i_n^-(\eta)$, carried by electrons in the conduction band are obtained from Eqns. 8–58 and 8–59 as shown in Eqns. 8–60 and 8–61:

$$i_n^+(\eta) = e\, k_{t,n}\, c_{RED}\, N_C \sqrt{\left(\frac{kT}{\pi\lambda}\right)} \exp\left(\frac{-(\varepsilon_C^s - \varepsilon_{RED})^2}{4\lambda k T} \right) , \tag{8-60}$$

$$i_n^-(\eta) = e\, k_{t,n}\, c_{OX}\, n_s \sqrt{\left(\frac{kT}{\pi\lambda}\right)} \exp\left(\frac{-(\varepsilon_C^s - \varepsilon_{OX})^2}{4\lambda k T} \right) . \tag{8-61}$$

Under the condition of band edge level pinning, where the interfacial electron level of electrodes relative to the redox electron level of redox particles is unchanged, the level differences of $\varepsilon_C^s - \varepsilon_{RED}$ and $\varepsilon_C^s - \varepsilon_{OX}$ remain constant irrespective of any change of the electrode potential. Consequently, the anodic transfer current of redox electrons, $i_n^+(\eta)$, in Eqn. 8–60 is independent of the overvoltage and remains equal to the exchange current $i_{n,0}$ as expressed in Eqn. 8–62:

$$i_n^+(\eta) = i_n^+(\eta) = i_{n,0} \quad . \tag{8-62}$$

On the other hand, the cathodic transfer current of redox electrons, $i_n^-(\eta)$, in Eqn. 8–61 is proportional to the concentration, n_s, of interfacial electrons in the electrode; n_s varies as an exponential function of the overvoltage η, $n_s = n_s^0 \exp(-e\eta/kT)$; hence, $i_n^-(\eta)$ is an exponential function of the overvoltage η as shown in Eqn. 8–63:

$$i_n^-(\eta) = i_{n,0} \exp\left(\frac{-e\eta}{kT}\right) . \tag{8-63}$$

Consequently, for the transfer reaction of the redox electrons via the conduction band mechanism, the anodic current is constant and independent of the electrode potential; whereas, the cathodic current increases with increasing cathodic overvoltage (decreasing electrode potential).

Similarly, for the transfer reaction of redox holes via the valence band mechanism the anodic and cathodic currents, $i_p^+(\eta)$ and $i_p^-(\eta)$, are obtained, respectively, as shown in Eqns. 8–64 and 8–65:

$$i_p^+(\eta) = i_{p,0} \exp\left(\frac{e\eta}{kT}\right) , \tag{8-64}$$

$$i_p^-(\eta) = i_p^-(\eta = 0) = i_{p,0} , \tag{8-65}$$

where $i_{p,0}$ is the exchange transfer current of redox holes. Thus, for the transfer reaction of redox holes via the valence band mechanism, the cathodic current is constant and independent of the electrode potential; whereas, the anodic current increases with increasing anodic overvoltage (increasing electrode potential).

The transfer currents of redox electrons and redox holes represented by Eqns. 8–63 and 8–64 are formally in agreement with the Tafel equation given by Eqns. 7–18 and 7–32. However, the Tafel constant (the transfer coefficient) α equals one or zero at semiconductor electrodes in contrast with metal electrodes at which α is close to 0.5. From Eqns. 8–64 and 8–65 for reaction currents, the *Tafel constants* is obtained as defined in Eqns. 8–66 and 8–67:

$$kT\frac{\partial \ln(i_n^+(\eta))}{\partial \eta} = \alpha_n^+ \rightarrow 0 , \quad kT\frac{\partial \ln(i_n^-(\eta))}{\partial \eta} = \alpha_n^- \rightarrow 1 , \tag{8-66}$$

$$kT\frac{\partial \ln(i_p^+(\eta))}{\partial \eta} = \alpha_p^+ \rightarrow 1 , \quad kT\frac{\partial \ln(i_p^-(\eta))}{\partial \eta} = \alpha_p^- \rightarrow 0 , \tag{8-67}$$

where the overvoltage is positive in the anodic direction and negative in the cathodic direction. Note that a Tafel constant $\alpha = 1$ corresponds to a Tafel slope $b = 60$ mV/decade at room temperature.

When the total overvoltage η is distributed not only in the space charge layer η_{SC} but also in the compact layer η_H, the Tafel constants of α_n^+ and α_p^- each becomes greater than zero; and the Tafel constants of α_p^+ and α_n^- each becomes less than one. In such cases, $i_n^+(\eta)$ and $i_p^-(\eta)$ do not remain constant but increase with increasing overvoltage. Further, if Fermi level pinning is established at the interface of semiconductor electrodes, the Tafel constant would be close to 0.5 for the transfer currents both of redox electrons and of redox holes as shown in the following table:

Tafel constant ($\eta = \eta_{SC}$)	Band edge level pinning ($\eta = \eta_H$)	Fermi level pinning
α_n^+	~ 0	~ 0.0
α_n^-	~ 1	~ 0.5
α_p^+	~ 1	~ 0.5
α_p^-	~ 0	~ 0.5

Fig. 8–24 illustrates schematically the transfer reaction currents of redox electrons and redox holes as functions of the electrode potential; these reaction current versus electrode potential curves are obtained from the formulation of the reaction currents in Eqns. 8–62 through 8–65.

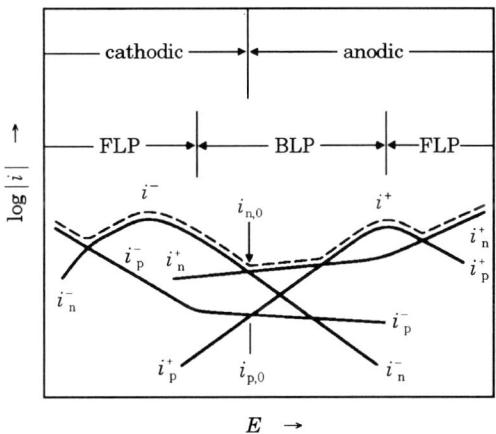

Fig. 8–24. Redox reaction currents via the conduction and the valence bands of semiconductor electrodes as functions of the electrode potential of semiconductors: anodic polarization corresponds to Figs. 8–20, 8–21 and 8–22. i^+ (i^-) = anodic (cathodic) current; i_n (i_p) = reaction current via the conduction (valence) band; BLP = band edge level pinning; FLP = Fermi level pinning.

266   ELECTRODE REACTIONS IN ELECTRON TRANSFER   CHAP. 8

In the discussion of this section, we have assumed that the transport current of charge carriers (electrons or holes) in semiconductor electrodes exceeds the interfacial reaction current; this occurs when the majority charge carriers are the main participants in the redox reaction. When the minority charge carriers participate mainly in the redox reaction, the transport current of the minority charge carriers in semiconductor electrodes may control the overall reaction current, as is discussed in Sec. 8.3.3.

8.3.3 The transport overvoltage of minority carriers

When the transport current of electrons or holes in semiconductor electrodes more or less influences the interfacial electron transfer current, the overvoltage η consists of the overvoltage of space charge layer η_{SC}, the overvoltage of compact layer η_H, and the *transport overvoltage* η_T in semiconductors as expressed in Eqn. 8–68:

$$\eta = \eta_{SC} + \eta_H + \eta_T . \tag{8–68}$$

The transport overvoltage, η_T, is usually zero when the transport current is carried by the majority charge carriers. However, when the transport current is carried by the minority charge carrier (holes in the n-type and electrons in the p-type semiconductors), the transport overvoltage becomes greater than zero. The transport overvoltages for minority charge carriers of electrons, $\eta_{T,n}$, and holes, $\eta_{T,p}$, are given by Eqn. 8–69:

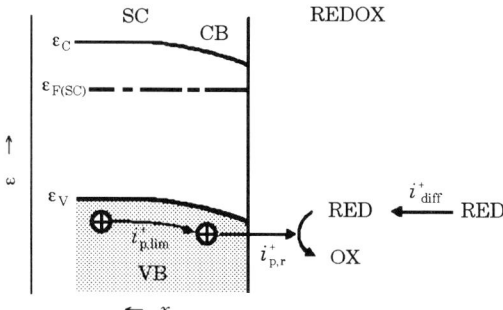

Fig. 8–25. Anodic transfer reaction of redox holes with the transport of minority charge carriers (holes) in n-type semiconductor electrodes: $i^+_{p,r}$ = anodic hole transfer current at the interface; $i^+_{p,\text{limit}}$ = limiting hole transport current; i^+_{diff} = limiting diffusion current of redox particles.

$$\eta_{T,n} = -kT\ln\left(1 - \frac{i_{n,r}}{i_{n,\lim}}\right), \quad \eta_{T,p} = kT\ln\left(1 - \frac{i_{p,r}}{i_{p,\lim}}\right), \quad (8\text{--}69)$$

where $i_{n,r}$ and $i_{p,r}$ are the transfer currents of electrons and holes at the electrode interface, and $i_{n,\lim}$ and $i_{p,\lim}$ are the limiting diffusion currents of electrons and holes in semiconductor electrodes, respectively [Sato, 1994; Myamlin-Pleskov, 1967].

As the polarization (the overvoltage $|\eta|$) increases for a redox reaction that requires the transport of *minority charge carriers* towards the electrode interface (anodic hole transfer at n-type and cathodic electron transfer at p-type electrodes), the transport overvoltage, $|\eta_T|$, increases from zero at small reaction currents to infinity; at this condition the reaction current is controlled by the limiting diffusion current ($i_{n,\lim}$ or $i_{p,\lim}$) of minority charge carriers as shown in Fig. 8–25.

8.3.4 Recombination of minority carriers

For redox reactions in which minority charge carriers are injected into the electrode (cathodic hole transfers to n-type and anodic electron transfers to p-type electrodes) as shown in Fig. 8–26, the reaction current is first controlled by the current due to the transfer (injection) of minority charge carriers at low overvoltages; and the reaction current increases up to the current equivalent to the limiting diffusion current of minority charge carriers; as the overvoltage increases the reaction current is controlled by the current due to the combination of the injected minority charge carriers with the majority charge carriers. Here, the injection current of minority charge carriers is much greater than the limiting diffusion current of minority charge carriers.

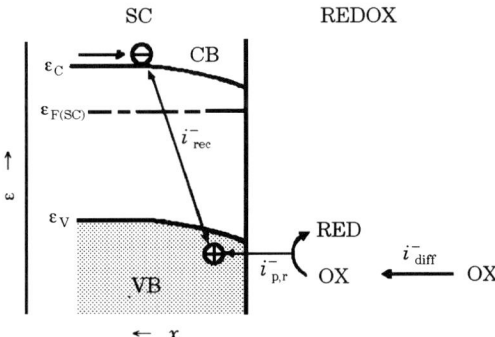

Fig. 8–26. Cathodic injection of minority charge carriers (holes) followed by the combination of minority charge carriers (holes) with majority charge carriers in n-type semiconductor electrodes: $i^-_{p,r}$ = cathodic current of the hole transfer at the interface; i^-_{rec} = current of the electron–hole recombination (cathodic hole transport current).

For n-type semiconductor electrodes in which the redox reaction due to the cathodic hole injection reaches its quasi-equilibrium state at the electrode interfaces, the current due to the recombination of injected holes (minority charge carriers) with electrons (the majority charge carriers), i_{rec}, is given by Eqn. 8–70 [Reineke-Memming, 1992]:

$$i_{rec} = -i_{p,\,lim} \exp\left(\frac{\varepsilon_{F(SC)} - \varepsilon_{F(REDOX)}}{\xi k T} \right) , \tag{8–70}$$

where $i_{p,\,lim}$ is the limiting diffusion current of thermally excited holes, and ξ is a parameter with values of $\xi = 2$ for the recombination in the space charge layer and $\xi = 1$ for the recombination in the electrode interior which is deeper than the space charge layer. The total transport current, i, is given in Eqn. 8–71 by the sum of the hole–electron recombination current, i_{rec}, and the limiting diffusion current of thermally excited holes, $i_{p,\,lim}$:

$$i = i_{p,\,lim} \left\{ 1 - \exp\left(\frac{\varepsilon_{F(SC)} - \varepsilon_{F(REDOX)}}{\xi k T} \right) \right\} . \tag{8–71}$$

Eqns. 8–70 and 8–71 reveal that the current of redox reactions determined by the recombination of minority charge carriers may be approximated by an equation of the Tafel type in which the Tafel slope is 120 mV/decade for $\xi = 2$ and 60 mV/decade for $\xi = 1$. Observations have shown that the redox reaction of hydrated complex iron ions at an n-type semiconductor electrode of tungsten selenide in sulfuric acid solutions proceeds with the Tafel slope of 120 mV/decade, suggesting that the recombination of the injected minority charge carriers (holes) with the majority charge carriers (electrons) in the space charge layer is the rate-determining step [Meissner-Memming, 1992].

8.3.5 Polarization curves of redox electron transfers

Based on the discussion in Sec. 8.3.4 we illustrate in Fig. 8–27 the polarization curves of redox reactions involving electron and hole transfer at n-type and p-type semiconductor electrodes. For n-type electrodes, the anodic current is carried by the minority charge carrier of holes in the valence band and is controlled at high overvoltages by the limiting current due to diffusion of holes in the electrode; whereas, the cathodic current is carried at low overvoltages by the majority charge carrier of electrons in the conduction band. The interfacial concentration of these majority charge carriers (electrons) increases with increasing cathodic overvoltage until the Fermi level is pinned in the conduction band at the electrode interface where the cathodic current becomes the hole injection current into the valence band of the electrode. As shown in Fig. 8–27, depending on whether the Fermi level of the redox electrons is close to the conduction band edge (marked by E_{REDOX} in Fig. 8–27) or to the valence band edge (marked by E'_{REDOX}), the anodic potential-independent current (solid line) or the cathodic potential-independent current (dotted line) equivalent to the exchange reaction current occurs near the equilibrium redox potential

For p-type electrodes, the cathodic current is carried at low overvoltages by the minority carriers (electrons) in the conduction band and is controlled at high overvoltages by the limiting current of electron diffusion; the anodic current is carried by the majority carriers (holes) in the valence band and the concentration of interfacial holes increases with increasing anodic overvoltage until the Fermi level is pinned in the valence band at the electrode interface where the anodic current becomes the current due to electron injection into the electrode.

The polarization curves shown in Fig. 8–27 correspond to Eqns. 8–62 to 8–65 while the electrode is controlled by band edge level pinning. As examples, Fig. 8–28 shows the cathodic polarization curves of several redox reactions at an

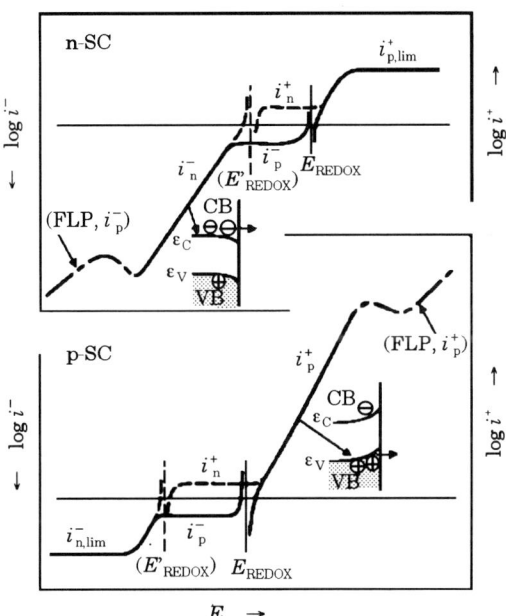

Fig. 8–27. Polarization curves of the redox electron transfer at n-type and p-type semiconductor electrodes: solid curve = reaction with the Fermi level of redox electrons close to the valence band edge; dotted curve = reaction with the Fermi level of redox electrons close to the conduction band edge; dot-dash curve = reaction in the state of Fermi level pinning (FLP).

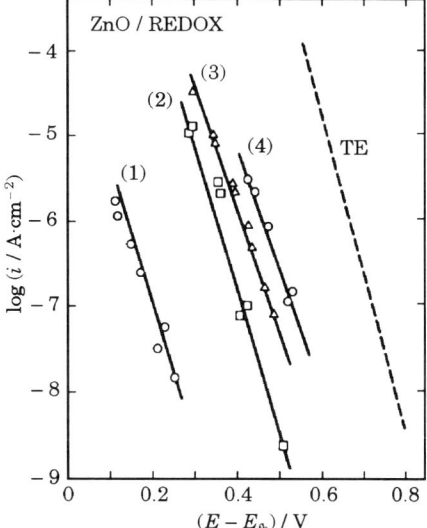

Fig. 8–28. Cathodic polarization curves for several redox reactions of hydrated redox particles at n-type semiconductor electrodes of zinc oxide in aqueous solutions: (1) = 1×10^{-2} M Ce^{4+} at pH 1.5; (2) = 1×10^{-2} M Ag(NH$_3$)$_2^{2+}$ at pH 12; (3) = 1×10^{-2} M Fe(CN)$_6^{3+}$ at pH 3.8; (4) = 1×10^{-2} M MnO$_4^-$ at pH 4.5; TE = thermal emission of electrons as a function of the potential barrier $E-E_{fb}$ of the space charge layer. [From Memming, 1987.]

n-type semiconductor electrodes of zinc oxide in aqueous solutions. The data of Fig. 8–28 shows that the polarization curves follow the Tafel equation with a Tafel slope equal to 60 mV/decade, indicating that the cathodic reaction current is proportional to the concentration of interfacial electrons at the electrode.

8.3.6 Redox Fermi level and band edge level

As discussed in Sec. 8.3.5, the redox reaction current due to electron or hole transfer depends not only on the concentration of interfacial electrons or holes at the electrode but also on the state density of the redox electrons or redox holes in the range of energy where the electron transfer takes place. Therefore, it is important in the kinetics of electron or hole transfer to determine the level of the band edge ε_C^s or ε_V^s of the electrode relative to the most probable level ε_{RED} or ε_{OX} of redox electrons or redox holes in the hydrated redox particles.

For redox reactions due to cathodic transfer of electrons via the conduction band, the cathodic current is expected to be maximum when the most probable vacant level ε_{OX} of the oxidant particle is in the same level as the conduction band edge ε_C^s; this cathodic current gradually decreases with increasing separation of ε_{OX} from ε_C^s as shown in Fig. 8–29. The same conclusion may also be drown from Eqn. 8–61.

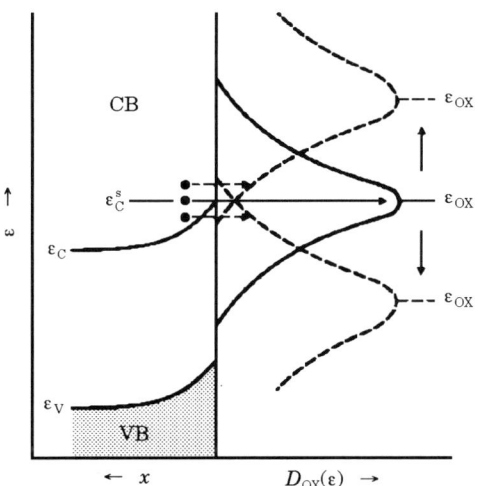

Fig. 8–29. Energy diagram for the most probable electron level, ε_{OX}, of oxidant particles and the conduction band edge level, ε_C^s, in a cathodic redox electron transfer via the conduction band: the cathodic current is maximum when ε_{OX} equals ε_C^s.

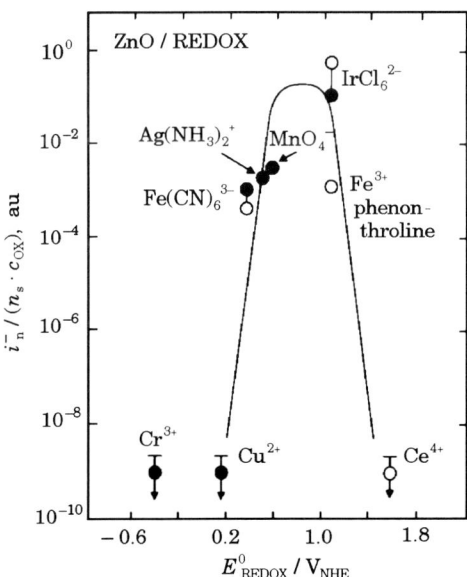

Fig. 8–30. Normalized cathodic current of redox reactions of hydrated redox particles as a function of the standard redox potential at n-type electrodes of zinc oxide: $i_n^- / (n_s c_{OX})$ = normalized cathodic reaction current; n_s = concentration of interfacial electrons; c_{OX} = concentration of oxidant particles; au = arbitrary unit. [From Morrison, 1969, 1980.]

Fig. 8–30 shows examples of the normalized cathodic transfer current of redox electrons for several redox reactions as a function of the standard redox potential E_{REDOX}^0 on n-type semiconductor electrodes of zinc oxide in aqueous solutions. The bell-like curve observed in Fig. 8–30 is in agreement with the forgoing conclusion that the maximum current occurs at the electrode potential at which ε_{OX} equals ε_C^s.

8.3.7 Electron transfer via the surface state

In addition to the redox reactions due to the direct transfer of electrons and holes via the conduction and valence bands, the transfer of redox electrons and holes via the surface states may also proceed at semiconductor electrodes on which surface states exist as shown in Fig. 8–31. Such transfer of redox electrons or holes involves the transition of electrons or holes between the conduction or valence band and the surface states, which can be either an exothermic or endothermic process occurring between two different energy levels. This transition of electrons or holes is followed by the transfer of electrons or holes across the interface of electrodes, which is an adiabatic process taking place at the same electron level between the surface states and the redox particles.

We now consider a cathodic transfer of electrons from the conduction band of the electrode to the vacant redox electron level in the hydrated oxidant particle to form a hydrated reductant particle in solution: $OX_{aq} + e_{CB} \rightarrow RED_{aq}$. Eqn. 8–72 expresses this reaction current, i_{CB}^-, due to the direct transfer of electrons from the conduction band to the oxidant particle based on Eqn. 8–61 as follows:

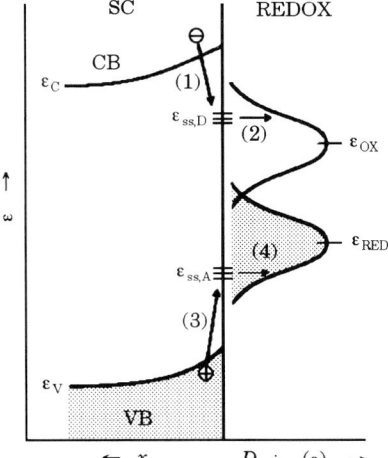

Fig. 8–31. Transfer reactions of redox electrons and holes via the surface states: (1) exothermic electron capture at the surface state $\varepsilon_{ss,D}$, (2) adiabatic transfer of electrons from the surface states to oxidant particles, (3) exothermic hole capture at the surface states $\varepsilon_{ss,A}$, (4) adiabatic transfer of holes from the surface states to reductant particles.

$$i_{CB}^- = e\, k_{t,n}\, c_{OX}\, N_C \sqrt{\left(\frac{kT}{\pi\lambda}\right)}\, \exp\left(\frac{-(\varepsilon_C^s - \varepsilon_{OX})^2}{4\lambda kT}\right) \exp\left(\frac{-(\varepsilon_C^s - \varepsilon_{F(SC)})}{kT}\right). \quad (8\text{--}72)$$

On the other hand, the cathodic reaction current, i_{ss}^-, via the surface states is given by Eqn. 8–73:

$$i_{ss}^- = e\, k_{t,ss}\, c_{OX}\, n_{ss} \sqrt{\left(\frac{kT}{\pi\lambda}\right)}\, \exp\left(\frac{-(\varepsilon_{ss} - \varepsilon_{OX})^2}{4\lambda kT}\right)$$

$$= e\, k_{t,ss}\, c_{OX}\, N_{ss} \sqrt{\left(\frac{kT}{\pi\lambda}\right)}\, \exp\left(\frac{-(\varepsilon_{ss} - \varepsilon_{OX})^2}{4\lambda kT}\right) \exp\left(\frac{-(\varepsilon_{ss} - \varepsilon_{F(SC)})}{kT}\right), \quad (8\text{--}73)$$

where $n_{ss} = N_{ss} \exp\{-(e_{ss} - e_{F(SC)})/kT\}$ is the concentration of electrons which occupy surface states, and N_{ss} is the concentration of surface states.

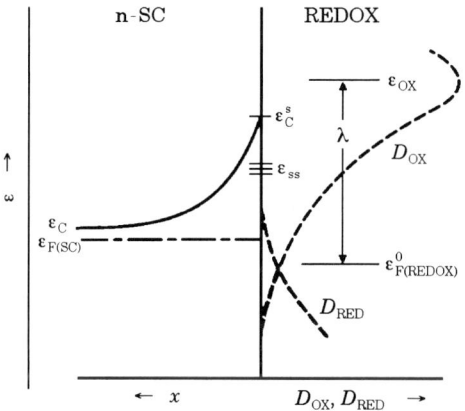

Fig. 8–32. Energy diagram for a redox electron transfer via the conduction band and via the surface states at n-type semiconductor electrodes: λ = reorganization energy of redox particles; ε_{ss} = surface state level.

Consequently, the ratio of i_{ss}^- to i_{CB}^- is given by Eqn. 8–74:

$$\frac{i_{ss}^-}{i_{CB}^-} = \frac{k_{t,ss}}{k_{t,n}} \frac{N_{ss}}{N_C} \exp\left(\frac{(\varepsilon_C^s - \varepsilon_{ss})(4\lambda - 2\varepsilon_{OX} + \varepsilon_C^s + \varepsilon_{ss})}{4\lambda kT}\right), \quad (8\text{--}74)$$

where λ is the reorganization energy of the redox particles (See Fig. 8–32.). Eqn. 8–74 is valid under the condition that $\varepsilon_{OX} \geq \varepsilon_C^s > \varepsilon_{ss}$ and hence that the term of

$(4\lambda - 2\varepsilon_{OX} + \varepsilon_C^s + \varepsilon_{ss})$ is positive as estimated in Fig. 8–32. The formulation of Eqn. 8–74 suggests that the ratio of the cathodic current of electron transfer via the surface states to the cathodic current via the conduction band increases with increasing concentration of the surface states N_{ss} and with an increasing energy gap between the conduction band edge ε_C^s and the surface state level ε_{ss}.

8.3.8 Electron tunneling through the space charge layer

As the anodic or cathodic polarization increases, the band level bending in the space charge layer (the depletion layer) becomes steeper and the electron tunneling through the space charge layer can occur particularly in semiconductor electrodes of high concentrations of donors or acceptors where the space charge layer is thin.

If the depletion layer is sufficiently thin (less than 3 nm) for the electron tunneling, the electron transfer between the conduction band and the redox electron level in hydrated redox particles would occur with a direct electron tunneling across both the space charge layer and the compact layer; this direct electron transfer can occur even at electron levels lower than the conduction band edge ε_C^s at the interface. Therefore, if electron tunneling through the *depletion layer* is involved, the reaction current due to transfer of redox electrons would be greater than the usual transfer current of electrons through the conduction band which can occur only at electron levels higher than the conduction band edge ε_C^s at the interface. Such an increase of anodic transfer current due to redox reactions associated with electron tunneling through the depletion layer at increased overvoltages has been observed, for example, with the oxygen electrode reactions at n-type tin oxide electrodes of high donor concentration (2×10^{19} cm^{-3}) in basic solutions [Memming, 1983].

8.4 Complexation and Adsorption in Electron Transfer Reactions

In addition to simple reactions of electron transfer (outer-sphere electron transfer) between the electrode and the hydrated redox particles, there are more complicated reactions of electron transfer in which *complexation* or *adsorption* of redox particles is involved. In such transfer reactions of redox electrons, the redox particles are coordinated with ligands in aqueous solution or contact-adsorbed on the electrode interface before the transfer of redox electrons; after the transfer of electrons, the particles are de-coordinated from ligands or desorbed from the electrode interface.

8.4.1 Complexation shifts the redox electron level.

We compare a simple electron transfer reaction of hydrated redox particles, $OX_{aq} + e_{(STD)} = RED_{aq}$, with an electron transfer reaction of complex redox particles, $L \bullet OX_{aq} + e_{(STD)} = L \bullet RED_{aq}$; where L is the ligand coordinated with redox particles and $e_{(STD)}$ is the electron in the gaseous standard state at the outer potential of the aqueous solution (Refer to Chap. 4.). The following reaction cycle may be used to obtain the energy relationship between the two redox reactions:

Reaction	Energy	
$L \bullet OX_{aq} + e_{(STD)} \to L \bullet RED_{aq}$	$\varepsilon^0_{F(comp)}$,	(8–75)
$L \bullet RED_{aq} \to L + RED_{aq}$	$-\Delta G^L_{RED}$,	(8–76)
$RED_{aq} \to OX_{aq} + e_{(STD)}$	$-\varepsilon^0_{F(aq)}$,	(8–77)
$L + OX_{aq} \to L \bullet OX_{aq}$	ΔG^L_{OX} ,	(8–78)

where ε^0_F is the standard Fermi level of redox reactions and ΔG^L is the standard free enthalpy of complexation. From the energy balance in the foregoing reaction cycle, the energy relationship in Eqn. 8–79 is obtained between the standard Fermi levels $\varepsilon^0_{F(aq)}$ and $\varepsilon^0_{F(comp)}$ of the two redox reactions:

$$\varepsilon^0_{F(comp)} - \varepsilon^0_{F(aq)} = -(\Delta G^L_{OX} - \Delta G^L_{RED}) .\qquad(8\text{–}79)$$

Therefore, complexation *raises* the standard Fermi level of redox electrons $\varepsilon^0_{F(REDOX)}$, provided that the affinity of ligand coordination is greater with the oxidant particle than with the reductant particle ($-\Delta G^L_{OX} > -\Delta G^L_{RED}$); whereas, the complexation *reduces* the standard Fermi level of redox electrons $\varepsilon^0_{F(REDOX)}$, provided that the affinity of ligand coordination is less with the oxidant particle than with the reductant particle ($-\Delta G^L_{OX} < -\Delta G^L_{RED}$). With the shift of the standard Fermi level of redox electrons due to the complexation, the most probable electron levels ε_{RED} and ε_{OX} of the redox particles are also shifted in the same direction.

Fig. 8–33 shows the shift of the level of redox electrons that would result from the complexation of redox particles. For simple cases, the reorganization energy, λ, of redox particles may be assumed nearly the same in the complex redox particles (coordinated with ligands) as in the simply hydrated redox particles (coordinated with water molecules).

The concentration of complex particles is obtained in Eqn. 8–80 by assuming that the complexation processes of Eqns. 8–76 and 8–78 are in equilibrium:

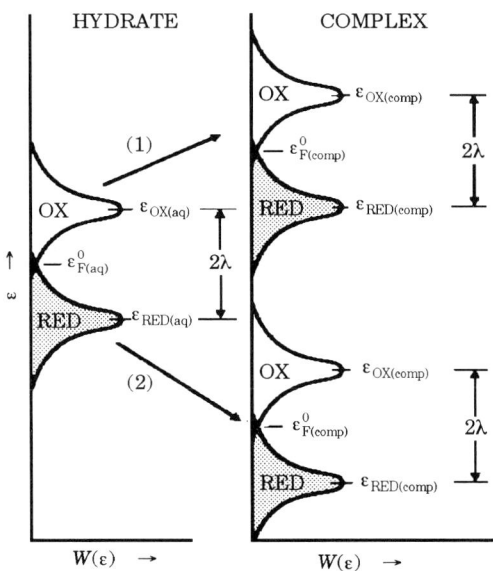

Fig. 8–33. Energy diagram showing a shift of redox electron levels due to the complexation of reductant and oxidant particles: (1) the affinity for complexation is larger with oxidants than with reductants, (2) the affinity for complexation is larger with reductants than with oxidants. COMPLEX = ligand-coordinated complex redox particles; HYDRATE = simply hydrated redox particles; W = probability density of the electron states; $\varepsilon^0_{F(aq)}$ = standard Fermi level of hydrated redox particles; $\varepsilon^0_{F(comp)}$ = standard Fermi level of ligand-coordinated complex redox particles.

$$\frac{c_{RED(comp)}}{c_L\, c_{RED(aq)}} = \exp\left(\frac{-\Delta G^L_{RED}}{kT}\right), \quad \frac{c_{OX(comp)}}{c_L\, c_{OX(aq)}} = \exp\left(\frac{-\Delta G^L_{OX}}{kT}\right). \qquad (8\text{--}80)$$

Here, we consider a redox reaction involving the transfer of anodic electrons in which the complexation of redox particles is involved at metal electrodes. The anodic reaction current, i^+, may be represented, to a first approximation, by the reaction current, $i^+(\varepsilon_{F(M)})$, at the Fermi level of the electrode $\varepsilon_{F(M)}$. From Eqns. 8–6 and 8–10 we obtain Eqn. 8–81 for the transfer current of anodic electrons, i_i^+, of simply hydrated redox particles or complex redox particles at the Fermi level of the electrode as follows:

$$i_i^+ \doteq i_i^+(\varepsilon_{F(M)}) = e\, k_t \frac{D_M(\varepsilon_{F(M)})}{2} D_{RED(i)}(\varepsilon_{F(M)}) = Z\, c_{RED(i)} \exp\left\{ \frac{-(\varepsilon_{F(M)} - \varepsilon_{RED(i)})^2}{4 \lambda_i k T} \right\}, \quad (8\text{–}81)$$

where Z is a constant, $c_{RED(i)}$ is the concentration of particle i (simply hydrated or ligand-coordinated), λ_i is the reorganization energy, and $\varepsilon_{RED(i)}$ is the most probable electron level of the hydrated or ligand-coordinated reductant particles. The ratio of anodic current, i_{comp}^+ / i_{aq}^+, of the reductants that are ligand-coordinated to simply hydrated is then expressed in Eqn. 8–82:

$$\frac{i_{comp}^+}{i_{aq}^+} = c_L \exp\left(\frac{-\Delta G_{RED}^L}{kT} \right) \exp\left\{ \frac{(\varepsilon_{F(M)} - \varepsilon_{RED(aq)})^2}{4 \lambda_{aq} k T} - \frac{(\varepsilon_{F(M)} - \varepsilon_{RED(comp)})^2}{4 \lambda_{comp} k T} \right\}, \quad (8\text{–}82)$$

indicating that the ratio increases with increasing concentration of ligand, c_L.

As shown in Fig. 8–34, when the most probable electron level of the reductant particle is higher in the ligand-coordinated state $\varepsilon_{RED(comp)}$ than in the simply hydrated state $\varepsilon_{RED(aq)}$, the transfer of anodic electrons would occur at higher energy levels (at less anodic potentials) with the ligand-coordinated reductant particle than with the simply hydrated reductant particle. In such a case the complexation of redox particles would favor the redox reaction involving the transfer of anodic electrons.

The effects of complexation of redox particles on the redox reaction kinetics are frequently more evident for semiconductor electrodes than for metal electrodes since the transfer of electrons takes place at the band edge levels rather than at the Fermi level of electrodes, respectively. For example, the anodic transfer of

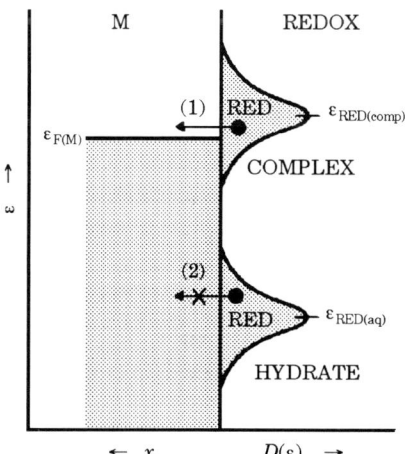

Fig. 8–34. Anodic transfer of redox electrons of simply hydrated reductants and of ligand-coordinated complex reductants at metal electrodes: (1) the anodic reaction of complex reductants takes place at less anodic potentials, (2) the anodic reaction of hydrated reductants can not occur unless at more anodic potentials where $\varepsilon_{RED(aq)}$ is close to $\varepsilon_{F(M)}$.

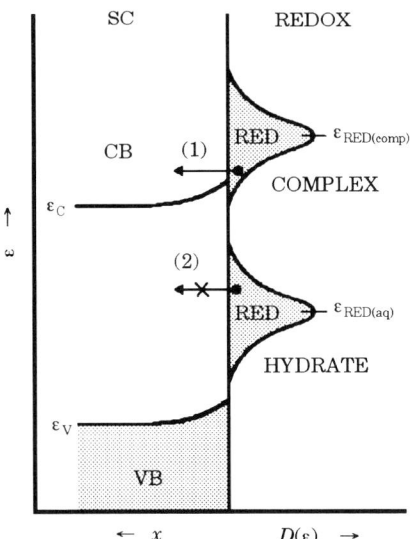

Fig. 8–35. Anodic transfer of redox electrons of hydrated reductants and of ligand-coordinated complex reductants at semiconductor electrodes: (1) the anodic reaction of complex reductants can proceed because their electron levels are located in the conduction band, (2) the anodic reaction of hydrated reductants can not proceed because their electron levels are located within the band gap.

redox electrons, which can not occur with the simply hydrated redox particle because its electron levels are located in the band gap of semiconductors, could proceed with the ligand-coordinated complex redox particle if the complexation raises the electron level of the reductant into the conduction band of semiconductors as shown in Fig. 8–35. In this case, the anodic electron transfer current, i^+, is given also by Eqn. 8–81 in which the conduction band edge ε_C^s is substituted for the Fermi level $\varepsilon_{F(M)}$ of the metal. The same discussion may also apply not only to the transfer of redox electrons via the conduction band but also to the transfer of redox holes via the valence band.

8.4.2 Contact adsorption shifts the redox electron level.

The kinetic treatment for the electron transfer of ligand-coordinated redox particles described in Sec. 8.4.1 may, in principle, apply also to the electron transfer of adsorbed redox particles (inner-sphere electron transfer). The *contact adsorption* on metal electrodes requires the dehydration of hydrated redox particles and hence inevitably shifts the standard Fermi level of redox electrons from $\varepsilon_{F(aq)}^0$ in the hydrated state to $\varepsilon_{F(ad)}^0$ in the adsorbed state. This shift of the Fermi level of redox electrons due to the *contact adsorption* of redox particles is expressed in Eqn. 8–83 similarly to Eqn. 8–79 of the complexation of redox particles (ligand coordination):

$$\varepsilon^0_{F(ad)} - \varepsilon^0_{F(aq)} = -(\Delta G_{OX(ad)} - \Delta G_{RED(ad)}), \tag{8-83}$$

where $\Delta G_{OX(ad)}$ and $\Delta G_{RED(ad)}$ are the standard free enthalpy for the contact adsorption of the oxidant and reductant particles, respectively. As shown in Fig. 8–36, the standard Fermi level of redox electrons, $\varepsilon^0_{F(REDOX)}$, would be *raised* by contact adsorption, if the adsorption affinity is greater with the oxidant particles than with the reductant particles ($-\Delta G_{OX(ad)} > -\Delta G_{RED(ad)}$); whereas, $\varepsilon^0_{F(REDOX)}$ would be *reduced* in the opposite case ($-\Delta G_{OX(ad)} < -\Delta G_{RED(ad)}$).

Further, contact adsorption may reduce the *reorganization* energy, λ, of redox particles in electron transfer; the distribution of electron levels of adsorbed redox particles may be more narrow than that of the simply hydrated redox particles as shown in Fig. 8–36. Further, contact adsorption of the redox particles produces: (1) an increase in the effective cross section for electron capture due to overlapping of the frontier orbital of the adsorbed particles with the band orbital of electrons

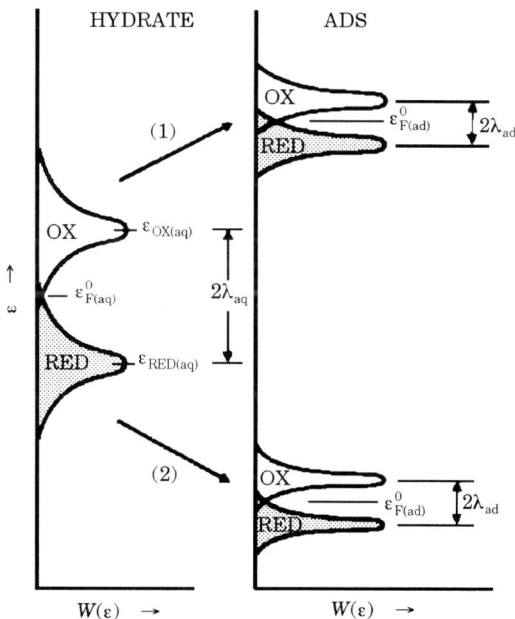

Fig. 8–36. Shift of redox electron levels due to the contact adsorption of reductant and oxidant particles: (1) adsorption affinity is larger with oxidants than with reductants, (2) adsorption affinity is larger with reductants than with oxidants. ADS = adsorbed particle; $\varepsilon^0_{F(ad)}$ = standard Fermi level of adsorbed particles; λ = reorganization energy of redox particles.

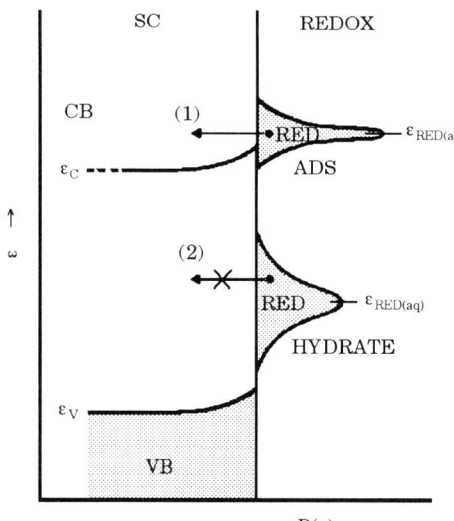

Fig. 8-37. Anodic electron transfer of hydrated reductants and of adsorbed reductants on semiconductor electrodes: (1) anodic reaction of adsorbed reductants can proceed because their electron levels are located in the conduction band, (2) anodic reaction of hydrated reductants can not proceed because their electron levels are located within the band gap.

of the electrode interface; (2) an increase in the effective concentration of redox particles at the electrode interface (10^{19} m^{-2} of adsorbed particles on the interface compared with 10^{17} m^{-2} of hydrated particles in the electron tunneling distance of 1.5 nm in 0.1 M solution); and (3) an increase of the electric field in the compact layer due to the contact adsorption of redox particles (Refer to Sec. 5.6.).

On account of these effects, the contact adsorption of the redox particles frequently accelerates the redox electron transfer, as compared with the direct electron transfer between the hydrated redox particle and the electrode. In other words, the reaction current due to redox electron transfer would be greater with the adsorbed redox particles than with the simply hydrated redox particles if the contact adsorption shifts the energy level of redox electrons in the favorable direction.

Fig. 8-37 shows the shift in the redox electron level of the reductant from the hydrated state, which can not anodically inject electrons into the conduction band because its electron levels are located within the band gap, to the adsorbed state, which then can inject electrons into the conduction band of semiconductor electrodes

Note that the redox reaction of electron transfer via adsorption intermediates requires the adsorption and desorption processes to occur as the preceding and following steps of the electron transfer; these adsorption and desorption are the ionic or atomic processes that obey the Boltzmann distribution function rather than the Fermi distribution function for electrons. The ionic adsorption and desorption processes involved in the transfer reactions of redox electrons are discussed in Chap. 9.

8.5 Electron Transfer at Film-Covered Metal Electrodes

Here, we examine reactions of redox electron transfer that occur at metal electrodes covered with *films* of semiconductors or insulators such as metal oxides. When the film is thin (1 to 2 nm thick), the *direct electron transfer* occurs easily by the tunneling mechanism between the metal electrode and the redox particles without passing through any definite electron levels in the film; hence, the thin-film-covered metal electrode may be regarded as being the same, in principal, as the metal electrode as shown in Fig. 8–38(a). As the film thickness increases, the direct electron transfer becomes more and more difficult, and the *indirect electron transfer* via electron levels in the film predominates. Accordingly, metal electrodes covered with thick semiconductor films may be regarded as being the same, in principal, as semiconductor electrodes as shown in Fig. 8–38(b). In general, electron transfer occurs between the electrode metal and the redox particles on thin-film-covered metal electrodes; whereas, electron transfer occurs between the film and the redox particles on thick-film-covered metal electrodes.

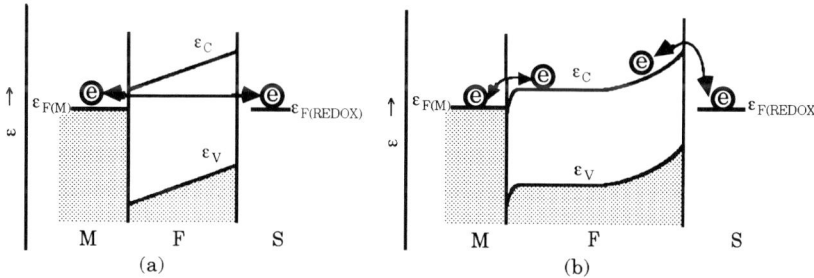

Fig. 8–38. Redox electron transfer at film-covered metal electrodes: (a) transfer of redox electrons across a thin semiconductor film (direct tunneling transfer of electrons), (b) transfer of redox electrons through a thick semiconductor film (indirect transfer of electrons through electron levels in the film).

8.5.1 Electron transfer between the electrode metal and the redox particles

For metal electrodes covered with a superficial thin film, electron transfer proceeds by direct tunneling between the electrode metal and the redox particles; the film influences only the barrier to the electron transfer. Usually, the transfer current of cathodic electrons, $i^-(\varepsilon)$, is given by Eqn. 8–84:

$$i^-(\varepsilon) = e\, k_t\, D_{M(e)}(\varepsilon)\, D_{OX}(\varepsilon) \ , \tag{8-84}$$

where $D_{M(e)}(\varepsilon)$ is the state density occupied by metal electrons, $D_{OX}(\varepsilon)$ is the state density vacant of redox electrons in the hydrated oxidant particles, and k_t is the rate constant of electron tunneling. The *tunneling probability* $\omega(\varepsilon)$ determining the tunneling rate $k_t = k_t^0\, \omega(\varepsilon)$ is a function of tunneling distance d as defined in Eqn. 8–85 [Schultze, 1978]:

$$\ln \omega(\varepsilon) = -\frac{2\, d\, \sqrt{2\, m_e\, \Delta\varepsilon}}{\hbar} \ , \tag{8-85}$$

where m_e is the effective mass of the electron, and $\Delta\varepsilon$ is the height of the energy barrier for electron tunneling. Simple calculation yields an electron tunneling probability of 0.01 for an energy barrier of 0.1 eV with a tunneling distance of 2 nm (about 15 atomic layers).

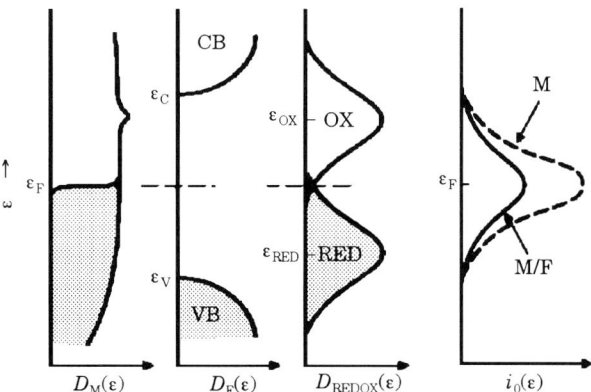

Fig. 8–39. Electron state density in electrode metals, D_M, semiconductor films, D_F, hydrated redox particles, D_{REDOX}, and exchange reaction current of redox electrons, i_0, in the electron transfer equilibrium: M = exchange current at bare metal electrodes, M/F = exchange current at a thin-film-covered metal electrode.

Fig. 8–39 illustrates, as a function of electron energy level, the electron state density $D(\varepsilon)$ in: the electrode metal, the thin superficial film, and the hydrated redox particles. Fig. 8–39 also illustrates the exchange current of electron transfer $i_0(\varepsilon)$ in the reaction equilibrium. This illustration shows that: (a) the energy-dependence of the exchange current of redox electrons is similar at both bare metal electrodes and thin-film-covered metal electrodes; (b) the thin superficial film reduces the exchange current because the electron tunneling distance is greater than the compact layer on the bare metal electrode.

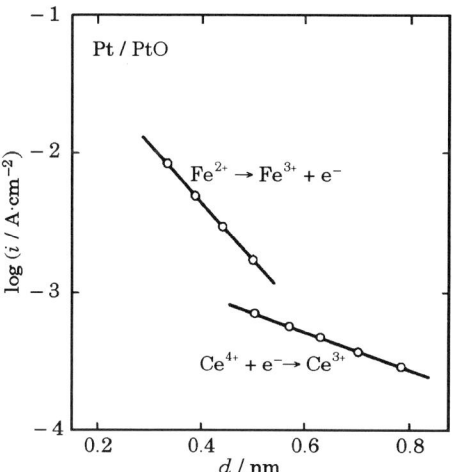

Fig. 8–40. Transfer currents of redox electrons observed on a platinum electrodes covered with thin platinum oxide films of PtO in acidic solutions as a function of the film thickness: hydrated redox particles Fe^{3+}/Fe^{2+} at 0.98 V_{NHE} in acidic solution at pH 0; hydrated redox particles Ce^{4+}/Ce^{3+} at 1.0 V_{NHE} in acidic solution at pH 3. [From Schultze, 1978; Schultze-Vetter, 1973.]

Fig. 8–40 shows the electron transfer current of two redox reactions (outer-sphere electron transfer) observed at constant potential for platinum electrodes covered with thin oxide films in acidic solutions as a function of the film thickness. As expected from Eqns. 8–84 and 8–85, a linear relationship is observed between the logarithm of the reaction currents and the thicknesses of the film.

For the transfer of redox electrons (inner-sphere electron transfer) in which redox particles are adsorbed on the thin superficial film that covers metal electrodes, the transfer current of redox electrons is not always decreased but rather increased by the presence of the thin film. Such an increase in the reaction current would occur, if the film acts as a reaction catalyst providing the adsorbed state of redox particles favorable for the redox electron transfer. For example, the anodic oxidation of carbon monoxide is catalyzed by the presence of thin anodic oxide-films on platinum electrodes in acidic solutions; of course, the reaction current decreases with increasing film thickness [Schultze, 1978].

8.5.2 Electron transfer between the film and the redox particles

As the thickness of a superficial film increases, the transfer of redox electrons through the electron levels in the film (indirect electron transfer) becomes predominant over the direct electron transfer between the electrode metal and the redox particles; the electron transfer takes place between the electron level in the film and the electron level in the redox particles.

We examine an electron transfer of hydrated redox particles (outer-sphere electron transfer) on metal electrodes covered with a thick film, as shown in Fig. 8–41, with an electron-depleted space charge layer on the film side of the film/solution interface and ohmic contact at the metal/film interface. It appears that no electron transfer may take place at electron levels in the band gap of the film since the film is sufficiently thick. Instead, electron transfer takes place at electron levels in the conduction and valence bands of the film.

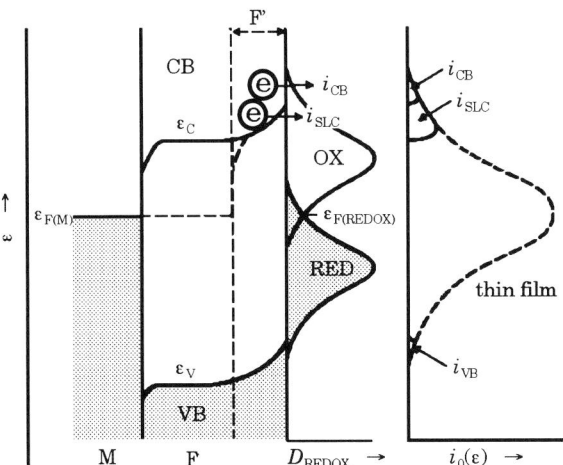

Fig. 8–41. Electron transfer reaction of hydrated redox particles in equilibrium on metal electrodes covered with a thick film (F, solid curve) and with a thin film (F', dotted curve): i_{CB} = electron transfer current via the conduction band; i_{SCL} = tunneling electron current through a depletion layer in the conduction band; i_{VB} = hole transfer current via the valence band.

As illustrated in Fig. 8–41, the usual exchange current of redox electrons, i_{CB}, occurs at electron levels higher than the conduction band edge; the exchange current, i_{SCL}, due to the electron tunneling through the space charge barrier occurs at electron levels slightly lower than the conduction band edge at the film/solution interface. On the other hand, the exchange current of redox holes, i_{VB}, takes place at electron levels lower than the valence band edge at the film/solution interface. Fig. 8–41 also compares the distribution of the exchange current of redox electrons at metal electrodes covered with thick films (solid curve) and with thin films (dashed curve).

Fig. 8–42 illustrates anodic and cathodic polarization curves observed for an outer-sphere electron transfer reaction with a typical thick film on metallic niobium electrodes. The thick film is anodically formed n-type Nb_2O_5 with a band gap of 5.3 eV; and the redox particles are hydrated ferric/ferrous cyano-complexes. The Tafel constant obtained from the observed polarization curve is $\alpha^+ \doteq 0$ for the anodic reaction and $\alpha^- \doteq 1$ for the cathodic reaction; these values agree with the Tafel constants for redox electron transfers via the conduction band of n-type semiconductor electrodes already described in Sec. 8.3.2 and shown in Fig. 8–27.

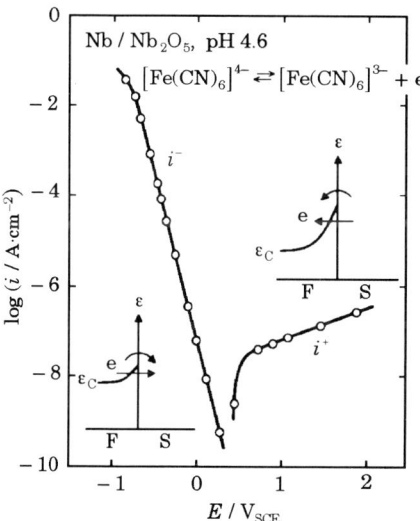

Fig. 8–42. Anodic and cathodic polarization curves observed for an electron transfer of hydrated redox particles at an electrode of metallic niobium covered with a thick niobium oxide Nb_2O_5 film (12 nm thick) in an acidic solution: the reaction is an electron transfer of hydrated redox particles, 0.25 M $Fe(CN)_6^{3-}$ /0.25 M $Fe(CN)_6^{4-}$, in 0.1 M acetic acid buffer solution of pH 4.6 at 25°C. i^+ = anodic reaction current; i^- = cathodic reaction current. [From Heusler-Schultze, 1975.]

8.5.3 Polarization curves observed

Fig. 8–43 shows polarization curves as a function of the film thickness observed for the anodic and cathodic transfer reactions of redox electrons of hydrated ferric/ferrous cyano-complex particles on metallic tin electrodes that are covered with an anodic tin oxide film of various thicknesses. The anodic oxide film of SnO_2 is an n-type semiconductor with a band gap of 3.7 eV; this film usually contains a donor concentration of 1×10^{-19} to 1×10^{-20} cm^{-3}. For the film thicknesses less than 2.5 nm, the redox electron transfer occurs directly between the redox particles and the electrode metal; the Tafel constant, α, is close to 0.5 both in the anodic and in the cathodic curves, indicating that the film-covered tin electrode behaves as a metallic tin electrode with the electron transfer current decreasing with increasing film thickness.

For the film thicknesses increasing more than 10 nm, the transfer of redox electrons via the conduction band of the film gradually predominates over the

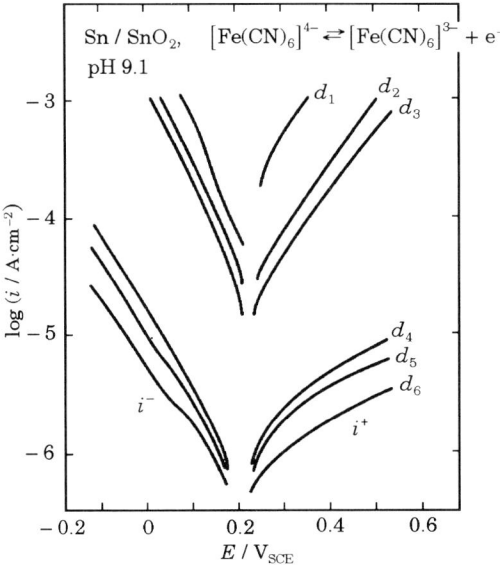

Fig. 8–43. Anodic and cathodic polarization curves observed for a redox electron transfer at metallic tin electrodes covered with anodic oxide SnO_2 films of various thicknesses d in an acidic solution: the reaction is a redox electron transfer of 0.25 M Fe(CN)$_6^{3-}$/0.25 M Fe(CN)$_6^{4-}$ in 0.2 M borate buffer solution of pH 9.1 at 25°C. d = film thickness; d_1 = 2 nm; d_2 = 2.3 nm; d_3 = 2.5 nm; d_4 = 20 nm; d_5 = 35.5 nm; d_6 = 55.5 nm. [From Kapusta-Hackerman, 1981.]

direct electron transfer between the Fermi level of the electrode metal and the Fermi level of the redox particles. This effect of increased film thickness causes the anodic Tafel constant to become less than 0.5, according to the participation of the conduction band electrons of the film in the redox reaction.

Fig. 8–44 shows the exchange currents of the redox electron transfer illustrated in Fig. 8–43 as a function of film thickness. It is obvious in this figure that the exchange current decreases steeply in the range of thin films, but much less in the range of thick films. The results in Fig. 8–44 indicates that the direct electron tunneling occurs only when the film is sufficiently thin.

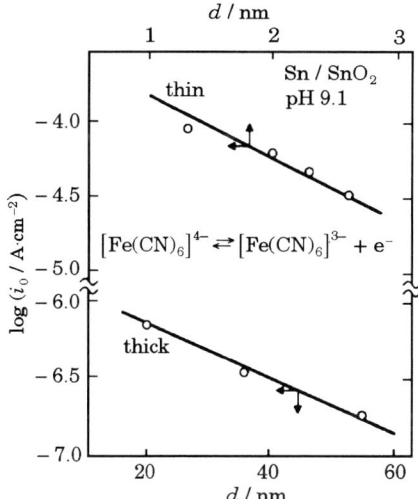

Fig. 8–44. Exchange current of a redox electron transfer observed in Fig. 8–43 as a function of the thickness of an oxide-film on metallic tin electrodes: the scale of film thickness differs for thin films and for thick films. [From Kapusta-Hackerman, 1981.]

References

[Bard-Faulkner, 1980]: A. Bard and L. R. Faulkner, *Electrochemical Methods, Fundamentals and Applications,* p. 136-212, John Wiley & Sons, New York, (1980).
[Gerischer, 1960]: H. Gerischer, *Z. Physik. Chem. N. F.,* **26**, 223, 325(1960).
[Gerischer, 1961]: H. Gerischer, *Z. Physik. Chem. N. F.,* **27**, 48(1961): *Adv. Electrochemistry and Electrochemical Engineering,* **Vol. 1**, (Edited by P. Delahay), p. 139, Interscience Publishers, New York, (1961).
[Gleria-Memming, 1975]: M. Gleria and R. Memming, *J. Electroanal. Chem. Interfacial Electrochem.,* **65**, 163(1975).

[Heusler-Schultze, 1975]: K. E. Heusler and M. Schultze, *Electrochimica Acta*, **20**, 237(1975).

[Kapusta-Hackerman, 1981]: S. Kapusta and N. Hackerman, *J. Electrochem. Soc.*, **128**, 327(1981).

[Meissner-Memming, 1992]: D. Meissner and R. Memming, *Electrochimica Acta*, **37**, 799(1992).

[Memming, 1983]: R. Memming, *Comprehensive Treatise of Electrochemistry*, **Vol. 7**, (Edited by B. E. Conway et al.), p. 529, Plenum Press, New York, (1983).

[Memming, 1987]: R. Memming, *Ber. Bunsenges. Phys. Chem.*, **91**, 353(1987).

[Morrison, 1969]: S. R. Morrison, *Surface Science*, **15**, 363(1969).

[Morrison, 1980]: S. R. Morrison, *Electrochemistry at Semiconductor and Oxidized Metal Electrodes,* p. 197, Plenum Press, New York, (1980).

[Myamlin-Pleskov, 1967]: V. A. Myamlin and Y. V. Pleskov, *Electrochemistry of Semiconductors,* p. 234, Plenum Press, New York, (1967).

[Reineke-Memming, 1992]: R. Reineke and R. Memming, *J. Phys. Chem.*, **96**, 1310, 1317(1992).

[Sato, 1994]: N. Sato, *Electrode Chemistry,* **Vol. 2**, p. 81, Japan Technical Information Service, Tokyo, (1994).

[Schultze, 1978]: J. W. Schultze, *Passivity of Metals*, (Edited by R. P. Frankenthal and J. Kruger), p. 82, The Electrochem. Soc. Inc., Princeton, (1978).

[Schultze-Vetter, 1973]: J. W. Schultze and K. J. Vetter, *Electrochimica Acta,* **18**, 899(1973).

[Vetter, 1967]: K. J. Vetter, *Electrochemical Kinetics, Theoretical and Experimental Aspects*, p. 465, Academic Press, New York, (1967).

CHAPTER 9

ELECTRODE REACTIONS IN ION TRANSFER

9.1 Metal Ion Transfer at Metal Electrodes

9.1.1 Metal ion transfer in a single elemental step

The transfer of metallic ions from the state of *metallic bonding* in metal electrodes to the state of *hydrated ions* in aqueous solutions (anodic metal dissolution) and its reverse reaction (cathodic metal deposition) belong to the category of *ionic processes*. Electrons in the metal participate in breaking and forming the metallic bond of metallic ions, but they *do not transfer* across the electrode interface where a high electric field exists. In general, the ionic processes of metallic dissolution–deposition are complicated and frequently involve hydrated anions as coordinated ligands with the transferring metal ions. To simplify the discussion, we assume a simple mechanism as expressed in Eqn. 9–1 in which the transfer of a monovalent metallic ion M^+ occurs in a single elemental step across the electrode interface:

$$M_s^+ \rightarrow M_{aq}^+ \quad , \tag{9-1}$$

where M_s^+ and M_{aq}^+ denote the metallic ions in the metallic bonding state at the interface of metal electrodes and in the state of hydrated ions at the plane of closest approach (OHP) to the electrode interface, respectively.

Fig. 9–1 illustrates the energy barrier to the transfer of metallic ions across the electrode interface; these energy barriers are represented by two potential energy curves, and their intersection, for surface metal ions in the metallic bond and for hydrated metal ions in aqueous solution. As described in Chaps. 3 and 4, the energy level (the *real potential*, $\alpha_{M_s^+}$) of interfacial metal ions in the metallic bonding state depends upon the electrode potential; whereas, the energy level (the *real potential*, $\alpha_{M_{aq}^+}$) of hydrated metal ions is independent of the electrode potential.

In the transfer equilibrium of metallic ions $\alpha_{M_s^+}$ equals $\alpha_{M_{aq}^+}$; and in the nonequilibrium state $\alpha_{M_s^+}$ deviates from $\alpha_{M_{aq}^+}$ to a certain level given by Eqn. 9–2:

$$\alpha_{M_s^+} = \alpha_{M_s^+(eq)} + e\,\eta = \alpha_{M_{aq}^+} + e\,\eta\,, \tag{9-2}$$

where $\alpha_{M_s^+(eq)}$ ($= \alpha_{M^+(aq)}$) is the real potential of the interfacial metallic ions in the transfer equilibrium (corresponding to the equilibrium potential E_{eq}), and η is the overvoltage. The intersection of the two potential energy curves of metallic ions in the metallic and hydrated states shown in Fig. 9–1 represents the activated state for the transfer of metallic ions.

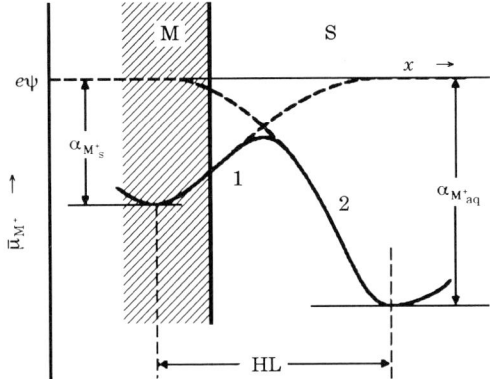

Fig. 9–1. Potential energy profile for the transfer of metal ions across the interface of metal electrode M/S: $\bar{\mu}_{M^+}$ = metal ion level (electrochemical potential); x = distance from the interface; $\alpha_{M_s^+}$ = real potential of interfacial metal ions; $\alpha_{M_{aq}^+}$ = real potential of hydrated metal ions; HL = compact layer (Helmholtz layer); ψ = outer potential of the solution, curve 1 = potential energy of interfacial metallic ions; curve 2 = potential energy of hydrated metal ions.

Fig. 9–2 shows the potential energy curves of metallic ions both in transfer equilibrium and in anodic polarization. The anodic and cathodic activation energies Δg^{*+} and Δg^{*-} are given as functions of the overvoltage η (positive in the anodic and negative in the cathodic direction), respectively, in Eqn. 9–3:

$$\Delta g^{*+} = \Delta g^{*+}_{(eq)} - (1-\beta)\,e\,\eta\,, \qquad \Delta g^{*-} = \Delta g^{*-}_{(eq)} + \beta\,e\,\eta \tag{9-3}$$

where $\Delta g^{*+}_{(eq)}$ and $\Delta g^{*-}_{(eq)}$ are the anodic and cathodic activation energies at the equilibrium potential E_{eq} of the reaction, respectively, and β is the symmetry factor introduced in Eqn. 7–27.

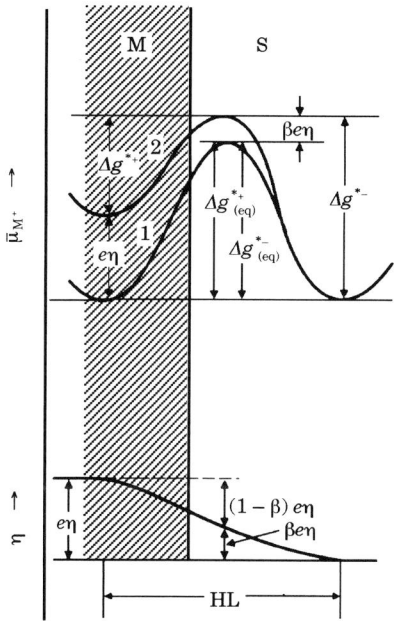

Fig. 9-2. Potential energy profile of transferring metallic ions across the interface of metal electrodes under anodic polarization: $\Delta g^{*+}_{(eq)} - (\Delta g^{*-}_{(eq)})$ = activation energy of anodic (cathodic) metal ion transfer in equilibrium; Δg^{*+} (Δg^{*-}) = activation energy of anodic (cathodic) metal ion transfer at anodic overvoltage η; curve 1 = potential energy profile in equilibrium; curve 2 = potential energy curve at an anodic overvoltage η.

Expressing the anodic and cathodic activation energies in the standard state as $\Delta g^{*+}_{(eq,\,0)}$ and $\Delta g^{*-}_{(eq,\,0)}$, respectively, and the standard equilibrium potential as $E_{eq,\,0}$, we obtain from Eqn. 9–3 the anodic and cathodic activation energies as shown in Eqn. 9–4:

$$\Delta g^{*+} = \Delta g^{*+}_{(eq,\,0)} - (1-\beta)\,e\,(\eta + \Delta E_{eq}) \quad , \quad \Delta g^{*-} = \Delta g^{*-}_{(eq,\,0)} + \beta\,e\,(\eta + \Delta E_{eq}) \quad , \quad (9\text{–}4)$$

where ΔE_{eq} is $\Delta E_{eq} = E_{eq} - E_{eq,\,0}$. From Nernst's equation in electrochemical thermodynamics, we obtain Eqn. 9–5 for ΔE_{eq}:

$$\Delta E_{eq} = E_{eq} - E_{eq,\,0} = \frac{kT}{e} \ln\left(\frac{a_{M^+_{aq}}}{a_{M^+_s}} \right) \quad , \tag{9–5}$$

where a denotes the activity of metallic ions.

Using Eqn. 9–4 for the activation energy, the anodic transfer current, i^+, of metallic ions is given by Eqn. 9–6:

$$i^+ = k^+\, a_{M^+_s} \exp\left(\frac{-\Delta g^{*+}}{kT} \right)$$

$$= k^+ a_{M_s^+} \exp\left(\frac{-\Delta g_{eq,0}^{*+}}{kT}\right) \exp\left(\frac{(1-\beta)e\Delta E_{eq}}{kT}\right) \exp\left(\frac{(1-\beta)e\eta}{kT}\right)$$

$$= k_0^+ (a_{M_s^+})^\beta (a_{M_{aq}^+})^{1-\beta} \exp\left(\frac{(1-\beta)e\eta}{kT}\right), \qquad (9\text{–}6)$$

Similarly, the cathodic transfer current of metallic ions, i^-, is given by Eqn. 9–7:

$$i^- = k_0^- (a_{M_s^+})^\beta (a_{M_{aq}^+})^{1-\beta} \exp\left(\frac{-\beta e\eta}{kT}\right). \qquad (9\text{–}7)$$

In the transfer equilibrium ($\eta = 0$), the anodic current equals the cathodic current, and the exchange current, i_0, can be defined by Eqn. 9–8:

$$i_0 = k_0 (a_{M_s^+})^\beta (a_{M_{aq}^+})^{1-\beta}, \qquad (9\text{–}8)$$

where $k_0 (= k^+, = k^-)$ is the rate constant.

The overall transfer current of metallic ions, i, is the algebraic sum of the anodic and cathodic current, i^+ and i^-, respectively, as defined in Eqn. 9–9:

$$i = i^+ - i^- = i_0 \left\{ \exp\left(\frac{(1-\beta)e\eta}{kT}\right) - \exp\left(\frac{-\beta e\eta}{kT}\right) \right\}. \qquad (9\text{–}9)$$

The plot of i versus η which results from Eqn. 9–9 is a polarization curve; this polarization curve is usually divided into two ranges of polarization as shown by estimated curves in Fig. 9–3: one is a range of polarization where a *linear rate* equation holds near the equilibrium potential ($\eta \to 0$); the other is a range of polarization (the Tafel range) where an *exponential rate* equation applies at potentials away from the equilibrium potential ($\eta \gg 0$).

The reaction current in the range of linear rates is described approximately by Eqn. 9–10:

$$i = \frac{i_0 e \eta}{kT} = \frac{\eta}{R_t}, \qquad R_t = \frac{kT}{i_0 e} = \frac{d\eta}{di}, \qquad (9\text{–}10)$$

where R_t is the *reaction resistance* to ion transfer which is also called the *polarization resistance*. The linear rate of Eqn. 9–10 occurs within a range of overvoltage less than 10 mV.

In the range of potential away from the equilibrium potential, the backward reaction current can be disregarded, and the anodic and cathodic reaction currents are expressed, respectively, by the Tafel equations described in Eqn. 9–11:

$$i^+ = i_0 \exp\left(\frac{(1-\beta)e\eta}{kT}\right), \qquad i^- = i_0 \exp\left(\frac{-\beta e\eta}{kT}\right). \qquad (9\text{–}11)$$

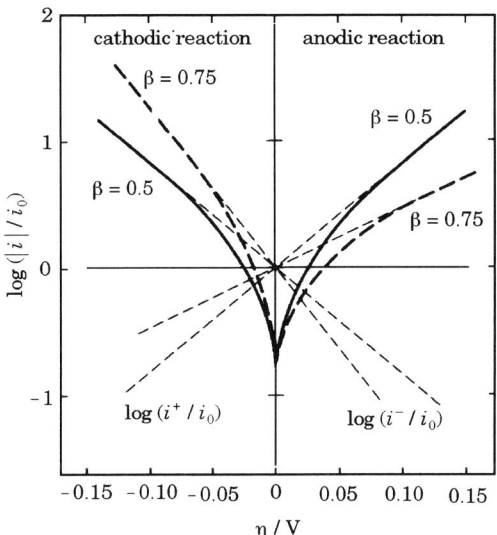

Fig. 9–3. Polarization curves estimated for a simple electrode reaction of metallic ion transfer: i = reaction current; i_0 = exchange reaction current in the reaction equilibrium; β = symmetric factor ($0 < \beta < 1$).

For the transfer of multivalent metallic ions M^{z+} the anodic and cathodic currents of ion transfer in the single elemental step of $M_s^{z+} \to M_{aq}^{z+}$ and $M_s^{z+} \leftarrow M_{aq}^{z+}$ are given in Eqn. 9–12 instead of Eqn. 9–11:

$$i^+ = i_0 \exp\left(\frac{(1-\beta)\,z\,e\,\eta}{k\,T} \right) \quad , \quad i^- = i_0 \exp\left(\frac{-\beta\,z\,e\,\eta}{k\,T} \right) . \tag{9–12}$$

As shown in Fig. 9–3, the anodic and cathodic polarization curves are symmetric if the symmetry factor is 0.5 ($\beta = 0.5$); whereas, they are unsymmetrical if the symmetry factor deviates from 0.5 ($\beta \neq 0.5$).

Fig. 9–4 shows the polarization curves observed for the transfer reaction of cadmium ions ($Cd_s^{2+} \rightleftarrows Cd_{aq}^{2+}$) at a metallic cadmium electrode in a sulfuric acid solution. It has been proposed in the literature that the transfer of cadmium ions is a single elemental step involving divalent cadmium ions [Conway-Bockris, 1958]. The Tafel constant, α, obtained from the observed polarization curves in Fig. 9–4 agrees well with that derived for a single transfer step of divalent ions; the Tafel constant is $\alpha^+ = z\,(1-\beta) \doteq 1$ in the anodic transfer and is $\alpha^- = z\,\beta \doteq 1$ in the cathodic transfer.

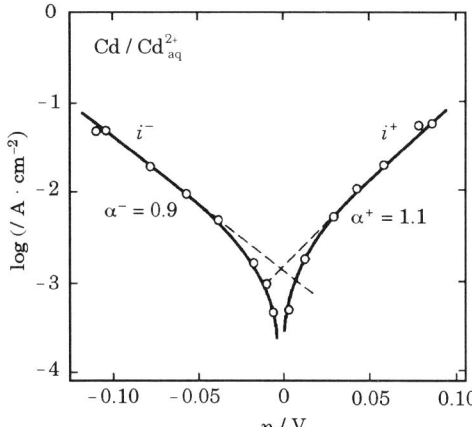

Fig. 9-4. Anodic and cathodic polarization curves measured for the transfer of divalent cadmium ions (dissolution–deposition) at a metallic cadmium electrode in a sulfate solution (0.005M Cd^{2+} + 0.4M SO$_4^{2-}$): i^+ (i^-) = anodic (cathodic) reaction current; α = Tafel constant (transfer coefficient). [From Lorenz, 1954.]

9.1.2 Metal ion transfer in a series of two elemental steps

For some metals the transfer of metallic ions involves a reaction intermediate of an adsorbed metallic ion complex which is coordinated with anionic ligands; hence, the overall reaction occurs in a series of two elemental steps rather than one Such a multistep transfer of ions can result, in the course of metallic ion transfer, from the reduction of the activation energy for ion transfer due to the formation of adsorbed intermediates. We examine a transfer reaction of divalent metallic ions via an adsorbed complex ion according to the steps in Eqn. 9–13:

$$A_{aq}^- + M_s^{2+} \rightarrow MA_{ad}^+ , \qquad (9\text{–}13a)$$
$$MA_{ad}^+ \rightarrow MA_{aq}^+ , \qquad (9\text{–}13b)$$
$$MA_{aq}^+ \rightleftarrows M_{aq}^{2+} + A_{aq}^- , \qquad (9\text{–}13c)$$

$$M_s^{2+} \rightarrow M_{aq}^{2+} , \qquad (9\text{–}13)$$

where A$^-$ is a monovalent anionic ligand; subscripts "aq", "ad", and "s" denote the hydrated particles, adsorbed particles, and surface sites, respectively. Eqn. 9–13a is the anodic transfer of anion A$^-$ across the outer part of the interfacial compact double layer (the outer Helmholtz layer) to form an adsorbed intermediate of a metallic ion complex which is coordinated with anionic ligands; Eqn. 9–13b is the anodic transfer of the adsorbed metallic ion complex across the outer Helmholtz layer to form a hydrated metallic ion complex (the desorption of metallic ion complex); and Eqn. 9–13c is the dissociation of the hydrated metal ion complex in aqueous solution which is assumed to be in equilibrium.

The anodic and cathodic currents i_1^+ and i_1^- for the formation of the adsorbed ion complex (Eqn. 9–13a) are given, respectively, in Eqns. 9–14 and 9–15:

$$i_1^+ = k_1^+ a_{A_{aq}^-} a_{M_s^{2+}} \exp\left(\frac{(1-\beta_1)e\eta}{kT} \right), \tag{9–14}$$

$$i_1^- = k_1^- a_{MA_{ad}^+} \exp\left(\frac{-\beta_1 e\eta}{kT} \right). \tag{9–15}$$

where k is the rate constant and a is the activity of particles. The activity of an adsorbed metallic ion complex in the adsorption equilibrium identified in Eqns. 9–14 and 9–15 is given by Eqn. 9–16:

$$a_{MA_{ad}^+(eq)} = a_{A_{aq}^-} a_{M_s^{2+}} \frac{k_1^+}{k_1^-} \exp\left(\frac{e\eta}{kT} \right). \tag{9–16}$$

Further, the anodic and cathodic currents, i_2^+ and i_2^-, for the desorption and adsorption of the metallic ion complex (Eqn. 9–13b) are given, respectively, in Eqns. 9–17 and 9–18:

$$i_2^+ = k_2^+ a_{MA_{ad}^+} \exp\left(\frac{(1-\beta_2)e\eta}{kT} \right), \tag{9–17}$$

$$i_2^- = k_2^- a_{MA_{aq}^+} \exp\left(\frac{-\beta_2 e\eta}{kT} \right). \tag{9–18}$$

In the rate equations, Eqn. 9–14 through Eqn. 9–18, we have assumed that Langmuir's adsorption isotherm holds; this assumption applies to the range of low adsorption coverages.

If the anodic anion transfer (anionic adsorption, Eqn. 9–13a) to form an adsorbed metallic ion complex is the rate–determining step, the Tafel constant, $\alpha^+ = 1 - \beta$, would be obtained from Eqn. 9–14. If the anodic transfer of the adsorbed metallic ion complex (desorption of complexes, Eqn. 9–13b) is the rate-determining step, the Tafel constant, $\alpha^+ = 2 - \beta$, would be obtained from Eqns. 9–16 and 9–17. Similarly, if the cathodic anion transfer (anionic desorption, Eqn. 9–13a) is determining the rate, the Tafel constant in the cathodic reaction, $\alpha^- = 1 + \beta$, would be obtained from Eqns. 9–15 and 9–16; and if the cathodic transfer of a metallic ion complex (adsorption of complexes, Eqn. 9–13b) is determining the rate, the Tafel constant, $\alpha^- = \beta$, would be obtained from Eqn. 9–18. In this discussion we have assumed $\beta_1 \doteq \beta_2 \equiv \beta$; therefore, Eqns. 9–19 and 9–20 follow:

$$\alpha^+ = 1 - \beta, \quad \alpha^- = 1 + \beta, \quad \text{anion transfer determining the rate,} \tag{9–19}$$

$$\alpha^+ = 2 - \beta, \quad \alpha^- = \beta, \quad \text{complex transfer determining the rate.} \tag{9–20}$$

Such a transfer reaction of metallic ions in a series connection of two elemental steps involving an intermediate of an adsorbed metallic ion complex may be illustrated by the anodic dissolution of metallic iron in acidic solution [Despic, 1983] as shown in Eqn. 9–21:

$$OH^-_{aq} + Fe^{2+}_s \rightarrow FeOH^+_{ad} \qquad (9\text{–}21a)$$
$$FeOH^+_{ad} \rightarrow FeOH^+_{aq} \qquad (9\text{–}21b)$$
$$FeOH^+_{aq} \rightarrow Fe^+_{aq} + OH^-_{aq} \qquad (9\text{–}21c)$$

$$Fe^{2+}_s \rightarrow Fe^{2+}_{aq} \qquad (9\text{–}21)$$

where Fe^{2+}_s is ferrous ion in the state of metallic bonding at the interface of iron electrodes, $FeOH^+_{ad}$ and $FeOH^+_{aq}$ are ferrous hydroxocomplexes in the state of adsorbed particles and in the state of hydrated particles, respectively.

Fig. 9–5 illustrates the anodic and cathodic polarization curves of the anodic dissolution and cathodic deposition of divalent iron ions in a weakly acidic solution. The anodic current of iron dissolution, i^+_{Fe}, has been found to obey an exponential rate equation (Tafel relation) with a Tafel constant of $\alpha^+ = 1.5$, suggesting that the desorption of the adsorbed ferrous hydroxocomplex, Eqn. 9–21b, is the rate-determining step. The observed cathodic current, i^-_{obs} (dotted curve), is the sum of the current of iron deposition, i^-_{Fe}, and the current of hydrogen ion reduction, i^-_H: the latter is controlled by the diffusion current of hydrogen ions. The real cathodic current of iron deposition, which is obtained by accounting the change in interfacial pH, also obeys an exponential rate equation with a Tafel constant of $\alpha^- = 0.5$; this suggests that the rate-determining step is the adsorption of the ferrous hydroxocomplex, Eqn. 9–21b.

For the mechanism of the anodic iron dissolution, which is described in this section, the formation process of the intermediate of ferrous hydroxocomplexes is in the quasi-equilibrium state so that the Nernst equation applies between the adsorption coverage, θ_{FeOH^+}, of the intermediate $FeOH^+_{ad}$ and the overvoltage, η. Accordingly, for the range of relatively low coverages of adsorption to which Langmuir's adsorption isotherm applies, we obtain Eqn. 9–22:

$$\theta_{FeOH^+} = K_{ad}\, c_{OH^-_{aq}} \exp\left(\frac{e\,\eta}{k\,T}\right). \qquad (9\text{–}22)$$

Further, for the range of relatively high coverages of adsorption where Temkin's adsorption isotherm applies [Temkin, 1947], we obtain Eqn. 9–23:

$$\exp(\kappa\, \theta_{FeOH^+}) = K_{ad}\, c_{OH^-_{aq}} \exp\left(\frac{e\,\eta}{k\,T}\right), \qquad \kappa = \frac{1}{k\,T}\frac{\partial \Delta G^0_{ad}}{\partial \theta_{FeOH^+}}, \qquad (9\text{–}23)$$

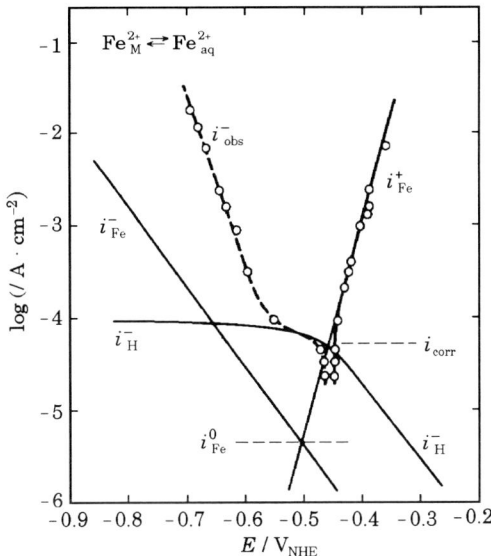

Fig. 9–5. Anodic and cathodic polarization curves observed for the transfer of divalent iron ions (dissolution–deposition) at an metallic iron electrode in a sulfuric acid solution at pH 4 (0.5 M Fe$_2$SO$_4$ + 0.5 M K$_2$SO$_4$): $i^+_{(Fe)}$ ($i^-_{(Fe)}$) = anodic iron dissolution (cathodic iron deposition) current; $i^0_{(Fe)}$ = exchange current of the iron ion transfer; $i^-_{(H)}$ = cathodic current of the hydrogen ion reduction (controlled by the hydrogen ion diffusion at high cathodic overvoltages); i_{corr} = corrosion current of iron (the coupled reaction of anodic iron dissolution and cathodic hydrogen ion reduction); i^-_{obs} = cathodic polarization current observed for the iron electrode (involving $i^-_{(Fe)}$, $i^-_{(H)}$ and a change in pH on the iron electrode); Tafel constant $\alpha^+ \doteq 1.5$; $\alpha^- \doteq 0.5$. [From Bockris-Drazig-Despic, 1961; Despic, 1983.]

where K_{ad} is the equilibrium constant for the formation of adsorbed intermediates at the equilibrium potential, and ΔG^0_{ad} is the standard free enthalpy for the formation of the adsorbed intermediates. Results for the coverage of adsorbed intermediate particles in the anodic iron dissolution, shown in Fig. 9–6, are in good agreement with Eqn. 9–22 in the range of low adsorption coverages and with Eqn. 9–23 in the range of high adsorption coverages.

The same two-step mechanism of metal dissolution has also been delivered for the anodic dissolution of nickel in acid solutions [Sato-Okamoto, 1964]. The mechanistic concepts for iron dissolution other than the two-step mechanism have also been presented in the literature [Heusler, 1958], and the mechanism of metal dissolution is still a subject of research [Plonski, 1996].

In Sec. 11.4.2, the anodic metal dissolution in the presence of a superficial oxide film is described.

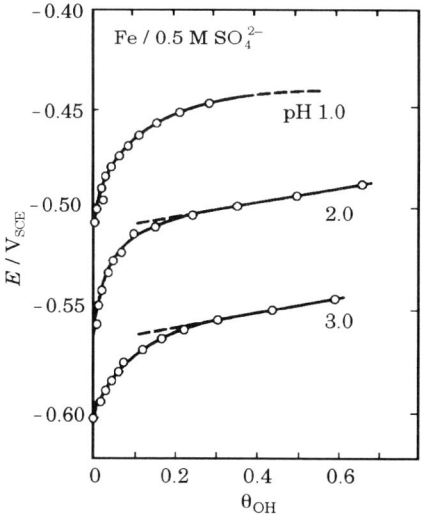

Fig. 9–6. Adsorption coverage of the reaction intermediate of hydroxocomplexes in the anodic dissolution of a metallic iron electrode as a function of electrode potential in acidic sulfate solutions at pH 1.0, 2.0 and 3.0: the solution is 0.5 M (Na$_2$SO$_4$ + H$_2$SO$_4$) at room temperature. θ_{OH} = adsorption coverage of the reaction intermediates of FeOH$_{ad}$ and FeOH$_{ad}^+$; V$_{SSE}$ = volt referred to the saturated silver–silver chloride electrode. [From Tsuru, 1991.]

9.2 Ion Transfer at Semiconductor Electrodes

9.2.1 Surface atom ionization of covalent semiconductor electrodes

We discuss the dissolution of surface atoms from simple substance semiconductor electrodes, which are covalent, such as silicon and germanium in aqueous solution. Generally, in covalent semiconductors, the bonding orbitals constitute the valence band and the antibonbing orbitals constitute the conduction band. Therefore, the accumulation of holes in the valence band or the accumulation of electrons in the conduction band at the electrode interface partially *breaks* the covalent bonding of the surface atom, S$_s$ (subscript "s" denotes the surface site).

As shown in Fig. 9–7, removing a pair of electrons from the covalent bonding orbital proceeds via an intermediate surface *radical*, S^{*+}, as indicated in Eqn. 9–24 [Gerischer-Mindt, 1968]:

$$S_s \rightarrow S_s^{*+} + e , \tag{9-24a}$$
$$S_s + h \rightarrow S_s^{*+} , \tag{9-24b}$$

and

$$S_s^{*+} \rightarrow S_s^{2+} + e , \tag{9-24c}$$
$$S_s^{*+} + h \rightarrow S_s^{2+} , \tag{9-24d}$$

where e is the conduction band electron, h is the valence band hole, and S_s^{2+} is the surface ion. Whether electrons in the conduction band or holes in the valence band predominate in the formation of surface radicals S^{*+} depends on both the electron level, ε_R, of the intermediate surface radical and the magnitude of the band gap, ε_g, as described in this section.

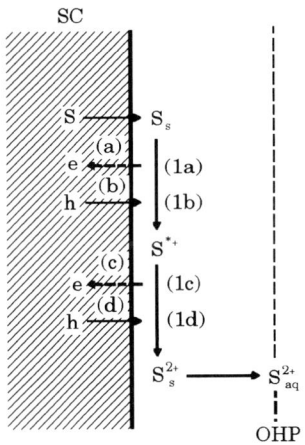

Fig. 9–7. Ionization of surface atoms followed ion transfer across the electrode interface in anodic dissolution of covalent semiconductors: S = covalently bonded atom in semiconductors; S_s = surface atom of semiconductors; S^{*+} = surface radical; S^{2+} = surface ion; S_{aq}^{2+} = hydrated ion; OHP = outer Helmholtz plane.

The activation energies of the respective elemental steps are denoted as Δg_a^*, Δg_b^*, Δg_c^*, and Δg_d^*; the corresponding anodic reaction rates are denoted as v_a, v_b, v_c, and v_d. Then, we obtain the rate equations shown in Eqns. 9–25a through 9–25d:

$$v_a = v_a c_{S_s} N_C \exp\left(\frac{-\Delta g_a^*}{kT}\right) , \tag{9-25a}$$

$$v_b = v_b c_{S_s} p_s \exp\left(\frac{-\Delta g_b^*}{kT}\right) , \tag{9-25b}$$

$$v_c = v_c c_{S_s^{*+}} N_C \exp\left(\frac{-\Delta g_c^*}{kT}\right) , \tag{9-25c}$$

$$v_d = v_d c_{S_s^{*+}} p_s \exp\left(\frac{-\Delta g_d^*}{kT}\right) , \tag{9-25d}$$

where v is the frequency factor, N_C is the effective electron state density of the conduction band ($\doteq N_C - n_s$, the effective state density of vacant electrons), p_s is the concentration of surface holes, c_{S_s} is the concentration of surface atoms, and $c_{S_s^{*+}}$ is the concentration of surface radicals.

In the two different steps, Eqns. 9–24a and 9–24b, for the formation of intermediate radicals, the initial state differs in energy by an amount equivalent to the band gap as shown in Fig. 9–8. Consequently, the activation energy differs in the two steps as shown in Eqn. 9–26:

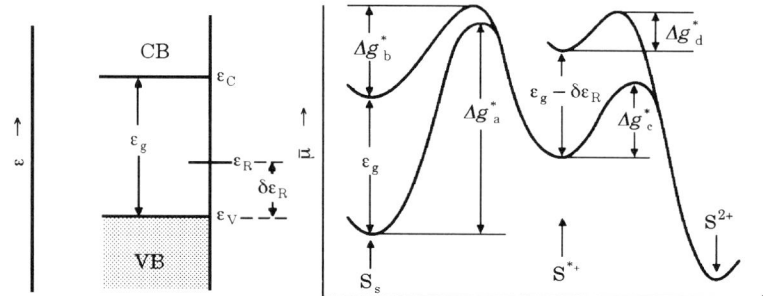

Fig. 9–8. Potential energy profile for the ionization of surface atoms in two steps on a covalent semiconductor electrode: ε_g = band gap energy; ε_R = electron level in an intermediate radical S^{*+}; Δg_a^* = activation energy for the first step of radical formation in the conduction band mechanism; Δg_b^* = activation energy for the first step of radical formation in the valence band mechanism; Δg_c^* = activation energy for the second step of radical ionization in the conduction band mechanism; Δg_d^* = activation energy for the second step of radical ionization in the valence band mechanism; $\delta\varepsilon_R = \varepsilon_R - \varepsilon_V^s$. [From Gerischer, 1970.]

$$\Delta g_a^* - \Delta g_b^* = \beta_{ab}\, \varepsilon_g \quad , \qquad 0 < \beta_{ab} < 1 \quad , \tag{9–26}$$

where β_{ab} is the symmetry factor of the activation barrier. From Eqns. 9–25a, 9–25b and 9–26, the reaction rates of the two steps of Eqns. 9–24a and 9–24b can be compared as shown in Eqn. 9–27:

$$\frac{v_b}{v_a} \doteq \frac{p_s}{N_C} \exp\left(\frac{\beta_{ab}\, \varepsilon_g}{kT} \right) . \tag{9–27}$$

The concentration of surface holes, p_s, is given as a function of the Fermi level ε_F^s at the surface, the valence band edge ε_V^s at the surface, and the effective state density of the valence band N_V ($\doteq N_C$) in Eqn. 9–28:

$$p_s = N_V \exp\left(\frac{-(\varepsilon_F^s - \varepsilon_V^s)}{kT}\right) . \tag{9-28}$$

Eqns. 9–27 and 9–28 yield Eqn. 9–29:

$$\frac{v_b}{v_a} \doteq \frac{N_V}{N_C} \exp\left(\frac{\beta_{ab}\,\varepsilon_g - (\varepsilon_F^s - \varepsilon_V^s)}{kT}\right) . \tag{9-29}$$

It follows that, when $\beta_{ab}\,\varepsilon_g$ is larger than $(\varepsilon_F^s - \varepsilon_V^s)$, the rate of capture of holes is greater than the rate of release of electrons ($v_b > v_a$). Namely, if the Fermi level ε_F^s at the surface is lower than the middle, $(\varepsilon_C^s + \varepsilon_V^s)/2$, of the band gap (the p-type surface) and if β_{ab} is close to 0.5, the valence band mechanism of Eqn. 9–24b would probably predominate over the conduction band mechanism of Eqn. 9–24a. As the band gap ε_g of the semiconductor electrode increases, the valence band mechanism become more predominant.

For the two different steps of the ionization of surface radicals, Eqns. 9–24c and 9–24d, which follow the radical formation, the activation energy also differs as shown in Eqn. 9–30:

$$\Delta g_c^* - \Delta g_d^* = \beta_{cd}\,(\varepsilon_C^s - \varepsilon_R) , \qquad 0 < \beta_{cd} < 1 , \tag{9-30}$$

where ε_R is the electron level of the surface radical and ε_C^s is the electron level of the conduction band edge. From Eqns. 9–25c, 9–25d and 9–30, the reaction rates of the two steps of Eqns. 9–24c and 9–24d can be compared in Eqn. 9–31:

$$\frac{v_d}{v_c} \doteq \frac{p_s}{N_C} \exp\left(\frac{\beta_{cd}\,(\varepsilon_C^s - \varepsilon_R)}{kT}\right) = \frac{N_V}{N_C} \exp\left(\frac{\beta_{cd}\,(\varepsilon_C^s - \varepsilon_R) - (\varepsilon_F^s - \varepsilon_V^s)}{kT}\right) . \tag{9-31}$$

Eqn. 9–31 indicates that the electron level, ε_R, of the intermediate radical is *decisive* in determining the ratio of rates v_d/v_c: if the electron level of ε_R is relatively close to the valence band edge ε_V^s, the valence band mechanism, Eqn. 9–24d, would predominate; whereas, if the electron level of ε_R is relatively close to the conduction band edge ε_C^s, the lone pair electron would be excited into the conduction band, and the conduction band mechanism, Eqn. 9–24c, would predominate. As the band gap of the semiconductor electrode decreases, the conduction band increasingly participates.

For germanium with small band gap (0.67 eV), both the valence band and the conduction band have been found to participate in anodic dissolution as shown in Eqn. 9–32:

$$Ge + \lambda h \rightarrow Ge_{aq}^{4+} + (4 - \lambda)\,e , \tag{9-32}$$

where the reaction order, λ, of holes shown in Eqn. 9–32 is in the range of $\lambda = 2$ to 4 [Gerischer, 1970].

In general, the activation energy for the release of electrons from surface atoms into the conduction band increases with increasing band gap of semiconductor electrodes; with this increase the capture of holes by the surface atoms and radicals predominates. Except for the germanium electrode, most covalent semiconductors have been found to dissolve anodically via this valence band mechanism [Memming, 1983].

9.2.2 Dissolution of covalent semiconductors

In the anodic dissolution of covalent bonded semiconductors, the transfer of surface ions across the compact layer (Helmholtz layer) occurs after the ionization of surface atoms S_s, illustrated in Eqn. 9–33, as described in Sec. 9.2.1:

$$S_s^{2+} \rightarrow S_{aq}^{2+} , \tag{9–33}$$

where S_s^{2+} is the surface ion of semiconductor electrodes, and S_{aq}^{2+} is the hydrated semiconductor ion in aqueous solution.

As shown in Fig. 9–9, the interfacial double layer of semiconductor electrodes consists of the space charge layer with the potential difference of $\Delta\phi_{SC}$ in the semiconductor and the compact layer with the potential difference of $\Delta\phi_H$ at the electrode interface. The potential difference $\Delta\phi_{SC}$ across the space charge layer controls the process of *ionization* of surface atoms (Eqn. 9–24); whereas, the potential difference $\Delta\phi_H$ across the compact layer controls the process of *transfer* of surface ions (Eqn. 9–33). The overvoltage η_{SC} across the space charge layer and the overvoltage η_H across the compact layer are expressed, respectively, in Eqn. 9–34:

$$\eta_H = \Delta\phi_H - \Delta\phi_H^0 , \qquad \eta_{SC} = \Delta\phi_{SC} - \Delta\phi_{SC}^0 , \tag{9–34}$$

where $\Delta\phi_H^0$ and $\Delta\phi_{SC}^0$ are the potential differences of the compact and space charge layers in the reaction equilibrium, respectively. The total overvoltage is, of course, the sum of the two overvoltages: $\eta = \eta_H + \eta_{SC}$.

When the electronic equilibrium is established in the space charge layer, the concentration of interfacial electrons is given by $n_s = n \exp(-e\,\Delta\phi_{SC}/kT)$ and the concentration of interfacial holes is given by $p_s = p \exp(e\,\Delta\phi_{SC}/kT)$ at the interface of semiconductor electrodes; n and p are the concentrations of electrons and holes, respectively, in the semiconductor interior. In general, the ionization of surface atoms (Eqn. 9–24) is in the quasi-equilibrium so that the concentration of surface ions $c_{S_s^{2+}}$ depends on the overvoltage of η_{SC} as shown in Eqn. 9–35:

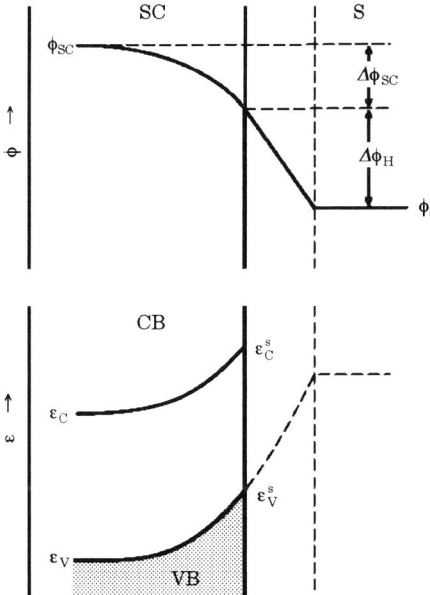

Fig. 9-9. Potential profile and band bending across the semiconductor electrode interface: φ = electrostatic inner potential; $\Delta\phi_{SC}$ ($\Delta\phi_H$) = potential difference in the space charge (the compact) layer.

$$\frac{c_{S_s^{2+}}}{c_{S_s^{2+}}^0} = \left(\frac{n_s^0}{n_s}\right)^2 = \left(\frac{p_s}{p_s^0}\right)^2 = \exp\left(\frac{2e\eta_{SC}}{kT}\right), \qquad (9\text{-}35)$$

where $c_{S_s^{2+}}^0$, n_s^0 and p_s^0 are the concentrations of the surface ions, surface electrons, and surface holes in the reaction equilibrium, respectively. It appears that the concentration of surface ions $c_{S_s^{2+}}$ is proportional to the square of the concentration of surface holes or inversely proportional to the square of the concentration of surface electrons.

The anodic current i of the transfer of surface ions, Eqn. 9-33, across the compact layer is given by Eqn. 9-36:

$$i = i_0 \left\{ \frac{c_{S_s^{2+}}}{c_{S_s^{2+}}^0} \exp\left(\frac{2e(1-\beta)\eta_H}{kT}\right) - \frac{c_{S_{aq}^{2+}}}{c_{S_{aq}^{2+}}^0} \exp\left(\frac{-2e\beta\eta_H}{kT}\right) \right\}$$

$$= i_0 \left\{ \exp\left(\frac{2e\beta\eta_{SC}}{kT}\right) \exp\left(\frac{2e(1-\beta)\eta_H}{kT}\right) - \exp\left(\frac{-2e\beta\eta_H}{kT}\right) \right\}, \qquad (9\text{-}36)$$

where, for simplicity, the concentration of hydrated ions $c_{S_{aq}^{2+}}$ in the solution may be assumed constant ($c_{S_{aq}^{2+}}/c_{S_{aq}^{2+}}^0 = 1$).

Eqn. 9–36 yields the anodic ion transfer current, i, in the state of band edge level pinning ($\eta = \eta_{SC}$, $\eta_H = 0$) as shown in Eqn. 9–37:

$$i = i_0 \left\{ \exp\left(\frac{2e\beta\eta}{kT} \right) - 1 \right\} . \tag{9-37}$$

Similarly, the anodic ion transfer current in the state of Fermi level pinning ($\eta = \eta_H$, $\eta_{SC} = 0$) can be obtained from Eqn. 9–36 as expressed in Eqn. 9–38:

$$i = i_0 \left\{ \exp\left(\frac{2e(1-\beta)\eta}{kT} \right) - \exp\left(\frac{-2e\beta\eta}{kT} \right) \right\} . \tag{9-38}$$

Generally, the band edge level pinning arises at low overvoltages at which the Fermi level at the interface is within the band gap; whereas, the Fermi level pinning arises at high overvoltages at which the Fermi level at the interface is in the valence band (Refer to Sec. 5.7.).

In discussion of this section, we have assumed that the limiting transport current of electrons or holes in semiconductor electrodes is much greater than the ion transfer current across the electrode interface. However, when the minority

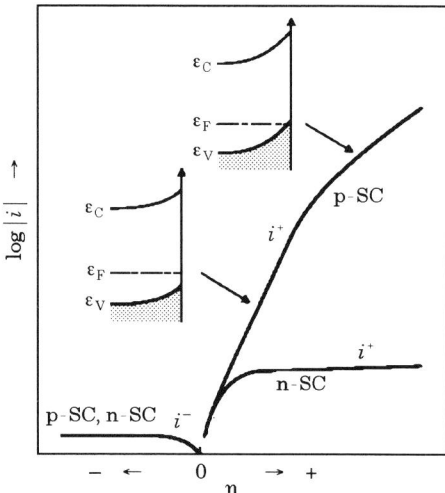

Fig. 9–10. Polarization curves of the anodic dissolution and cathodic deposition of n-type and p-type covalent semiconductor electrodes: n-SC (p-SC) = n-type (p-type) semiconductor electrode; i^+ (i^-) = anodic dissolution (cathodic deposition) current; ε_F = Fermi level.

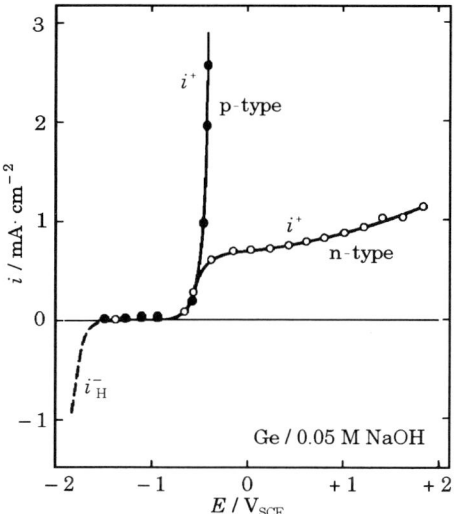

Fig. 9–11. Polarization curves observed for the anodic dissolution of n-type and p-type semiconductor electrodes of germanium in 0.05 M NaOH solution: $i^+(i^-)$ = current of the anodic dissolution of germanium (the cathodic reduction of hydrogen ions). [From Brattain-Garrett, 1955.]

charge carrier transports charge in the electrode, the limiting transport current of the minority charge carriers finally controls the anodic ion transfer across the electrode interface (Refer to Sec. 8.3.). For example, the anodic dissolution of semiconductors involving valence band holes proceeds readily at p-type semiconductor electrodes; but this dissolution rate levels off at n-type semiconductor electrodes as shown in Fig. 9–10. Fig. 9–11 illustrates the anodic dissolution current, as a function of the electrode potential, observed for p-type and n-type germanium electrodes in basic solutions.

The same discussion may apply to the anodic dissolution of covalently bonded electrodes of compounds such as gallium arsenide. In general, covalent semiconductors of compounds contain varying ionic polarity, in which the component atoms of positive polarity are likely to become surface cations; and the component atoms of negative polarity are likely to become surface radicals. For such semiconductors of compounds in the anodic dissolution, the valence band mechanism predominates over the conduction band mechanism with increasing band gap and increasing polarity of the compounds.

9.2.3 Dissolution of ionic semiconductors

For the covalent semiconductors described in Sec. 9.2.2, holes in the valence band or electrons in the conduction band are required for initially breaking the covalent bonding of the surface atoms to form the surface ions; thus, the dissolution of covalent semiconductor ions is accompanied by an anodic current of holes or electrons in the electrode, and this current can be observed in the circuit of the polarization cell. In contrast, for *ionic* semiconductors surface ions exist initially and they can transfer across the electrode interface without requiring any currents of holes and electrons in the electrode; thus, no dissolution current can occur in the circuit of the polarization cell.

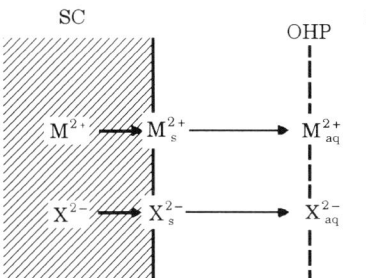

Fig. 9–12. Simultaneous transfer of cations and anions at a semiconductor electrode of ionic compound MX: M^{2+} = cation; X^{2-} = anion.

We examine semiconductors of the ionic compound MX composed of cation M^{2+} and anion X^{2-}. As shown in Fig. 9–12, the dissolution of ionic compound MX involves the transfer of M^{2+} and X^{2-} ions at the electrode interface as shown in Eqn. 9–39 and 9–40:

$$M_s^{2+} \to M_{aq}^{2+} \; , \tag{9–39}$$

$$X_s^{2-} \to X_{aq}^{2-} \; , \tag{9–40}$$

where subscripts "s" and "aq" denote a surface site on electrodes and a hydrated ion in aqueous solution, respectively. Transfer of cations in Eqn. 9–39 is the anodic reaction and transfer of anions in Eqn. 9–40 is the cathodic reaction.

From neutrality requirement, the anodic current of cation transfer equals the cathodic current of anion transfer. The transfer rate of cations v_M^+ and the transfer rate of anions v_X^-, depend on the potential difference $\Delta\phi_H$ across the compact layer and are given, respectively, by Eqn. 9–41 and Eqn. 9–42:

$$v_M^+ = k_M^+ c_{M_s} \exp\left(\frac{e \, \alpha_M^+ \, \Delta\phi_H}{k T} \right) , \tag{9–41}$$

$$v_X^- = k_X^- c_{X_s} \exp\left(\frac{-e \, \alpha_X^- \, \Delta\phi_H}{k T} \right) , \tag{9–42}$$

where k is the rate constant, c is the concentration of surface ions, and α is the transfer coefficient (Tafel constant). Therefore, the dissolution rate, v_{MX}, of ionic solid MX is expressed by Eqn. 9–43:

$$v_{MX} = v_M^+ = v_X^- = \sqrt{k_M^+ k_X^-}\sqrt{c_{X_s} c_{M_s}}\exp\left(\frac{e(\alpha_M^+ - \alpha_X^-)\Delta\phi_H}{2kT}\right). \tag{9–43}$$

When the interface of semiconductor electrodes is in the state of band edge level pinning, the potential difference $\Delta\phi_H$ across the compact layer remains constant and independent of the electrode potential; however, this $\Delta\phi_H$ depends on the composition of the solution. Thus, the dissolution rate of the ionic compound v_{MX}, which depends on $\Delta\phi_H$, is a function of the solution composition. For example, it is known that the rate of dissolution of metal oxides depends on the pH of the solution.

Furthermore, in the dissolution of ionic compounds where $v_M^+ = v_X^-$, we obtain from Eqns. 9–41 and 9–42 the potential difference $\Delta\phi_H$ of the compact layer as shown in Eqn. 9–44:

$$\Delta\phi_H = \frac{kT}{e(\alpha_M^+ + \alpha_X^-)}\ln\left(\frac{k_X^- c_{X_s}}{k_M^+ c_{M_s}}\right), \tag{9–44}$$

where the rate of backward reactions (deposition of ionic compounds) is disregarded.

As a reference state for the surface composition, an electrode surface of stoichiometric composition of MX ($c_{X_s} = c_{M_s}$) is chosen; at this surface composition the dissolution rate is denoted by v_{MX}^0 and the potential difference of compact layer is denoted by $\Delta\phi_H^0$. Thus, we obtain the dissolution rate v_{MX} of the interface of nonstoichiometric composition of ionic compounds as a function of $\Delta\phi_H$ or $\delta\Delta\phi_H (= \Delta\phi_H - \Delta\phi_H^0)$ as shown in Eqn. 9–45 [Vermilyea, 1966]:

$$v_{MX} = \frac{2 v_{MX}^0}{\exp\left(\dfrac{-e\alpha_M^+ \delta\Delta\phi_H}{kT}\right) + \exp\left(\dfrac{e\alpha_X^- \delta\Delta\phi_H}{kT}\right)}. \tag{9–45}$$

When $\delta\Delta\phi_H$ is negative ($\delta\Delta\phi_H \ll 0$, $\Delta\phi_H$ is lower than $\Delta\phi_H^0$), the second term of the denominator in Eqn. 9–45 is disregarded to obtain Eqn. 9–46:

$$v_{MX} = 2 v_{MX}^0 \exp\left(\frac{e\alpha_M^+ \delta\Delta\phi_H}{kT}\right). \tag{9–46}$$

When $\delta\Delta\phi_H$ is positive ($\delta\Delta\phi_H \gg 0$, $\Delta\phi_H$ is higher than $\Delta\phi_H^0$), the first term of the denominator in Eqn. 9–45 is disregarded to obtain Eqn. 9–47:

$$v_{MX} = 2 v_{MX}^0 \exp\left(\frac{-e\alpha_X^- \delta\Delta\phi_H}{kT}\right). \tag{9–47}$$

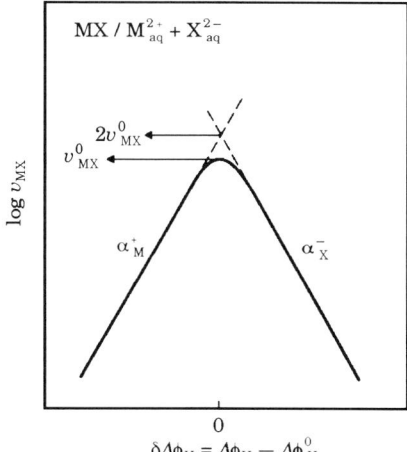

Fig. 9–13. Reaction rate of simultaneous dissolution of surface cations and anions from a semiconductor electrode of ionic compound as a function of the potential difference of the compact layer: $\Delta\phi_H$ ($\Delta\phi_H^0$) = potential difference of the compact layer (at the stoichiometric surface composition); v_{MX} (v_{MX}^0) = dissolution rate of compound MX (at the stoichiometric surface composition); α_M^+ = Tafel constant of the cation dissolution; α_X^- = Tafel constant of the anion dissolution. [From Vermilyea, 1966.]

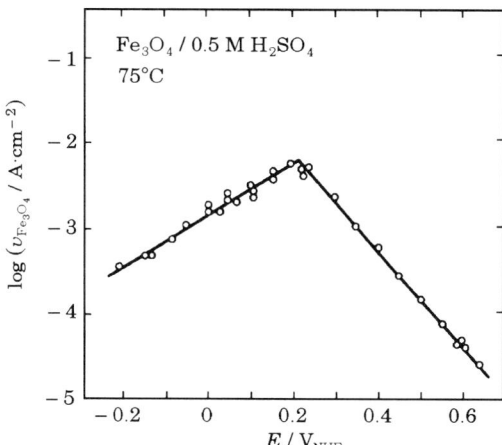

Fig. 9–14. Dissolution rate of a magnetite electrode observed as a function of the electrode potential in sulfuric acid solution: $v_{Fe_3O_4}$ = dissolution rate; V_{NHE} = volt referred to the normal hydrogen electrode. [From Engell, 1956.]

It follows from Eqns. 9–46 and 9–47 that the dissolution rate, v_{MX}, is determined by the transfer of cations at less anodic potentials ($\delta\Delta\phi_H \ll 0$) and by the transfer of anions at more anodic potentials ($\delta\Delta\phi_H \gg 0$) as schematically shown in Fig.

9–13. For example, Fig. 9–14 illustrates the dissolution rate of a magnetite electrode as a function of the electrode potential in a sulfuric acid solution; the interface of the magnetite electrode is in the state of Fermi level pinning so that all the change in the electrode potential occurs in the potential difference $\Delta\phi_H$ of the compact layer.

From Eqns. 9–41 and 9–42, we obtain in Eqn. 9–48 the ratio c_{M_s}/c_{X_s} of concentrations of surface cations to surface anions in the dissolution of the compound ($v_M^+ = v_X^-$):

$$\frac{c_{M_s}}{c_{X_s}} = \frac{k_X^-}{k_M^+} \exp\left(\frac{-e(\alpha_M^+ + \alpha_X^-)\Delta\phi_H}{kT} \right). \tag{9-48}$$

It appears that the concentration of surface cations increases with increasing cathodic polarization (decreasing $\Delta\phi_H$); whereas, the concentration of surface anions increases with increasing anodic polarization (increasing $\Delta\phi_H$). The dependence of concentrations of surface constituents on $\Delta\phi_H$ suggested that the dissolution rate of MX is determined by the transfer of cations at less anodic potentials and by the transfer of anions at more anodic potentials.

Note that the potential difference $\Delta\phi_H$ across the compact layer can not be changed by changing the electrode potential unless the electrode interface is in the state of Fermi level pinning. In the state of band edge level pinning $\Delta\phi_H$ remains independent of the electrode potential; $\Delta\phi_H$ depends on the concentration of potential determining ions in aqueous solution.

9.2.4 Oxidative and reductive dissolution of ionic semiconductors

Semiconductor electrodes of ionic compounds can also dissolve with the *oxidation* of surface anions or with the *reduction* of surface cations as shown schematically in Fig. 9–15.

The *oxidative dissolution* may be expressed by Eqns. 9–49 and 9–50:

$$M_s^{2+} \rightarrow M_{aq}^{2+}, \tag{9-49}$$

$$X_s^{2-} + 2h \rightarrow X_{ad} \rightarrow \frac{1}{2} X_{2,\,ad} \rightarrow \frac{1}{2} X_{2,\,aq}. \tag{9-50}$$

Eqn. 9–49 is the anodic transfer of surface cation M_s^{2+} into aqueous solution (cation dissolution); and Eqn. 9–50 is the anodic oxidation (hole capture) of surface anion X_s^{2-} producing molecules of $X_{2,\,aq}$ (e.g. gaseous oxygen molecules from oxide ions). Electric neutrality requires that the rate of cation dissolution equals the rate of anion oxidation; hence, the rate of the oxidative dissolution of semiconductor electrodes can be represented by the anodic hole current for the oxidation of surface anions.

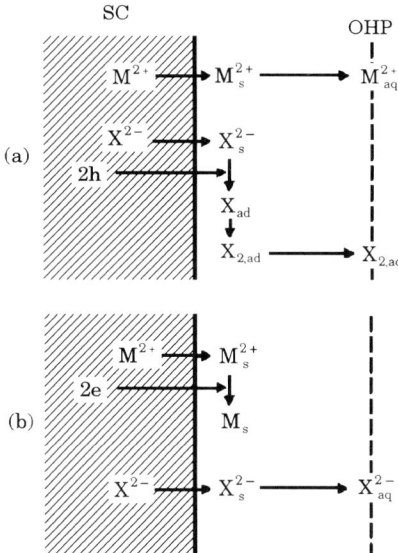

Fig. 9–15. Oxidative and reductive dissolution reactions of semiconductor electrodes of ionic compounds: (a) cation dissolution coupled with an anodic hole oxidation of surface anions, (b) anion dissolution coupled with a cathodic electron reduction of surface cations.

The rate v_M^+ of the cation transfer of Eqn. 9–49 is given by Eqn. 9–51 as a function of the potential difference $\Delta\phi_H$ of the compact layer:

$$v_M^+ = k_M^+ c_{M_s} \exp\left(\frac{e\,\alpha_M^+\,\Delta\phi_H}{kT}\right) \quad . \tag{9-51}$$

The rate v_X^+ of the oxidation of surface anions is given by Eqn. 9–52 as a function of the potential difference $\Delta\phi_{SC}$ of the space charge layer:

$$v_X^+ = k_X^+ c_{X_s} p_s^2 = k_X^+ c_{X_s} p_b^2 \exp\left(\frac{2e\,\Delta\phi_{SC}}{kT}\right) \quad , \tag{9-52}$$

where k is the rate constant, and p_s and p_b are the concentrations of holes at the interface and in the interior of the semiconductor electrode, respectively.

The rate of oxidative dissolution, v_{MX} (= $v_M^+ = v_X^+$), can be observed by measuring the anodic hole current, $i_{MX,p}$, of the oxidation of surface anions in Eqn. 9–50 as shown in Eqn. 9–53:

$$i_{MX,p} = 2e\,v_{MX} = 2e\,v_M^+ = 2e\,v_X^- = 2e\,k_X^- c_{X_s} p_b^2 \exp\left(\frac{2e\,\Delta\phi_{SC}}{kT}\right) \quad . \tag{9-53}$$

In the dissolution ($v_M^+ = v_X^+$), the ratio of concentrations of surface ions, c_{M_s}/c_{X_s}, can be obtained from Eqns. 9–51 and 9–52 as expressed in Eqn. 9–54:

$$\frac{c_{M_s}}{c_{X_s}} = \frac{k_X^+}{k_M^+} p_b{}^2 \exp\left(\frac{e(2\,\Delta\phi_{SC} - \alpha_M^+ \Delta\phi_H)}{kT} \right). \tag{9-54}$$

Eqn. 9–54 indicates that when the electrode interface is in the state of band edge level pinning ($\Delta\phi_H$ is constant), the concentration of surface cations increases with increasing anodic polarization.

In the state of band edge level pinning where all the change in the electrode potential occurs in the space charge layer, $\Delta\phi_{SC}$, the anodic polarization curve of the oxidative dissolution follows Eqn. 9–53. As the anodic polarization increases, the electrode interface enters a state of Fermi level pinning in which all the change in the electrode potential occurs in the compact layer, $\Delta\phi_H$, and the concentration of surface cations in Eqns. 9–54 then decreases with increasing anodic polarization.

Next, we discuss the *reductive dissolution* of ionic compounds shown in Fig. 9–15(b). The reductive dissolution is composed of the cathodic reduction of surface cations and the cathodic transfer of surface anions as expressed in Eqns. 9–55 and 9–56:

$$M_s^{2+} + 2\,e \rightarrow M_s\ , \tag{9-55}$$

$$X_s^{2-} \rightarrow X_{aq}^{2-}\ , \tag{9-56}$$

where M_s is the surface atom produced by the reduction of surface cations. The rate of the cathodic reduction of surface cations in Eqn. 9–55 is a function of the potential difference $\Delta\phi_{SC}$ of the space charge layer; the rate of cathodic anion transfer in Eqn. 9–56 is a function of the potential difference $\Delta\phi_H$ of the compact layer. Further, the rate of the reductive dissolution, where the rate of the reduction of surface cations equals the rate of anion transfer, can be observed by measuring the cathodic electron current of the reduction of surface cations in Eqn. 9–55.

Fig. 9–16 illustrates the polarization curves for the anodic oxidative and the cathodic reductive dissolution of semiconductors of ionic compounds. The anodic oxidative dissolution proceeds readily at p-type semiconductor electrodes in which the majority charge carriers are holes; whereas, the cathodic reductive dissolution proceeds readily at n-type semiconductor electrodes in which the majority charge carriers are electrons.

In order that the oxidative and reductive dissolution reactions may proceed, the affinity for the reaction is required to be positive; the reaction affinity is represented by the difference between the Fermi level, $\varepsilon_{F(SC)}$, of the electrode and the *equivalent* Fermi level $\varepsilon_{F(dec)}$ of the ion transfer reaction (Refer to Sec. 4.4.).

The overall reactions of the oxidative and reductive dissolution processes are represented by Eqn. 9–57 and Eqn. 9–58, respectively:

$$MX + 2\,h \rightarrow M^{2+}_{aq} + \frac{1}{2} X_2 \, , \qquad \varepsilon_{F(p,\,dec)} \qquad (9\text{--}57)$$

$$MX + 2\,e \rightarrow M + X^{2-}_{aq} \, , \qquad \varepsilon_{F(n,\,dec)} \qquad (9\text{--}58)$$

where $\varepsilon_{F(p,\,dec)}$ and $\varepsilon_{F(n,\,dec)}$ are the equivalent Fermi levels of the oxidative and reductive dissolution reactions (compound decomposition), respectively.

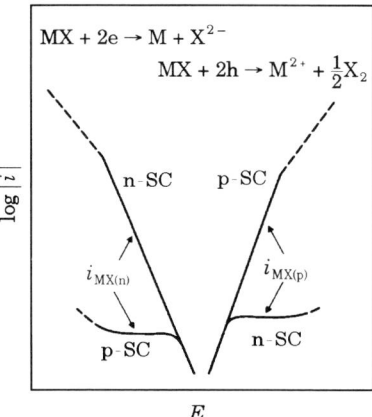

Fig. 9–16. Polarization curves of the anodic oxidative dissolution and cathodic reductive dissolution of semiconductor electrodes of an ionic compound MX: $i_{MX(p)}$ ($i_{MX(n)}$) = anodic oxidative (cathodic reductive) dissolution current; solid curve = band edge level pinning at the electrode interface, dotted curve = Fermi level pinning at the electrode interface.

The affinity for the reaction of Eqn. 9–57 is positive when $\varepsilon_{F(SC)}$ is lower than $\varepsilon_{F(p,\,dec)}$; the affinity for the reaction of Eqn. 9–58 is positive when $\varepsilon_{F(SC)}$ is higher than $\varepsilon_{F(n,\,dec)}$. Since the Fermi level of p-type semiconductors can not be lower than the valence band edge ε^s_V, the affinity ($\varepsilon_{F(p,\,dec)} - \varepsilon_{F(SC)}$) for the oxidative dissolution can not be positive (the oxidative dissolution can not occur) when the equivalent Fermi level $\varepsilon_{F(p,\,dec)}$ of the oxidative dissolution is within the valence band as in Fig. 9–17(a). However, when $\varepsilon_{F(p,\,dec)}$ is within the band gap as shown in Fig. 9–17(b), the oxidative dissolution is thermodynamically possible. Similarly, when the equivalent Fermi level $\varepsilon_{F(n,\,dec)}$ for the reductive dissolution is within the conduction band, the affinity ($\varepsilon_{F(SC)} - \varepsilon_{F(n,\,dec)}$) for the reductive dissolution is negative and hence the reductive dissolution is thermodynamically impossible as shown in Fig. 9–17(c). However, the reductive dissolution becomes thermodynamically possible, when $\varepsilon_{F(n,\,dec)}$ is within the band gap as shown in Fig. 9–17(d).

Ion Transfer at Semiconductor Electrodes   313

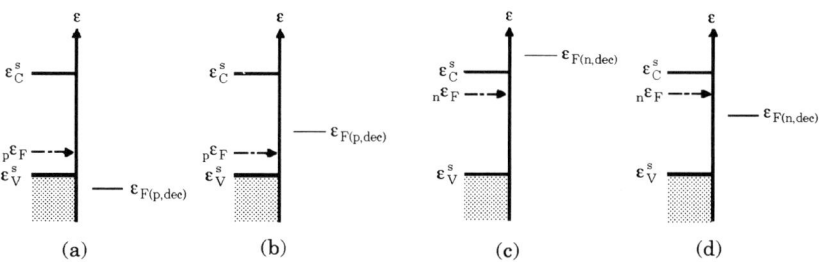

Fig. 9–17. Thermodynamic stability of electrodes of compound semiconductors relative to the oxidative and reductive dissolution in the state of band edge level pinning: (a) the oxidative dissolution is thermodynamically impossible ($\varepsilon_{F(p,\,dec)} < \varepsilon_V^s$), (b) the oxidative dissolution may occur ($\varepsilon_{F(p,\,dec)} > \varepsilon_V^s$), (c) the reductive dissolution is thermodynamically impossible ($\varepsilon_{F(n,\,dec)} > \varepsilon_C^s$), (d) the reductive dissolution may occur ($\varepsilon_{F(n,\,dec)} < \varepsilon_C^s$). $\varepsilon_{F(p,\,dec)}$ ($\varepsilon_{F(n,\,dec)}$) = equivalent Fermi level of the oxidative (reductive) dissolution.

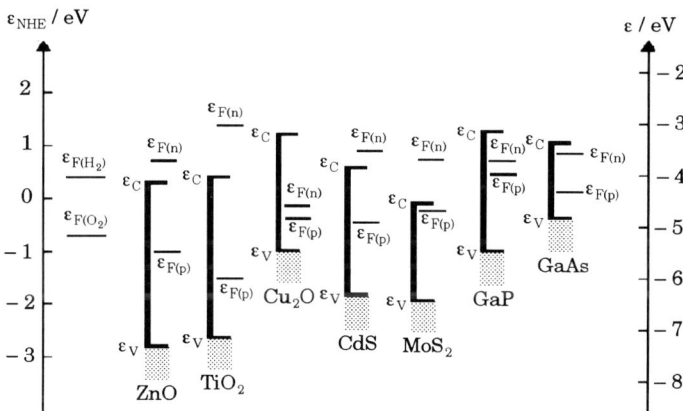

Fig. 9–18. Band edge levels and equivalent Fermi levels of oxidative and reductive dissolution reactions of compound semiconductors in aqueous solutions at pH 7: $\varepsilon_{F(n)}$ = $\varepsilon_{F(n,\,dec)}$; $\varepsilon_{F(p)} = \varepsilon_{F(p,\,dec)}$; $\varepsilon_{F(H_2)}$ ($\varepsilon_{F(O_2)}$) = electron level of the hydrogen (oxygen) reaction; ε_{NHE} = electron level relative to the normal hydrogen electrode; ε = electron level relative to the standard gaseous electron. [From Gerischer, 1978.]

Fig. 9–18 illustrates the band edge levels of compound semiconductor electrodes in aqueous solutions and the equivalent Fermi levels of the following oxidative and reductive dissolution reactions:

$$\text{ZnO} + 2\,\text{Cl}^-_{\text{aq}} + 2\,\text{h} \to \text{ZnCl}_{2,\,\text{aq}} + (1/2)\,\text{O}_2$$
$$\text{ZnO} + 2\,\text{H}^+_{\text{aq}} + 2\,\text{e} \to \text{Zn} + \text{H}_2\text{O}_{\text{aq}}$$
$$\text{TiO}_2 + 4\,\text{Cl}^-_{\text{aq}} + 4\,\text{h} \to \text{TiCl}_{4,\,\text{aq}} + \text{O}_2$$
$$\text{TiO}_2 + 4\,\text{H}^+_{\text{aq}} + 4\,\text{e} \to \text{Ti} + 2\,\text{H}_2\text{O}_{\text{aq}}$$
$$\text{CdS} + 2\,\text{Cl}^-_{\text{aq}} + 2\,\text{h} \to \text{CdCl}_{2,\,\text{aq}} + \text{S}$$
$$\text{CdS} + 2\,\text{H}^+_{\text{aq}} + 2\,\text{e} \to \text{Cd} + \text{H}_2\text{S}$$
$$\text{GaP} + 6\,\text{H}_2\text{O} + 6\,\text{h} \to \text{Ga(OH)}_3 + \text{H}_3\text{PO}_{3,\,\text{aq}} + 6\,\text{H}^+_{\text{aq}}$$
$$\text{GaP} + 3\,\text{H}^+_{\text{aq}} + 6\,\text{e} \to \text{Ga} + \text{PH}_3\ .$$

It appears in Fig. 9–18 that both oxidative and reductive dissolutions are thermodynamically possible with Cu_2O, GaP and GaAs. On the other hand, with ZnO, TiO_2, CdS and MoS_2, the reductive dissolution is thermodynamically impossible but the oxidative mode of dissolution is allowed to occur.

9.3 Ion Adsorption on Metal Electrodes

9.3.1 Ion adsorption equilibrium

The process in which a hydrated ion is dehydrated to form an adsorbed ion on the electrode interface, and its reverse process, are *ion transfer* processes across the compact layer on the electrode interface. As an example, we consider the proton adsorption (cathodic proton transfer) and proton desorption (anodic proton transfer) on metal electrodes represented by Eqn. 9–59:

$$\text{H}^+_{\text{aq}} \rightleftarrows \text{H}^+_{\text{ad}}\ , \tag{9–59}$$

In the adsorption–desorption equilibrium, the adsorption coverage of protons, $\theta_{\text{H}^+}\ (= c_{\text{H}^+_{\text{ad}}}/c^{\max}_{\text{H}^+_{\text{ad}}})$, is derived from the Nernst equation of electrochemical equilibria to produce Eqn. 9–60 in the range of low absorption coverages where Langmuir's adsorption isotherm applies:

$$\frac{c_{\text{H}^+_{\text{ad}}}}{c^{\max}_{\text{H}^+_{\text{ad}}}} \doteq \frac{c_{\text{H}^+_{\text{ad}}}}{c^{\max}_{\text{H}^+_{\text{ad}}}(1-\theta_{\text{H}^+})} = K\,c_{\text{H}^+_{\text{aq}}}\exp\left(\frac{-e\,\Delta\phi_\text{H}}{k\,T}\right), \tag{9–60}$$

where $c_{\text{H}^+_{\text{ad}}}$ is the concentration of adsorbed protons, $c^{\max}_{\text{H}^+_{\text{ad}}}$ is the maximum possible concentration of adsorbed protons, K is the equilibrium constant, and $\Delta\phi_\text{H}$ is the potential difference of the compact layer. Strictly, $\Delta\phi_\text{H}$ in Eqn. 9–60 should be replaced by the difference of potential between the plane of adsorbed dehydrated protons (the inner Helmholtz layer) and the pane of closest approach of hydrated protons (the outer Helmholtz layer)

As the proton coverage increases, the interaction among adsorbed protons increases leading to Temkin's adsorption isotherm [Temkin, 1947], and Eqn. 9–61 is obtained:

$$\exp\left(\kappa \frac{c_{H^+_{ad}}}{c^{max}_{H^+_{ad}}}\right) = K\, c_{H^+_{aq}} (1 - \theta_{H^+}) \exp\left(\frac{-e\,\Delta\phi_H}{kT}\right), \quad \kappa = \frac{1}{kT}\frac{\partial \Delta G^0_{ad}}{\partial \theta_{H^+}}. \quad (9\text{–}61)$$

where ΔG^0_{ad} is the standard free enthalpy of adsorption.

The change in the electrode potential of the metal equals the change in the potential difference $\Delta\phi_H$ across the compact layer on the electrode interface. Consequently, a linear relation is expected to apply between the logarithm of the proton coverage, θ_{H^+}, and the electrode potential, E, at low proton coverages as shown in Eqn. 9–60; whereas, to a first approximation, a linear relation applies between the coverage, θ_{H^+}, and the electrode potential, E, at high proton coverages as shown in Eqn. 9–61. Such relationships between θ_{H^+} and E have been observed with the adsorption of hydroxide ions on the interface of metallic iron electrodes as already shown in Fig. 9–6.

9.3.2 Electron levels of adsorbed ions

Adsorbed ions on metal electrodes may change their ionic valence by donating or accepting electrons to or from the metal electrode to be oxidized or reduced. For example, as described in Sec. 5.7, the adsorbed proton constitutes an interfacial redox system in Eqn. 9–62:

$$H^+_{ad} + e = H_{ad}, \quad (9\text{–}62)$$

in which the vacant electron level of the oxidized state of H^+_{ad} (the most probable electron level, $\varepsilon_{H^+_{ad}}$) is higher than the occupied electron level of the reduced state of H_{ad} (the most probable electron level, $\varepsilon_{H_{ad}}$) by an amount equal to the reorganization energy of the adsorbed particles.

The ratio of concentrations of adsorbed hydrogen atoms to adsorbed protons, $c_{H_{ad}}/c_{H^+_{ad}}$, is derived from Eqn. 5–55 as a function of the Fermi level, $\varepsilon_{F(H^+/H)}$, of the *interfacial redox electrons* as shown in Eqn. 9–63:

$$\frac{c_{H_{ad}}}{c_{H^+_{ad}}} = \exp\left(\frac{\varepsilon_{F(H^+/H)} - \varepsilon^0_{F(H^+/H)}}{kT}\right), \quad (9\text{–}63)$$

where $\varepsilon^0_{F(H^+/H)}$ is the standard Fermi level at the ratio of concentrations equal to unity $c_{H_{ad}}/c_{H^+_{ad}} = 1$.

When equilibrium is established between the electrode electrons and the interfacial redox electrons in the adsorbed protons, the Fermi level of the metal electrode $\varepsilon_{F(M)}$ equals the Fermi level of the interfacial redox particles $\varepsilon_{F(H^+/H)}$: $\varepsilon_{F(M)} = \varepsilon_{F(H^+/H)}$. It appears from Fig. 9–19 that when the Fermi level $\varepsilon_{F(M)}$ of the electrode is higher than the standard Fermi level $\varepsilon^0_{F(H^+/H)}$ of the adsorbed redox particles, the concentration of adsorbed hydrogen atoms, $c_{H_{ad}}$, exceeds the concentration of adsorbed protons, $c_{H^+_{ad}}$: i.e. $c_{H_{ad}} > c_{H^+_{ad}}$. On the other hand, when $\varepsilon_{F(M)}$ is lower than $\varepsilon^0_{F(H^+/H)}$, the concentration of adsorbed protons, $c_{H^+_{ad}}$, exceeds the concentration of adsorbed hydrogen atoms, $c_{H_{ad}}$: i.e. $c_{H^+_{ad}} > c_{H_{ad}}$.

If the difference of potential is constant between the plane of adsorbed protons and the plane of the electrode surface (the inner Helmholz layer), the electron level of the adsorbed protons ($\varepsilon^0_{F(H^+/H)}$) relative to the electron level of the electrode ($\varepsilon_{F(M)}$) remains unchanged with changing electrode potential ($\varepsilon_{F(M)} - \varepsilon^0_{F(H^+/H)} = \varepsilon_{F(H^+/H)} - \varepsilon^0_{F(H^+/H)} =$ constant); hence, the concentration ratio of adsorbed hydrogen

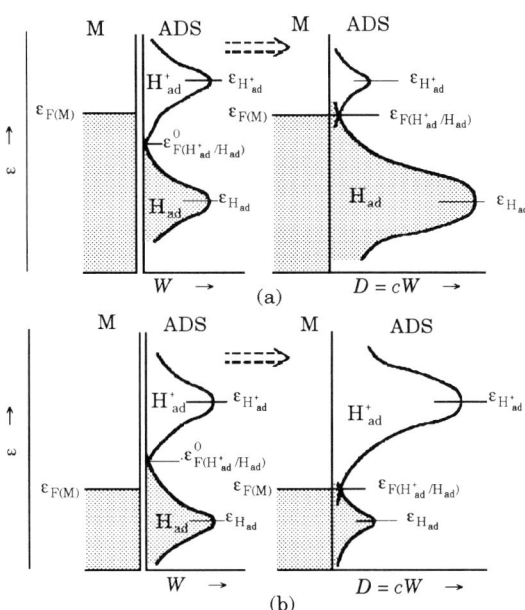

Fig. 9–19. Electron state density of adsorbed proton/hydrogen redox particles (H^+_{ad}/H_{ad}) on metal electrodes: (a) the relative concentration of adsorbed reductant hydrogen atoms (H_{ad}) would be higher if the Fermi level $\varepsilon_{F(M)}$ of electrode is higher than the standard Fermi level $\varepsilon^0_{F(H^+/H)}$ of adsorbed redox particles, (b) the relative concentration of adsorbed oxidant protons (H^+_{ad}) would be higher if the Fermi level $\varepsilon_{F(M)}$ of electrode is lower than the standard Fermi level $\varepsilon^0_{F(H^+/H)}$. ADS = adsorbed layer; $\varepsilon_{H^+_{ad}}$ ($\varepsilon_{H_{ad}}$) = most probable vacant electron level of adsorbed protons (most probable occupied electron level of adsorbed hydrogen atoms); W = probability density; c = concentration, D = electron state density.

atoms to adsorbed protons is constant and independent of the electrode potential. Assuming that the difference of potential across the inner Helmholtz layer is unchanged, we obtain from Eqns. 9–60 and 9–63 the concentration of adsorbed hydrogen atoms, $c_{H_{ad}}$, as a function of the electrode potential (the difference of potential across the compact layer $\Delta\phi_H$) as shown in Eqn. 9–64:

$$c_{H_{ad}} = K\, c_{H_{ad}^+}^{max}\, c_{H_{aq}^+} \exp\left(\frac{\varepsilon_{F(M)} - \varepsilon_{F(H^+/H)}^0}{kT}\right) \exp\left(\frac{-e\,\Delta\phi_H}{kT}\right). \tag{9–64}$$

Eqn. 9–64 indicates that the concentration of adsorbed hydrogen atoms is an exponential function of the electrode potential.

Even if we assume that the difference of potential between the plane of adsorbed protons and the surface plane of the electrode changes with the electrode potential, we could also derive the concentration of adsorbed hydrogen atoms, $c_{H_{ad}}$, as an exponential function of the electrode potential (Refer to Sec, 5.7.4).

9.4 Ion Adsorption on Semiconductor Electrodes

9.4.1 Ion adsorption equilibrium

We consider the dehydration–adsorption of hydrated protons (cathodic proton transfer) and the desorption–hydration of adsorbed protons (anodic proton transfer) on the interface of semiconductor electrodes. Since these adsorption and desorption of protons are *ion transfer* processes across the compact layer at the interface of semiconductor electrodes, the adsorption–desorption equilibrium is expressed as a function of the potential difference $\Delta\phi_H$ of the compact layer in the same way as Eqns. 9–60 and 9–61. In contrast to metal electrodes where $\Delta\phi_H$ changes with the electrode potential, semiconductor electrodes in the state of band edge level pinning maintain the potential difference $\Delta\phi_H$ of the compact layer constant and independent of the electrode potential. Consequently, the concentration of adsorbed protons, $c_{H_{ad}^+}$, is determined not by the electrode potential but by the concentration of hydrated protons in aqueous solutions.

When the interface of the semiconductor electrode is in the state of Fermi level pinning, the potential difference $\Delta\phi_H$ of the compact layer changes with the electrode potential; hence, the equilibrium of adsorption–desorption of protons on semiconductor electrodes depends on the electrode potential in the same way as that on metal electrodes.

9.4.2 Electron levels of adsorbed ions

Adsorbed protons on the interface of semiconductor electrodes constitute an *interfacial redox reaction* system as shown in Eqn. 9–65:

$$H_{ad}^+ + e = H_{ad}, \qquad H_{ad}^+ = H_{ad} + h \; . \tag{9-65}$$

Electrons in the conduction band is predominantly involved in the redox reaction when the electron level of adsorbed protons is close to the conduction band edge. On the other hand, holes in the valence band participate predominantly in the redox reaction when the electron level of adsorbed protons is close to the valence band edge.

In the equilibrium of interfacial redox reactions of adsorbed protons and hydrogens, the Fermi level $\varepsilon_{F(SC)}^s$ of the semiconductor electrons at the interface of electrodes equals the Fermi level $\varepsilon_{F(H^+/H)}$ of the interfacial redox electrons in adsorbed protons and hydrogens. The Fermi level $\varepsilon_{F(SC)}^s$ of the interface of the semiconductor electrode depends on the potential difference $\Delta\phi_{SC}$ of the space charge layer as shown in Eqn. 9–66:

$$\varepsilon_{F(SC)}^s = \varepsilon_{F(SC), \, fb} - e \, \Delta\phi_{SC} \quad , \tag{9-66}$$

where $\varepsilon_{F(SC), \, fb}$ is the Fermi level of the semiconductor at the flat band potential. Hence, we obtain from Eqn. 9–63 the ratio of concentrations of adsorbed hydrogen atoms to adsorbed protons in the adsorption–desorption equilibrium as shown in Eqn. 9–67:

$$\frac{c_{H_{ad}}}{c_{H_{ad}^+}} = \exp\left(\frac{\varepsilon_{F(H^+/H)} - \varepsilon_{F(H^+/H)}^0}{kT} \right) = \exp\left(\frac{\varepsilon_{F(SC)}^s - \varepsilon_{F(H^+/H)}^0}{kT} \right)$$

$$= \exp\left(\frac{\varepsilon_{F(SC), \, fb} - \varepsilon_{F(H^+/H)}^0}{kT} \right) \exp\left(\frac{-e \, \Delta\phi_{SC}}{kT} \right) . \tag{9-67}$$

Consequently, it follows that the ratio of concentrations $c_{H_{ad}}/c_{H_{ad}^+}$ in the adsorption equilibrium on semiconductor electrodes in the state of band edge level pinning changes with changing electrode potential as shown in Fig. 9–20. This differs from the case of metal electrodes where the ratio of concentrations remains constant irrespective of the electrode potential (Refer to Sec. 9.3.). Combining Eqns. 9–60 and 9–67 to produce Eqn. 9–68 describes the concentration of adsorbed hydrogen atoms as an exponential function of the electrode potential ($\Delta\phi_{SC}$) for low adsorption coverages:

$$c_{H_{ad}} = K \, c_{H_{aq}^+}^{max} \, c_{H_{aq}^+} \exp\left(\frac{\varepsilon_{F(SC), \, fb} - \varepsilon_{F(H^+/H)}^0}{kT} \right) \exp\left(\frac{-e \, \Delta\phi_{SC}}{kT} \right) . \tag{9-68}$$

For the state of Fermi level pinning of the electrode interface, we may also derive the concentration of adsorbed hydrogen atoms which can be represented by the same equation as Eqn. 9–64 for metal electrodes.

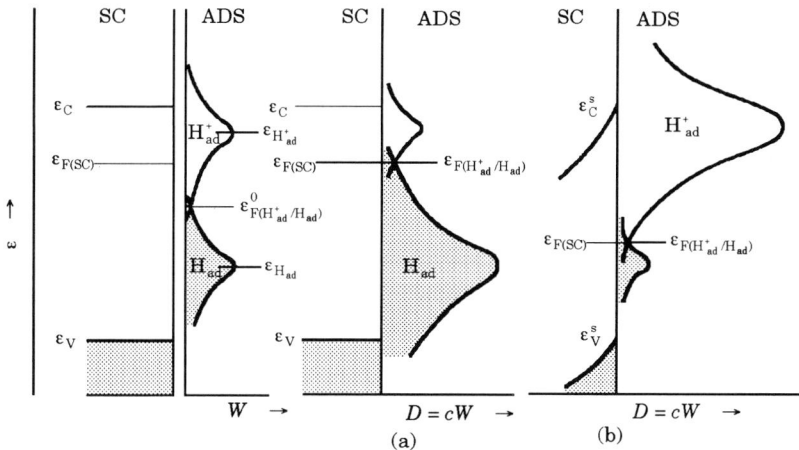

Fig. 9–20. Electron state density D of adsorbed redox particles (H^+_{ad}/H_{ad}) on semiconductor electrodes: (a) at the flat band potential, (b) under the anodic polarization which increases the concentration of adsorbed oxidants (H^+_{ad}) if the band edge level is pinned. W = probability density of the electron states.

9.4.3 Proton levels on electrode surfaces

In general, semiconductor electrodes in aqueous solutions adsorb water molecules, hydronium ions, and hydroxide ions in addition to various solute ions. As a result, the dissociation–association equilibria of the adsorbed hydronium ions and water molecules produce, in the proton dissociation–association reactions of Eqns. 9–69 and 9–70, the acidic and basic *proton levels*, respectively, on the interface of semiconductor electrodes as shown in Fig. 9–21:

$$S \cdot H^+_{ad}(H_2O) = S \cdot H_2O_{ad} + H^+_{aq} \, , \qquad K_1 \qquad (9\text{--}69)$$

$$S \cdot H_2O_{ad} = S \cdot OH^-_{ad} + H^+_{aq} \, , \quad (S \cdot H^+_{ad} + OH^-_{aq} = S \cdot H_2O_{ad}) \, , \quad K_2 \, , \qquad (9\text{--}70)$$

where S denotes the surface atom of semiconductors, K is the equilibrium constant, and subscripts "ad" and "aq" denote the adsorbed and hydrated states, respectively.

Reactions in Eqns. 9–69 and 9–70 are the dissociation processes of (1) the acidic protons from adsorbed hydronium ions and (2) the basic protons from adsorbed water molecules on the electrode interface, respectively. Eqn. 9–71 gives the equilibrium constants, K_1 and K_2, of these proton dissociation reactions:

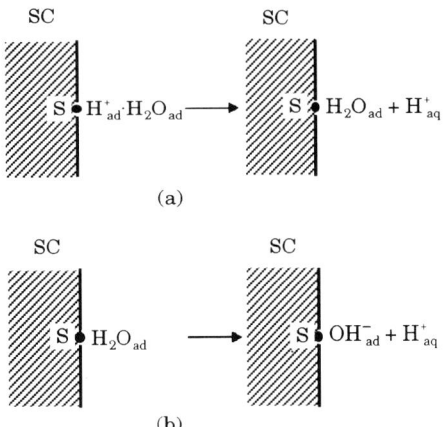

Fig. 9-21. Proton levels of adsorbed hydronium ions and of adsorbed water molecules on semiconductor electrodes: (a) acidic proton dissociation of adsorbed hydronium ions, (b) basic proton dissociation of adsorbed water molecules. S• = semiconductor surface atom.

$$K_1 = \frac{[\text{S·H}_2\text{O}_{ad}][\text{H}^+_{aq}]}{[\text{S·H}^+_{ad}(\text{H}_2\text{O})]} \quad , \quad K_2 = \frac{[\text{S·OH}^-_{ad}][\text{H}^+_{aq}]}{[\text{S·H}_2\text{O}_{ad}]} \quad , \tag{9-71}$$

where [i] denotes the concentration of particle i. These equilibrium dissociation constants correspond, respectively, to the standard electrochemical potential of acidic protons (acidic occupied proton level) in adsorbed hydronium ions and to the electrochemical potential of basic protons (basic occupied proton level) in adsorbed water molecules. To the extent that the state of the electrode interface is characterized by band edge level pinning, these acidic and basic proton levels are independent of the electrode potential but depend on the pH of the solution.

In aquatic chemistry, the *unitary proton level* of the proton dissociation reaction is expressed by the logarithm of the reciprocal of the proton dissociation constant i.e. $pK = -\log K$; here, a higher level of proton dissociation corresponds with a lower pK. When the pK_1 of the adsorbed protons is lower than the pH of the solution, the protons in the adsorbed hydronium ions desorb, leave acidic vacant proton levels in adsorbed water molecules, and form hydrated protons in the aqueous solution. Fig. 9-22 shows the occupied and vacant proton levels for the acidic and basic dissociations of adsorbed hydronium ions and of adsorbed water molecules on the interface of semiconductor electrodes.

Since the two proton dissociation reactions of Eqns. 9-69 and 9-70 can proceed simultaneously on the electrode interface, Eqn. 9-71 yields Eqn. 9-72:

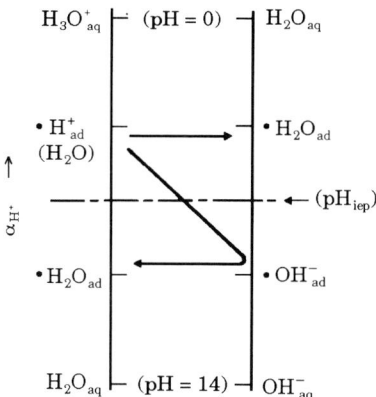

Fig. 9–22. Unitary proton levels of hydrated and adsorbed hydronium ions (acidic proton) and of hydrated and adsorbed water molecules (basic proton): the left side is the occupied proton level α_{H^+} (the real potential of acidic protons), and the right side is the vacant proton level. •$H^+_{ad}(H_2O)$ = unitary occupied proton level of adsorbed hydronium ions (acidic proton level); •H_2O_{ad} = unitary vacant proton level of adsorbed hydronium ions (acidic proton level) and unitary occupied proton level of adsorbed water molecules (basic proton level); •OH^-_{ad} = unitary vacant proton level of adsorbed water molecules (basic proton level); (pH_{iep}) = hydrated proton level at the iso-electric point pH.

$$pH = \frac{1}{2}(pK_1 + pK_2) + \frac{1}{2}\log\left(\frac{[S\cdot OH^-_{ad}]}{[S\cdot H^+_{ad}(H_2O)]}\right). \quad (9\text{–}72)$$

Eqn. 9–72 indicates that the logarithm of the ratio of concentrations of adsorbed protons (acidic occupied proton level) to adsorbed hydroxide ions (basic vacant proton level) depends linearly on the pH of the solution.

The pH at which the concentration of acidic occupied proton levels of adsorbed hydronium ions equals the concentration of basic vacant proton levels of adsorbed water molecules is called the *iso-electric point* pH_{iep}; here, the net interfacial charge of adsorbed ions at the interface is zero. The iso-electric point pH_{iep} is expressed in Eqn. 9–73:

$$pH_{iep} = \frac{1}{2}(pK_1 + pK_2). \quad (9\text{–}73)$$

For compound semiconductors, the adsorbed proton level differs with different constituents in the semiconductor; thus, the distinction between the acidic and basic proton levels, pK_1 and pK_2, is greater than in the case simple substance semiconductors. For example, on metal oxide electrodes the acidic proton dissoci-

occurs at the surface sites of oxide ions as shown in Eqns. 9–74 and 9–75:

$$M^{2+} \cdot H_2O_{ad} = M^{2+} \cdot OH^-_{ad} + H^+_{aq} \quad , \tag{9-74}$$

$$O^{2-} \cdot H_2O_{ad} = O^{2-} \cdot H^+_{ad} + OH^-_{aq} \quad . \tag{9-75}$$

For metal oxide electrodes, the iso-electric point pH_{iep} is also located midway between the unitary acidic proton level and the unitary basic proton level. Table 9–1 shows the iso-electric point pH_{iep} of several metal oxides in aqueous solutions.

TABLE 9–1. Iso-electric point pH_{iep} of metal oxides in aqueous solution. [From Aikawa, 1992.]

MO_x	pH_{iep}	MO_x	pH_{iep}
MgO	12.4	Fe_3O_4	6.5
NiO	10.3	TiO_2	5.3
CuO	9.5	SnO_2	4.5
Al_2O_3	9.2	SiO_2	2.2
ZnO	8.8	WO_3	0.5
Fe_2O_3	8.6		

References

[Aikawa, 1992]: Y. Aikawa, *Zairyou-to-Kankyo*, **41**, 111(1992).
[Bockris-Drazig-Despic, 1961]: J. O'M. Bockris, D. Drazig and A. Despic, *Electrochimica Acta*, **4**, 325(1961).
[Brattain-Garret, 1955]: W. H. Brattain and C. G. B. Garret, *Bell System Tech. J.*, **34**, 129(1955).
[Conway-Bockris, 1958]: B. E. Conway and J. O'M. Bockris, *Proc. Roy. Soc.*, **A248**, 394(1958).
[Despic, 1983]: A. Despic, *Comprehensive Treatise of Electrochemistry*, **Vol. 7**, (Edited by B. E. Conway et al.), p. 451, Plenum Press, New York, (1983).
[Engell, 1956]: H. J. Engell, *Z. Phys. Chem. N. F.*, **7**, 158(1956)
[Gerischer, 1970]: H. Gerischer, *Physical Chemistry, An Advanced Treatise*, **Vol. IX A**, *Electrochemistry*, (Edited by H. Eyring), p. 463, Academic Press, New York, (1970).

[Gerischer, 1978]: H. Gerischer, *J. Vac. Sci. Technology,* **15**, 1422(1978).
[Gerischer-Mindt, 1968]: H. Gerischer and W. Mindt, *Electrochimica Acta*, **13**, 1329(1968).
[Heusler, 1958]: K. E. Heusler, *Z. Elektrochem.*, **62**, 582(1958).
[Lorenz, 1954]: W. Lorenz, *Z. Elektrochem.*, **58**, 912(1954).
[Memming, 1983]: R. Memming, *Comprehensive Treatise of Electrochemistry*, **Vol. 7**, (Edited by B. E. Conway et al.), p. 529, Plenum Press, New York, (1983).
[Plonski, 1996]: I-H. Plonski, Modern Aspects of Electrochemistry, **No. 29**, (Edited by J. O'M. Bockris et al.), p. 203, Plenum Press, New York, (1996).
[Sato-Okamoto, 1964]: N. Sato and Go Okamoto, *J. Electrochem. Soc.*, **111**, 897(1964).
[Temkin, 1947]: M. Temkin, *Zh. Fiz. Khim.*, **15**, 296(1941); **21**, 517(1947).
[Tsuru, 1991]: T. Tsuru, *Materials Science and Engineering*, **A146**, 1(1991).
[Vermilyea, 1966]: D. A. Vermilyea, *J. Electrochem. Soc.*, **113**, 1067(1966).

CHAPTER 10

SEMICONDUCTOR PHOTOELECTRODES

10.1 Quasi-Fermi Level of Excited Electrons and Holes

10.1.1 Quasi-Fermi level

Under light illumination, semiconductor electrodes absorb the energy of photons to produce excited electrons and holes in the conduction and valence bands. Compared with the photoelectrons in metals, the photoexcited electrons and holes in semiconductors are relatively stable so that the photo-effect on electrode reactions can be manifested more distinctly with semiconductor electrodes than with metal electrodes.

When a semiconductor electrode is irradiated by a photon beam (light) with the energy greater than the band gap energy of the semiconductor, the electrode absorbs the photon energy to excite electrons from the valence band to the conduction band, thereby producing photoexcited pairs of electrons in the conduction band and holes in the valence band as shown in Fig. 10–1. For n-type semiconductors, the photoexcited electrons hardly influence the concentration of electrons in the conduction band because electrons are the majority charge carriers, but the photoexcited holes significantly increase the concentration of holes (minority charge carriers) in the valence band. Generally, in semiconductor electrodes, photoexcitation significantly increases the concentration of minority charge carriers, but influences little the concentration of majority charge carriers.

The concentrations of electrons and holes in the conduction and valence bands, n^* and p^*, in the *photostationary* state under irradiation of the photon beam are expressed, respectively, in Eqn. 10–1:

$$n^* = n + \Delta n^* , \qquad p^* = p + \Delta p^* , \qquad (10\text{--}1)$$

where n and p are the concentrations of electrons and holes in the dark, Δn^* and Δp^* are the increases of concentration of electrons and holes produced by photoexcitation ($\Delta n^* = \Delta p^*$), respectively.

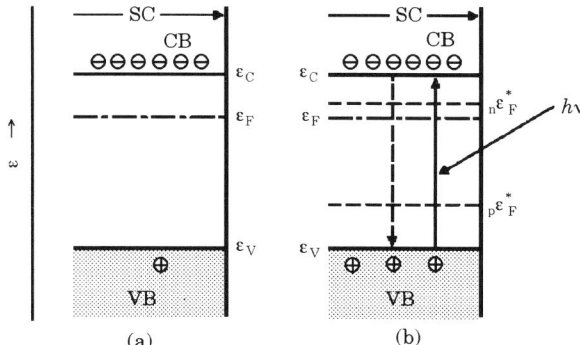

Fig. 10–1. Splitting of the Fermi level, $\varepsilon_{F(SC)}$, into both the quasi-Fermi level of electrons, $_n\varepsilon_F^*$, and the quasi-Fermi level of holes, $_p\varepsilon_F^*$, in photoexcited semiconductors: (a) in the dark, (b) in the photon irradiation. SC = semiconductor; $h\nu$ = photon energy.

In the case in which the photoexcited pairs of electrons and holes are relatively stable so that the thermal equilibrium is established between phonons and electrons in the conduction band as well as between phonons and holes in the valence band, we can define individually the electrochemical potentials of photoexcited electrons and of holes in the photostationary state. Here, thermal equilibrium is not established between the photoexcited electrons in the conduction band and the holes in the valence band. The electrochemical potential, thus defined, for the photoexcited electrons and holes is called the *quasi-Fermi level of electrons*, $_n\varepsilon_F^*$, and the *quasi-Fermi level of holes*, $_p\varepsilon_F^*$. [Schockley, 1950; Gerischer, 1990].

In the dark, thermal equilibrium is established between electrons in the conduction band and holes in the valence band so that both the quasi-Fermi level of electrons and the quasi-Fermi level of holes equal the original Fermi level of the semiconductor ($_n\varepsilon_F^* = {_p\varepsilon_F^*} = \varepsilon_F$). However, under the condition of photoexcitation, the quasi-Fermi level of electrons is higher and the quasi-Fermi level of holes is lower than the original Fermi level of the semiconductor ($_n\varepsilon_F^* > \varepsilon_F > {_p\varepsilon_F^*}$). Consequently, photoexcitation splits the Fermi level of semiconductors into two quasi-Fermi levels: the quasi-Fermi level of electrons for the conduction band and the quasi-Fermi level of holes for the valence band as shown in Fig. 10–1.

Such a concept of quasi-Fermi level is valid under the condition that the time constant for the establishment of thermal equilibrium of electrons or holes in the conduction or valence band (the reciprocal of the rate of establishing equilibrium between electrons and phonons) is much smaller than the time constant for the recombination of excited electron–hole pairs (the reciprocal of the recombination rate of electrons and holes); this condition is satisfied in usual semiconductors.

As is described in Sec. 2.4, the Fermi levels of n-type and p-type semiconductors are given, respectively, in Eqn. 10–2:

$$\varepsilon_{F(SC)} = \varepsilon_C - kT \ln\left(\frac{N_C}{n}\right), \quad \varepsilon_{F(SC)} = \varepsilon_V + kT \ln\left(\frac{N_V}{p}\right). \quad (10\text{--}2)$$

Accordingly, the quasi-Fermi levels of electrons and holes are expressed, respectively, in Eqns. 10–3 and 10–4:

$$_n\varepsilon_F^* = \varepsilon_C - kT \ln\left(\frac{N_C}{n^*}\right) = \varepsilon_{F(SC)} + kT \ln\left(\frac{n + \Delta n^*}{n}\right), \quad (10\text{--}3)$$

$$_p\varepsilon_F^* = \varepsilon_V + kT \ln\left(\frac{N_V}{p^*}\right) = \varepsilon_{F(SC)} - kT \ln\left(\frac{p + \Delta p^*}{p}\right). \quad (10\text{--}4)$$

For n-type semiconductors ($n \gg p$, and $n \gg \Delta n^*$), the quasi-Fermi level of electrons (Eqn. 10–3) approximately equals the original Fermi level (Eqn. 10–2); whereas, the quasi-Fermi level of holes (Eqn. 10–4) is lower than the original Fermi level (Eqn. 10–2) because the concentration of photoexcited holes, Δp^*, exceeds the concentration of holes, p, in the dark ($p \ll \Delta p^*$). In general, under the condition of photoexcitation, the quasi-Fermi level of the majority charge carriers remains close to the original Fermi level; but the quasi-Fermi level of the minority charge carriers shifts away from the original Fermi level.

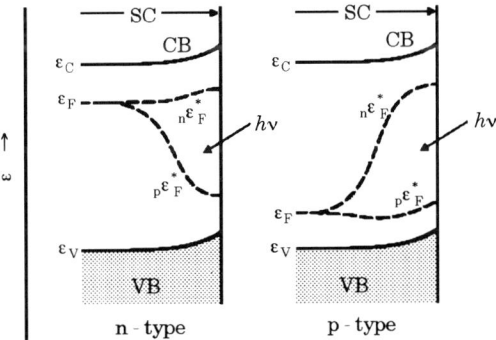

Fig. 10–2. Splitting of the Fermi level of electrodes, $\varepsilon_{F(SC)}$, into the quasi-Fermi levels of electrons, $_n\varepsilon_F^*$, and of holes, $_p\varepsilon_F^*$, respectively, in a surface layer of photoexcited n-type and p-type semiconductors: the shift of quasi-Fermi levels from the original Fermi level are greater for the minority charge carrier than for the majority charge carrier.

Since photoexcited electron–hole pairs are formed only within a limited depth from the semiconductor surface to which the irradiating photons can penetrate, the photon-induced split of the Fermi level into the quasi-Fermi levels of electrons and holes occurs only in a surface layer of semiconductors of limited depth as shown in Fig. 10–2.

10.1.2 Quasi-Fermi levels and electrode reactions

Charge transfer reactions on semiconductor electrodes proceed under the condition of anodic and cathodic polarization in which the Fermi level $\varepsilon_{F(SC)}$ is different either from the Fermi level $\varepsilon_{F(REDOX)}$ of redox electron transfer reactions or from the equivalent Fermi level $\varepsilon_{F(ION)}$ of ion transfer reactions. For redox electron transfer reactions, the thermodynamic requirement for the anodic and cathodic reactions to proceed is given by the following inequalities:

$$\varepsilon_{F(SC)} > \varepsilon_{F(REDOX)}, \quad \text{cathodic reaction proceeds}, \tag{10-5}$$

$$\varepsilon_{F(SC)} < \varepsilon_{F(REDOX)}, \quad \text{anodic reaction proceeds}. \tag{10-6}$$

Under the condition of photoexcitation, the quasi-Fermi level, instead of the original Fermi level, determines the direction of redox electron transfer reactions. The thermodynamic requirement for the transfer of cathodic electrons to proceed from the conduction band to the oxidant particles is then given by the inequality of Eqn. 10–7:

$$_n\varepsilon_F^* > \varepsilon_{F(REDOX)}. \tag{10-7}$$

Similarly, the thermodynamic requirement for the transfer of anodic holes to proceed from the valence band to the reductant particles is given by Eqn. 10–8:

$$_p\varepsilon_F^* < \varepsilon_{F(REDOX)}. \tag{10-8}$$

The cathodic current of the electron transfer is proportional to the concentration of interfacial electrons, n^*, and the anodic current of the hole transfer is proportional to the concentration of interfacial holes, p^*, in semiconductor electrodes as described in Sec. 8.3. Since the concentration of interfacial electrons or holes depends on the quasi-Fermi level of interfacial electrons or holes in the electrode, as shown in Eqn. 10–3 or 10–4, the transfer current of cathodic electrons or anodic holes under the condition of photoexcitation depends on the quasi-Fermi level of interfacial electrons, $_n\varepsilon_F^*$, or the quasi-Fermi level of interfacial holes, $_p\varepsilon_F^*$. It also follows from Sec. 8.3 that the anodic current of the electron transfer (the injection of electrons into the conduction band) or the cathodic current of the hole transfer (the injection of holes into the valence band) does not depend on the interfacial concentration of electrons or holes and hence appears independent of the quasi-Fermi level of interfacial electrons or holes.

Fig. 10–3 juxtaposes the Fermi levels of the following redox reactions in aqueous solutions and the quasi-Fermi levels of interfacial electrons and holes in the electrode: $\varepsilon_{F(H_2O/H_2)}$ of the hydrogen redox reaction; $\varepsilon_{F(O_2/H_2O)}$ of the oxygen redox reaction; $\varepsilon_{F(SC)}$ of the n-type semiconductor; $_n\varepsilon_F^*$ and $_p\varepsilon_F^*$ of the n-type semiconductor in the photoexcitation. In the dark, the Fermi level $\varepsilon_{F(SC)}$ of the semiconductor is located between $\varepsilon_{F(H_2O/H_2)}$ and $\varepsilon_{F(O_2/H_2O)}$ so that both the electron transfer of the cathodic hydrogen reaction and the hole transfer of the anodic oxygen reaction are thermodynamically impossible. In photoexcitation, the Fermi level of the semiconductor shifts by an energy equivalent to the photopotential (Refer to Sec. 10.2.) and also splits into the two quasi-Fermi levels, $_n\varepsilon_F^*$ and $_p\varepsilon_F^*$, for interfacial electrons and holes. Therefore, the transfer reactions of both cathodic electrons of the hydrogen reaction and anodic holes of the oxygen reaction become thermodynamically possible provided that the energy conditions, Eqns. 10–7 and 10–8, are satisfied as shown in Fig. 10–3 and expressed in the following:

$_n\varepsilon_F^* > \varepsilon_{F(O_2/H_2O)}$,   cathodic electron transfer reaction proceeds,

$_p\varepsilon_F^* < \varepsilon_{F(O_2/H_2O)}$,   anodic hole transfer reaction proceeds.

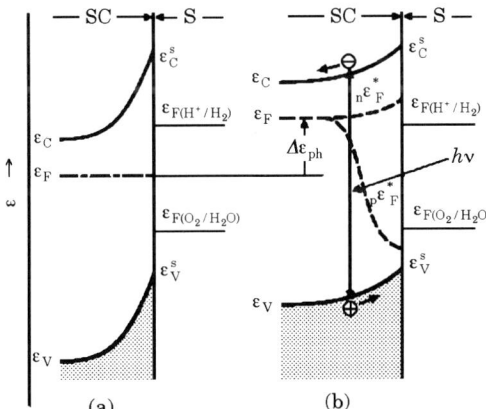

Fig. 10–3. Energy diagrams for an n-type semiconductor electrode (a) in the dark and (b) in the photoexcited state: S = aqueous solution; ε_C^s = conduction band edge level at the interface; ε_V^s = valence band edge level at the interface; $\varepsilon_{F(O_2/H_2O)}$ = Fermi level of the oxygen electrode reaction; $\varepsilon_{F(H^+/H_2)}$ = Fermi level of the hydrogen electrode reaction.

Thus, photoexcitation permits electron transfer reactions to occur which are thermodynamically impossible in the dark provided that the photon irradiation shifts the energy level of interfacial electrons (or holes) in the anodic (or cathodic) direction and produces the thermodynamic affinity for the reactions.

10.2 Photopotential

When a semiconductor electrode is at the flat band potential, photoexcited electrons and holes are soon annihilated by their recombination. However, in the presence of a space charge layer, the photoexcited electrons and holes are separated, with each moving in the opposite direction under the electric field in the space charge layer as shown in Fig. 10–4.

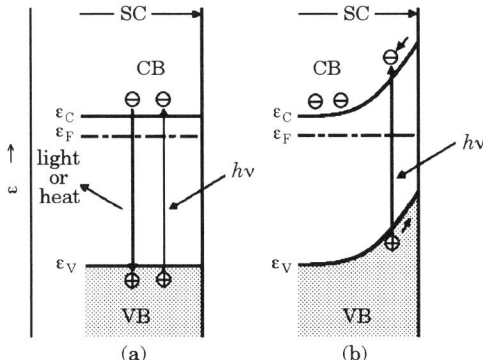

Fig. 10–4. Generation of electron–hole pairs by photoexcitation and their recombination or separation in semiconductors: (a) the generation and recombination of photoexcited electron–hole pairs in the flat band state, (b) the generation and separation of photoexcited electron–hole pairs in the space charge layer.

Such a migration of photoexcited electrons and holes induces in the electrode an inverse potential which reduces the difference of potential across the space charge layer and retards the migration of electrons and holes in the opposite direction as shown in Fig. 10–5. This inverse potential, induced by photoexcitation, is called the *photopotential*. Since the photopotential, $\Delta E_{ph} = -\Delta \varepsilon_{ph}/e$, arises in a direction to reduce the difference of potential across the space charge layer, the Fermi level of the semiconductor interior rises by an energy of $\Delta \varepsilon_{ph}$ (the electrode potential decreases by a magnitude of $\Delta E_{ph} = -\Delta\varepsilon_{ph}/e$) when the band edge level bends upward (See Fig. 10–5.); whereas, the Fermi level decreases by an energy of $\Delta\varepsilon_{ph}$ (the electrode potential rises by a magnitude of $\Delta E_{ph} = -\Delta\varepsilon_{ph}/e$) when the band edge level bends downward in the space charge layer.

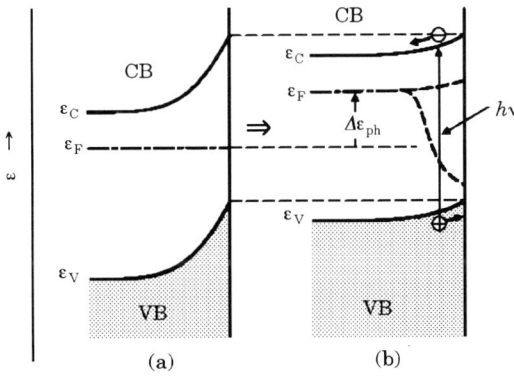

Fig. 10–5. Photopotential and band bending in the space charge layer of semiconductor electrodes (a) in the dark and (b) in the photoexcited state: $\Delta E_{ph} = -\Delta\varepsilon_{ph}/e$ = photopotential.

It follows that the greatest possible photopotential, ΔE_{ph}^0, of semiconductor electrodes is the difference between the dark electrode potential E and the flat band potential E_{fb} at which the band bending disappears as shown in Fig. 10–6 and expressed in Eqn. 10–9:

$$|\Delta E_{ph}| \le |\Delta E_{ph}^0| = |E - E_{fb}| \quad . \tag{10-9}$$

For semiconductor electrodes in which the concentration of impurities is relatively high (N_D, $N_A \ge 10^{17}$ cm^{-2}), the photopotential has been derived as a function of the concentration of electrons and holes to obtain Eqns. 10–10 and 10–11 [Myamlin-Pleskov, 1967]:

$$\Delta E_{ph} = -\frac{kT}{e}\ln\left(\frac{\Delta p^*}{p}\right), \quad \text{for n-type semiconductor electrodes,} \tag{10-10}$$

$$\Delta E_{ph} = -\frac{kT}{e}\ln\left(\frac{\Delta n^*}{n}\right), \quad \text{for p-type semiconductor electrodes.} \tag{10-11}$$

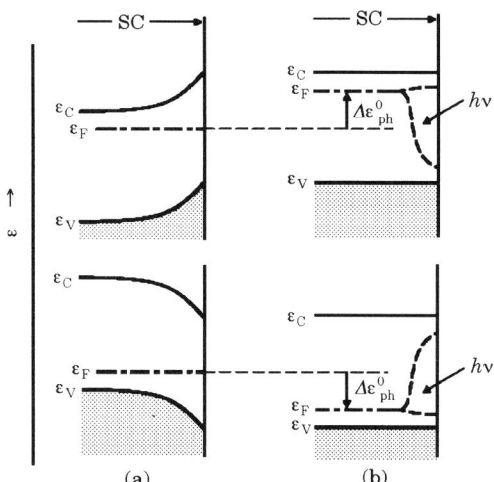

Fig. 10–6. Generation of the greatest possible photopotential in semiconductor electrodes: (a) in the dark, (b) in the photoexcited state. $\Delta E_{ph}^0 = -\Delta \varepsilon_{ph}^0/e$ = greatest possible photopotential.

These are approximate equations that are applicable to the case in which no reactions proceed at the electrode interface.

Fig. 10–7 illustrates the photopotential observed with n-type and p-type semiconductor electrodes of silicon in a sulfate solution. For n-type silicon electrodes, the photopotential is relatively high in the range of anodic potential (more positive potentials) where a depletion or inversion layer of space charge is formed; whereas, the photopotential is small in the cathodic range of potential (more negative potentials) where an accumulation layer of space charge is formed. Contrastively, for p-type silicon electrodes, the photopotential is high in the range of cathodic potentials where a depletion or inversion layer of space charge is formed; the photopotential is small in the range of anodic potentials where an accumulation layer of space charge is formed. These results agree with the photopotentials expected from Eqns. 10–10 and 10–11 and from Figs. 10–5 and 10–6.

The fact that the photopotential is smaller with the accumulation layer than with the depletion layer is due to the maximum potential barrier that is possible with the space charge layer which is smaller in the accumulation layer than in the depletion layer as expected from the energy difference between the Fermi level and the band edge levels; this energy difference is small in the accumulation layer but high in the depletion layer.

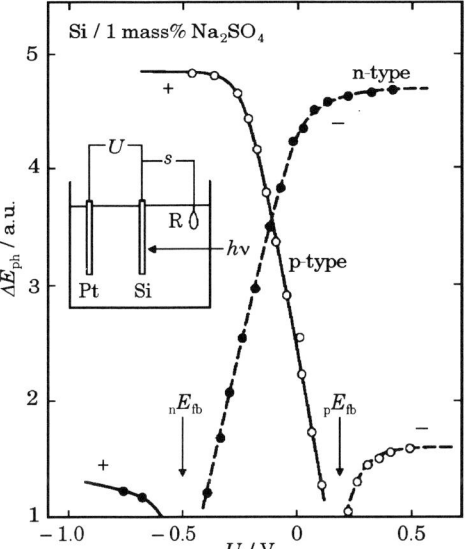

Fig. 10–7. Photopotential as a function of the electrode potential of n-type and p-type semiconductor silicon electrodes in an aqueous sulfate solution: U = cell voltage before photoexcitation; ΔE_{ph} = photopotential in arbitrary units; $_nE_{fb}$ = flat band potential of n-type electrodes; $_pE_{fb}$ = flat band potential of p-type electrodes; au = arbitrary unit; solid curve = positive photopotential; dotted curve = negative photopotential. [From Harten, 1960.]

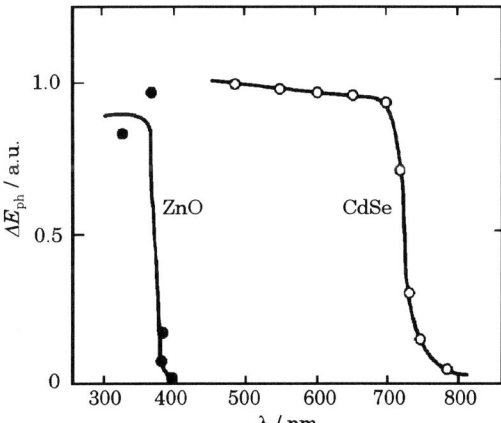

Fig. 10–8. Photopotential as a function of the wave length of photons for zinc oxide and cadmium selenide electrodes in aqueous solutions: λ = photon wave length. [From Williams, 1960.]

It is shown in Fig. 10–7 that the sign of the photopotential changes at a certain potential which corresponds to the flat band potential of electrodes (See Fig. 10–6.). This provides a way of estimating the flat band potential by measuring the potential at which the photopotential changes its sign.

The photon energy needed to produce excited electron–hole pairs must be higher than the band gap energy ε_g of semiconductors. Fig. 10–8 shows the photopotential observed in semiconductor electrodes of zinc oxide and cadmium selenide as a function of the wave length of irradiating photons; the threshold wave length corresponds to the band gap energy ε_g.

10.3 Photoexcited Electrode Reactions

10.3.1 Photoexcited electrode reaction current (Photocurrent)

Photoexcited electrons or holes in the space charge layer migrate towards the electrode interface where they participate in transfer reactions of cathodic electrons or anodic holes to provide the reaction current as shown in Fig. 10–9. Such a reaction current of photoexcited electrons or holes is called the *photoexcited reaction current* or simply the *photocurrent*.

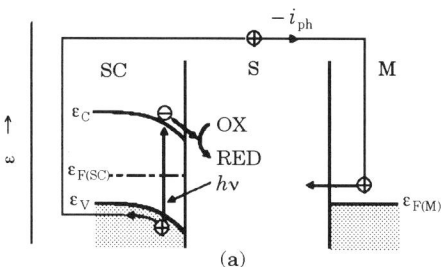

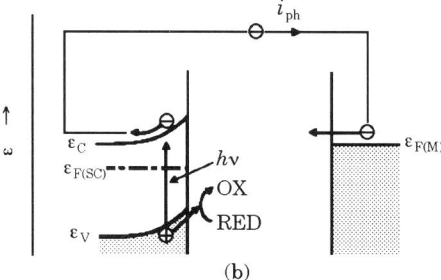

Fig. 10–9. Photoexcited reaction current (photocurrent) at semiconductor electrodes: (a) photoexcited reaction of cathodic electron transfer (OX + e → RED) at p-type semiconductor electrodes, (b) photoexcited reaction of anodic hole transfer (RED → OX + e) at n-type semiconductor electrodes. i_{ph} = photocurrent.

The photocurrent of electrons or holes is insignificant when the reacting particles are the majority charge carriers of which the concentration is hardly increased by photoexcitation; whereas, the same photocurrent is relatively great when the reacting particles are the minority charge carriers of which the concentration is increased greatly by photoexcitation compared with the concentration of the minority charge carriers in the dark. Consequently, the photocurrent can be observed *only* with the reaction in which the minority charge carriers participate; namely, the transfer reaction of anodic holes at n-type electrodes and the transfer reaction of cathodic electrons at p-type electrodes as schematically illustrated in Fig. 10–10.

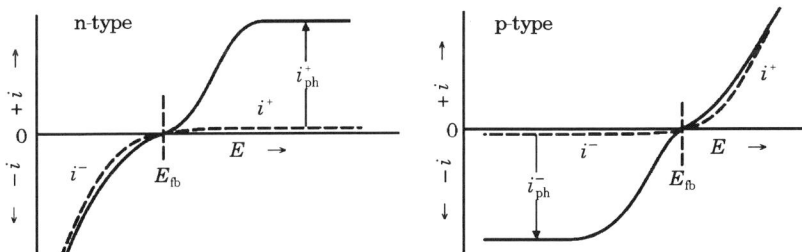

Fig. 10–10. Polarization curves for electrode reactions at n-type and p-type semiconductor electrodes in the dark and in the photoexcited state: dotted curve = dark; solid curve = photoexcited; i^+ (i^-) = anodic (cathodic) current in the dark; i_{ph}^+ (i_{ph}^-) = anodic (cathodic) current in the photoexcited state.

As shown in Fig. 10–9, the photoexcited reaction current occurs only when an appreciable electric field exists in the space charge layer. No photocurrent occurs at the flat band potential because the electric field that is required to separate the photoexcited electron–hole pairs is absent. The photocurrent occurs at any potentials different from the flat band potential; hence, the flat band potential may be regarded as the potential for the onset of the photocurrent. It follows, therefore, that photoexcited electrode reactions may occur at potentials at which the same electrode reactions are thermodynamically impossible in the dark.

The rate of formation of photoexcited electron–hole pairs, $G(x)$, is given as a function of the intensity of the photon beam I_0, the absorption coefficient of photons α, and the depth of photon-penetration x as shown in Eqn. 10–12 [Butler, 1977]:

$$G(x) = I_0 \exp(-\alpha x) \quad . \tag{10-12}$$

Integration of Eqn. 10–12 yields the total photocurrent i_{ph} in Eqn. 10–13 [Memming, 1983]:

$$\begin{aligned} i_{ph} &= e\, I_0 \left\{ 1 - \frac{\exp(-\alpha\, \delta_{SC})}{1 + \alpha L} \right\} \\ &= e\, I_0 \left[1 - \frac{\exp\left(-2\, \alpha\, L_{D,\,eff} \sqrt{\dfrac{e\, \Delta\phi_{SC}}{kT} - 1}\right)}{1 + \alpha L} \right], \end{aligned} \tag{10-13}$$

where δ_{SC} is the thickness of the space charge layer, $L_{D,\,eff}$ is the effective Debye length, $\Delta\phi_{SC}$ is the difference of potential across the space charge layer, and L is the diffusion distance of minority charge carriers.

It follows from Eqn. 10–13 that, if $\alpha\,\delta_{SC}$ is much larger than 1 ($\alpha\,\delta_{SC} \gg 1$, both α and δ_{SC} being great), all the photoexcited minority charge carriers would be consumed in the interfacial reaction ($i_{ph} = e\, I_0$). In such a case, the photocurrent is constant at potentials away from the flat band potential as shown in Fig. 10–11; this figure plots the anodic current of photoexcited dissolution for a gallium arsenide electrode as a function of the electrode potential.

If $\alpha\,\delta_{SC}$ is much smaller than one ($\alpha\,\delta_{SC} \ll 1$) and $\alpha\, L_{D,\,eff} \sqrt{(e\, \Delta\phi_{SC}/kT) - 1}$ is also much smaller than one, Eqn. 10–13 would be expressed by Eqn. 10–14:

$$i_{ph} = e\, I_0\, \alpha\, (L + \delta_{SC}) = e\, I_0\, \alpha \left[L + 2\, L_{D,\,eff} \sqrt{\frac{e\, \Delta\phi_{SC}}{kT} - 1} \right]. \tag{10-14}$$

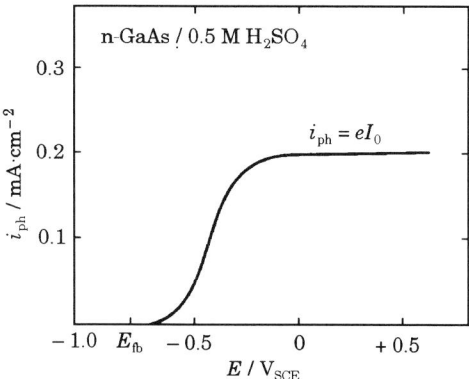

Fig. 10–11. Anodic photoexcited dissolution current of an n-type semiconductor electrode of gallium arsenide as a function of the electrode potential in a 0.5 M sulfuric acid solution: I_0 = photon intensity; i_{ph} = photocurrent. [From Memming-Kelly, 1981.]

Eqn. 10–14 indicates that if the diffusion length L of the minority charge carrier is much longer than the thickness δ_{SC} of the space charge layer ($L \gg \delta_{SC}$), the photocurrent would be constant and independent of the electrode potential.

On the other hand, if the diffusion length of the minority charge carrier is much shorter than the thickness of the space charge layer ($L \ll \delta_{SC}$), Eqn. 10–13 would yield Eqn. 10–15:

$$(i_{ph})^2 = (2 e I_0 \alpha L_{D, eff})^2 \left(\frac{e \Delta\phi_{SC}}{kT} - 1 \right) \doteq (2 e I_0 \alpha L_{D, eff})^2 \frac{e}{kT}(E - E_{fb}) \ , \quad (10\text{–}15)$$

where E is the electrode potential, E_{fb} is the flat band potential, and $\Delta\phi_{SC} = E - E_{fb}$. Eqn. 10–15 indicates that the square of the photocurrent depends linearly on the electrode potential. Fig. 10–12 illustrates this dependence for the photoexcited current due to the anodic oxidation of water molecules to produce oxygen gas on an n-type tungsten oxide (WO_3) electrode in which $L \ll \delta_{SC}$.

The point at which the straight line of $(i_{ph})^2$ versus E intersects the coordinate of potential represents the flat band potential. Eqn. 10–15 holds when the reaction rate at the electrode interface is much greater than the rate of formation of photoexcited electron–hole pairs; here, the interfacial reaction is in the state of quasi-equilibrium and the interfacial overvoltage η_H is close to zero.

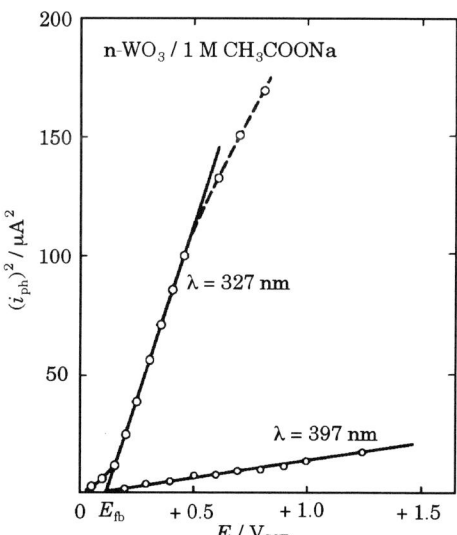

Fig. 10–12. Anodic photoexcited current of the oxidation of water molecules to produce oxygen gas as a function of the electrode potential of an n-type tungsten oxide electrode in an acidic solution: λ = wave length of photons. [From Butler, 1977.]

10.3.2 The range of electrode potential for photoelectrode reactions

Electrochemical thermodynamics predicts that the anodic reaction may proceed only at electrode potentials more positive than the equilibrium potential of the reaction; and the cathodic reaction may proceed only at electrode potentials more negative than the equilibrium potential. In other words, relative to the Fermi level of the reaction, the Fermi level of the electrode must be higher for the cathodic reaction to proceed and lower for the anodic reaction to proceed.

For metal electrodes, the concentrations of both electrons and holes in the electrode are sufficiently high at the Fermi level of the electrode that the energy of electrons and holes, which participate in the electrode reaction, may be represented by the Fermi level of the electrode; namely by the electron level corresponding to the electrode potential. Contrastively, for semiconductor electrodes, the electrons and holes that participate in the electrode reaction are not at the Fermi level of the electrode, but at the levels of the conduction and valence bands which are different from the Fermi level of the electrode, i. e. the electron level corresponding to the electrode potential. For example, as shown in Fig. 10–13, the anodic oxygen reaction at n-type semiconductor electrodes proceeds with interfacial holes in the valence band of which the energy $-\varepsilon_V^s$ ($-\varepsilon_V^s = e\, E_V^s$) is higher than the hole energy $-\varepsilon_F$ ($-\varepsilon_F = e\, E$) at the Fermi level of the electrode: note that the sign of the hole energy level is opposite to the electron energy level.

Since the Fermi level, ε_F, of n-type semiconductor electrodes is close to the conduction band edge ε_C^s, the energy level $-\varepsilon_V^s$ of interfacial holes in the valence band is higher than the hole energy $-\varepsilon_F$ at the Fermi level of the electrode by an energy nearly equal to the band gap ε_g. Therefore, the energy $-\varepsilon_V^s$ of interfacial holes that participate in the anodic oxygen reaction is higher than the hole energy $-\varepsilon_{F(SC)}$ at the Fermi level of the electrode: $-\varepsilon_V^s \doteqdot -\varepsilon_{F(SC)} + \varepsilon_g$. The energy of interfacial holes participating in the anodic reaction is at the potential E_V^s of the valence band edge which is more anodic than the observed electrode potential E.

For n-type semiconductor electrodes, the anodic oxygen reaction (anodic hole transfer) will not occur in the dark because the concentration of interfacial holes in the valence band is extremely small; whereas, the same reaction will occur in the photon irradiation because the concentration of interfacial holes in the valence band is increased by photoexcitation and the quasi-Fermi level $_p\varepsilon_F^*$ of interfacial holes becomes lower than the Fermi level $\varepsilon_{F(O_2/H_2O)}$ of the oxygen redox reaction.

For illustrations, we compare the transfer of anodic holes at metal electrodes and the transfer of anodic photoexcited holes at n-type semiconductor electrodes for the oxygen redox reaction shown in Eqn. 10–16:

$$2\,H_2O_{aq} + 4\,h \rightarrow O_{2,\,aq} + 4\,H_{aq}^+ \quad , \tag{10–16}$$

where h is the hole in the valence band of the electrode.

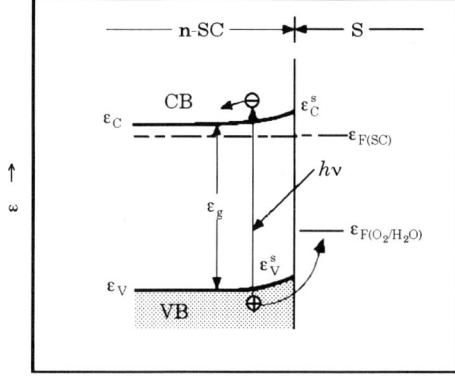

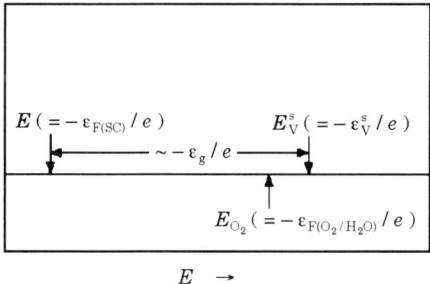

Fig. 10–13. Anodic transfer of photoexcited holes (minority charge carrier) at n-type semiconductor electrodes: $E\,(=-\varepsilon_{F(SC)}/e)$ = electrode potential; $E_V^s\,(=-\varepsilon_V^s/e)$ = potential of the valence band edge; $E_{O_2}\,(=-\varepsilon_{F(O_2/H_2O)})$ = equilibrium electrode potential of the oxygen electrode reaction.

For metal electrodes, the anodic oxygen reaction proceeds at electrode potentials which are more anodic than the equilibrium potential E_{O_2} of the reaction as shown in Fig. 10–14. For n-type semiconductor electrodes, the anodic photoexcited oxygen reaction proceeds at electrode potentials where the potential E_V^s of the valence band edge (precisely, the potential $_pE_F^*$ of the quasi-Fermi level of interfacial holes, $_p\varepsilon_F^* = -e\,_pE_F^*$) is more anodic than the equilibrium oxygen potential E_{O_2}, *even if* the observed electrode potential E is less anodic than the equilibrium oxygen potential E_{O_2}. Accordingly, the anodic hole transfer of the oxygen reaction at photoexcited n-type semiconductor electrodes occurs *even* in the range of potential which is *less anodic* than the equilibrium potential E_{O_2} of the reaction as shown in Fig. 10–14.

The potential $E_{O_2(ph)}$ at which the anodic photoexcited oxygen reaction occurs is lower by an amount ΔE_{ph} than the equilibrium potential E_{O_2} which is the potential for the onset of the anodic oxygen reaction without photoexcitation. This potential difference ΔE_{ph} is nearly equivalent to the band gap ε_g as indicated in Eqn. 10–17:

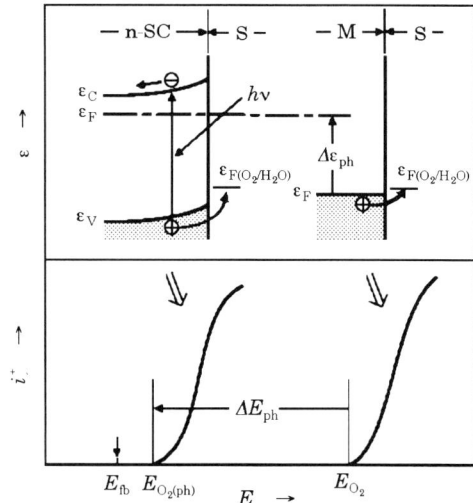

Fig. 10–14. Energy levels and polarization curves (current vs. potential) for the anodic transfer of photoexcited holes in the oxygen reaction ($2\,H_2O_{aq} + 4\,h \rightarrow O_2 + 4\,H^+_{aq}$) on a metal electrode and on an n-type semiconductor electrode: i^+ = anodic reaction current; $\varepsilon_{F(O_2/H_2O)}$ = Fermi level of the oxygen electrode reaction; $\Delta\varepsilon_{ph}$ = gain of photoenergy; $E_{O_2(ph)}$ = potential for the onset of the anodic photoexcited oxygen reaction; $\Delta E_{ph}\,(=-\Delta\varepsilon_{ph}/e)$ = shift of the potential fir the onset of the anodic oxygen reaction from the equilibrium oxygen potential in the negative direction due to the gain of photoenergy in the n-type electrode; E_{fb} = flat band potential of the n-type electrode.

$$\Delta E_{ph} = E_{O_2(ph)} - E_{O_2} \doteq -\frac{\Delta\varepsilon_{ph}}{e} \leq -\frac{\varepsilon_g}{e}\,, \tag{10-17}$$

where $\Delta\varepsilon_{ph}\,(=\varepsilon_{F(SC)} - \varepsilon_{F(M)})$ is the difference in the Fermi level (electrode potential) between the metal electrode ($\approx$ the equilibrium oxygen potential) and the photoexcited n-type semiconductor electrode ($\approx$ the flat band potential). Since upward band bending is required for the anodic photoexcited hole reaction to occur at n-type semiconductor electrodes, the lowest possible potential for the onset of the anodic photoexcited reaction is assigned at the flat band potential E_{fb}. Therefore, Eqn. 10–18 is obtained:

$$|\Delta E_{ph}| < |E_{fb} - E_{O_2}|\,. \tag{10-18}$$

It appears from Eqns. 10–17 and 10–18 that the potential for the *onset* of the anodic photoexcited hole transfer in the oxygen reaction is shifted from the equilibrium electrode potential of the oxygen reaction toward the less anodic direction (cathodic direction) by a magnitude nearly equivalent to the absorbed photon energy that is the energy of the band gap. This shift of the potential for the onset of the reactions results from the energy that is absorbed in semiconductor electrodes by photoexcitation to increase the concentration of interfacial holes in the valence band; this phenomenon does *not contradict* the thermodynamic prediction.

Similarly, the cathodic photoexcited electron transfer of the hydrogen reaction shown in Eqn. 10–19 can occur at p-type semiconductor electrodes at which the cathodic hydrogen reaction is thermodynamically impossible in the dark.

$$2\,H^+_{aq} + 2\,e \rightarrow H_2 \quad . \tag{10-19}$$

For metal electrodes this reaction can proceed only at potentials more cathodic than the equilibrium hydrogen electrode potential E_{H_2}; but for photoexcited p-type semiconductor electrodes the reaction can proceed *even* at potentials *less cathodic* (more anodic) than E_{H_2} as shown in Fig. 10–15. The potential for the onset of the

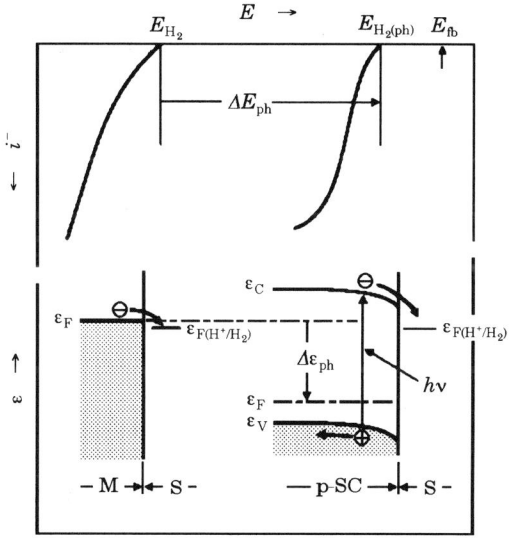

Fig. 10–15. Energy levels and polarization curves of the cathodic hydrogen reaction at a metal electrode and at a photoexcited p-type semiconductor electrode: i^- = cathodic current; $\varepsilon_{F(H^+/H_2)}$ = Fermi level of the hydrogen reaction; $\Delta\varepsilon_{ph}$ = gain of photoenergy; $E_{H_2(ph)}$ = onset potential for the cathodic photoexcited hydrogen reaction at a p-type electrode; ΔE_{ph} = shift of the onset potential for the cathodic hydrogen reaction from the equilibrium potential due to the gain of photoenergy in a p-type electrode; E_{fb} = flat band potential.

cathodic electron transfer of the hydrogen reaction is shifted toward the less cathodic direction (anodic direction) from the equilibrium hydrogen electrode potential E_{H_2} to the potential $E_{H_2(ph)}$ for the onset of the hydrogen reaction at photoexcited p-type semiconductor electrodes; this shift of the potential for the onset of the hydrogen electrode reaction ΔE_{ph} (= $E_{H_2(ph)} - E_{H_2}$) is nearly equivalent to the absorbed photon energy, i.e. the band gap energy.

The potential, E_{ph}, for the onset of the photoexcited reaction relative to the equilibrium electrode potential E_{eq} of the same reaction can also be derived from the reaction kinetics [Memming, 1987]. Here, we consider the transfer of anodic holes (minority charge carriers) at an n-type semiconductor electrode at which the hole transfer at the interface is in the quasi-equilibrium; then, the anodic reaction rate is controlled by the photogeneration and transport of holes in the n-type semiconductor electrode. The current of hole transport, i_{tp}, has been given by Eqn. 8–71 as a function of polarization ($E - E_{eq}$) as shown in Eqn. 10–20:

$$i_{tp} = i_{p,lim} \left\{ 1 - \exp\left(\frac{-e(E - E_{eq})}{\xi k T} \right) \right\} , \qquad (10\text{–}20)$$

where $i_{p,lim}$ is the limiting diffusion current of holes and ξ (=1 or 2) is a constant.

In the photostationary state, the current of hole transfer, i, across the electrode interface is the sum of the current of hole transport, i_{tp}, and the current of photoexcited hole generation, i_{ph}, as shown in Eqn. 10–21:

$$i = i_{tp} + i_{ph} . \qquad (10\text{–}21)$$

Therefore, the potential for the onset of the cathodic redox reaction corresponds to the potential $E_{ph,i=0}$ at which the overall cathodic reaction current is zero ($i = 0$). Then, Eqns. 10–20 and 10–21 yield Eqn. 10–22:

$$E_{ph,i=0} - E_{eq} = -\frac{\xi k T}{e} \ln\left(\frac{i_{ph}}{i_{p,lim}} + 1 \right) . \qquad (10\text{–}22)$$

The left hand side of Eqn. 10–22 represents the shift of the potential for the onset of the reaction, ΔE_{ph}, due to the photoexcitation ($\Delta E_{ph} = E_{ph,i=0} - E_{eq}$). If the diffusion distance of minority charge carriers is sufficiently long, the photoexcited reaction current i_{ph} would not depend on the electrode potential as shown in Eqn. 10–14. Consequently, the potential for the onset of the photoexcited reaction $E_{ph,i=0}$ shifts in the cathodic (negative) direction toward the flat band potential as the photon intensity increases. Usually, therefore, the potential for the onset of photoexcited reactions at high intensities of photon irradiation is close to the *flat band* potential. Note that few cases occur in which the flat band potential itself changes with the photon irradiation as is described in Sec. 10.3.3.

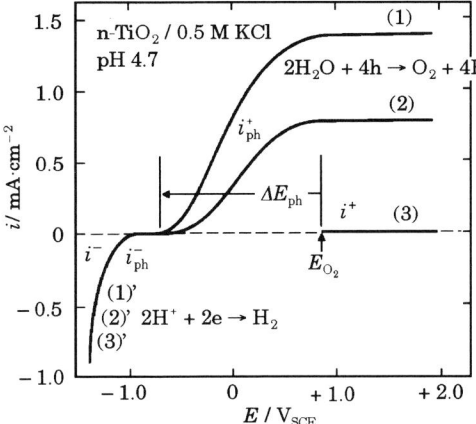

Fig. 10-16. Polarization curves for the anodic oxygen and cathodic hydrogen redox reactions on an n-type semiconductor electrode of titanium oxide in the dark and in the photoexcited state: i^+ = anodic current in the dark (zero); i_{ph}^+ = anodic current in the photoexcited state; $i_{ph}^+ \doteqdot i_{ph}^-$ = cathodic current; (1)/(2)/(3) = anodic polarization curves at the relative photon intensity ratio 1.00/0.48/0.00. [From Fujishima-Honda-Kikuchi, 1969.]

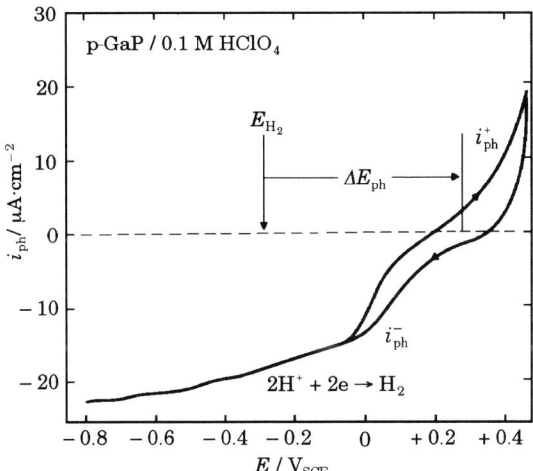

Fig. 10-17. Polarization curves for the cathodic hydrogen redox reaction on a photoexcited p-type semiconductor electrode of gallium phosphide: E_{H_2} = equilibrium potential of the hydrogen reaction. [From David-Butler, 1988.]

Fig. 10-16 shows polarization curves for the anodic oxygen reaction (anodic hole transfer) and for the cathodic hydrogen reaction (cathodic electron transfer) on an n-type semiconductor electrode of titanium oxide. The data in Fig. 10-16 shows that the anodic current due to transfer of holes (minority charge carriers) in the oxygen reaction is insignificant in the dark but increases with increasing photon intensity at potentials more anodic than the potential for the onset of the anodic photoexcited oxygen reaction but less anodic than the equilibrium electrode potential of the reaction (See Fig. 10–14.). Contrastively, the cathodic hydrogen reaction due to transfer of electrons (majority charge carriers) proceeds following almost the same polarization curve in the dark and in the photon irradiation.

Fig. 10–17 shows the polarization curves for the cathodic hydrogen reaction (cathodic electron transfer) on a p-type semiconductor electrode of gallium phosphide. The potential for the onset of the cathodic photoexcited hydrogen reaction shifts significantly from the equilibrium electrode potential of the hydrogen reaction toward the flat band potential of the p-type electrode (See Fig. 10–15.).

10.3.3 The flat band potential of photoexcited electrodes

The flat band potential of semiconductor electrodes is determined by the potential difference across the compact layer at the electrode interface and is characteristic of individual semiconductor electrodes. For semiconductor electrodes in the state of band edge level pinning, the potential difference across the compact layer remains constant and independent of the electrode potential. However, for some semiconductor electrodes, photon irradiation changes the potential difference across the compact layer on the electrode and, hence, shifts the flat band potential of the electrode.

An example of the effect of photon irradiation on the flat band potential is shown in Fig. 10–18; this figure compares a Mott-Schottky plot with the anodic polarization curves of the dissolution reactions of a semiconductor anode of n-type molybdenum selenide in an acidic solution in the dark and in the photoexcited conditions. In this example photoexcitation shifts the flat band potential from E_{fb} in the dark to $E_{fb(ph)}$ in the photoexcited state; $E_{fb(ph)}$ is about 0.75 V more positive than E_{fb}. This *photo-shift* of the flat band potential, $E_{fb(ph)} - E_{fb}$, corresponds to the change in the potential, $\delta\Delta\phi_H$, of the compact layer due to the photoexcitation as defined in Eqn. 10–23:

$$E_{fb(ph)} - E_{fb} = \delta\Delta\phi_H . \tag{10-23}$$

As shown in Fig. 10–18, the flat band potential that characterizes the potential for the onset of the photocurrent shifts from the dark flat band potential E_{fb} to the photoexcited flat band potential $E_{fb(ph)}$ as photoexcitation continues

A shift of the flat band potential due to the photoexcitation of the type shown in Fig. 10–18 can be produced by the capture of holes in the surface state level, ε_{ss}, on the electrode as shown in Fig. 10–19. We now consider a dissolution reaction involving the transfer of anodic ions of a simple substance semiconductor electrode according to Eqns. 10–24 and 10–25:

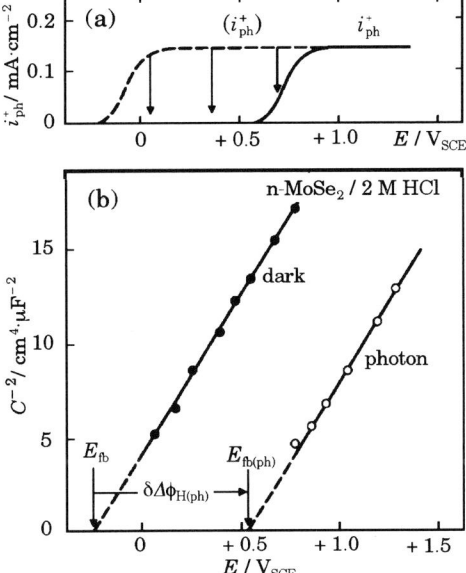

Fig. 10–18. (a) Polarization curves of anodic dissolution and (b) Mott-Schottky plots of an n-type semiconductor electrode of molybdenum selenide in the dark and in the photoexcited state in an acidic solution: C = electrode capacity; (i_{ph}^+) = anodic dissolution current immediately after photoexcitation (dotted curve); i_{ph}^+ = anodic dissolution current in the photostationary state (solid curve); $E_{fb(ph)}$ = flat band potential in the photostationary state. [From McEvoy-Etman-Memming, 1985.]

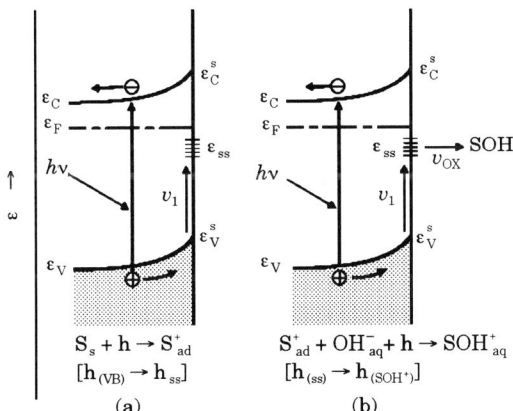

Fig. 10–19. (a) Capture of photogenerated holes in the surface state to form surface ions and (b) anodic dissolution of surface ions to form hydrated ions on an n-type semiconductor electrode: v_1 = rate of hole capture in the surface state; v_{OX} = rate of anodic dissolution of surface ions; ε_{ss} = surface state level; S_s = surface atom of the semiconductor electrode; $h_{(VB)}$ = hole in the valence band; $h_{(ss)}$ = hole captured in the surface state; $h_{(SOH^+)}$ = hole in dissolved semiconductor ions.

$$S + h(h\nu) \rightarrow S^+_{ad} \, , \tag{10-24}$$

$$S^+_{ad} + OH^-_{aq} + h(h\nu) \rightarrow SOH^+_{ad} \, , \tag{10-25}$$

where $h(h\nu)$ is the photoexcited hole, and S^+_{ad} is the surface ion with the surface state level, ε_{ss}. If the rate of capture of holes, v_1, in the surface state is greater than the rate of dissolution of surface ions, v_{ox}, the concentration of holes in the surface state would increase and the positive surface charge would increase; hence, the potential of the compact layer would change and the flat band potential would shift. Conversely, if the rate of dissolution, v_{ox}, is greater than the rate of capture of holes, v_1, holes would not accumulate in the surface state and the flat band potential would not shift. The former is illustrated by an n-type semiconductor electrode of molybdenum selenide in acidic solutions (See Fig. 10–18.), in which v_1 is greater than v_{ox}; the latter is illustrated by the same n-type semiconductor electrode of molybdenum selenide, but in basic solutions, in which v_{ox} is greater than v_1 [Sinn-Meissnere-Memming, 1990].

Similarly, the flat band potential also would shift at photoexcited n-type semiconductor electrodes on which a transfer reaction involving anodic redox holes occurs via the surface state level ε_{ss}, if the rate of hole capture at the surface state is greater than the rate of hole transfer across the compact layer as shown in Fig. 10–20(a).

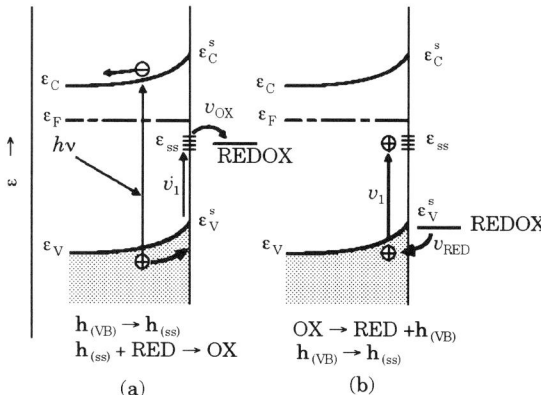

Fig. 10–20. Capture of photogenerated holes and cathodically injected holes in the surface state on n-type semiconductor electrodes: (a) the surface state captures the photogenerated holes at the rate v_1 followed by the anodic hole transfer to redox particles, (b) the surface state captures the cathodically injected holes.

Furthermore, the flat band potential can shift even in the dark: for example, if cathodic holes are injected into n-type semiconductor electrodes from oxidant particles, the injected holes would be captured in the surface state thereby increasing the interfacial positive charge and shifting the flat band potential as shown in Fig. 10–20(b). For instance, the injection of cathodic holes from hydrated cerous ions of the (Ce^{4+}/Ce^{3+}) redox particles shifts the flat band potential of an n-type semiconductor electrode of molybdenum selenide in the anodic (positive) direction; and the magnitude of the shift of the flat band potential due to the redox hole injection is the same as that caused by the formation of photoexcited holes [McEvoy-Etman-Memming, 1985]. Therefore, it follows that if the concentration of holes in the surface state exceeds the equilibrium concentration in the dark, the excess positively charged holes would cause a change of the potential difference in the compact layer and a shift of the flat band potential, regardless whether the excess holes are photogenerated or cathodically injected.

10.4 The Rate of Photoelectrode Reactions

10.4.1 Anodic transfer reactions of photoexcited holes

We consider a redox reaction involving the transfer of anodic holes at semiconductor electrodes as shown in Eqn. 10–26:

$$RED + h \to OX \ . \tag{10-26}$$

Here, the reaction rate is proportional to the concentration of interfacial holes in the electrode as described in Sec. 8.3. Since the concentration of interfacial holes depends on the Fermi level $\varepsilon_{F,s}$ of the electrode interface, the reaction current due to anodic holes depends on the Fermi level $\varepsilon_{F,s}$ of the interface in the dark and on the quasi-Fermi level of interfacial holes $_p\varepsilon^*_{F,s}$ in the photoexcited state.

This conclusion is valid regardless whether the electrode is n-type or p-type. Consequently, if the quasi-Fermi level of interfacial holes $_p\varepsilon^*_{F,s}$ in a photoexcited n-type semiconductor electrode equals the quasi-Fermi level of interfacial holes $_p\varepsilon^*_{F,s}$ (equal to the Fermi level $_p\varepsilon_{F,s}$ of the interface) in a p-type electrode of the same semiconductor in the dark, the current due to anodic holes would be the same on the two electrodes and, hence, the curves of the anodic reaction current as a function of the quasi-Fermi level of interfacial holes would be the same for the two electrodes as suggested in Fig. 10–21. However, the curves of the anodic reaction current represented as a function of the electrode potential (the Fermi level of the electrode), instead of the quasi-Fermi level of interfacial holes, are not the same for the two electrodes.

The quasi-Fermi level $_p\varepsilon_{F,s}^*$ of interfacial holes nearly equals the Fermi level $_p\varepsilon_{F,s}$ ($\fallingdotseq\ _p\varepsilon_{F,s}^*$) in both dark and photoexcited p-type electrodes, but the quasi-Fermi level $_p\varepsilon_{F,s}^*$ of interfacial holes is lower than the Fermi level $_n\varepsilon_{F,s}$ ($>\ _p\varepsilon_{F,s}^*$) of the photoexcited n-type electrode as shown in Fig. 10–21. Eventually, it follows that the range of electrode potential, where the anodic reaction occurs on the photoexcited n-type electrode, shifts from the range of potential for the same anodic reaction on the dark p-type electrode toward the cathodic (more negative) direction by an energy equivalent to $(_n\varepsilon_{F,s} -\ _p\varepsilon_{F,s}^*)$.

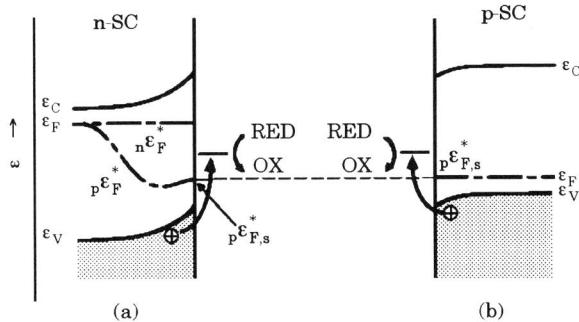

Fig. 10–21. Quasi-Fermi levels of holes in the transfer reaction of anodic holes from the valence band (a) of a photoexcited n-type electrode and (b) of a dark p-type electrode of the same semiconductor to redox particles: $_p\varepsilon_{F,s}^*$ = quasi-Fermi level of interfacial holes in the photoexcited n-type electrode where $_p\varepsilon_{F,s}^*$ is lower than the Fermi level ε_F and in the dark n-type electrode where $_p\varepsilon_{F,s}^*$ equals the Fermi level ε_F.

The transfer of anodic holes is associated with the three processes: the generation and transport of holes in the electrode; the hole transfer across the compact layer; and the diffusion of redox particles in aqueous solution. The total overvoltage, η, is the sum of the three overvoltages; $\eta_{p,SC}$ for the generation and transport of holes in the electrode, η_H for the transfer of holes across the electrode interface, and η_{diff} for the diffusion of redox particles in the solution as defined in Eqn. 10–27:

$$\eta = \eta_{p,SC} + \eta_H + \eta_{diff} . \tag{10–27}$$

Further, the total overvoltage, η, is the difference of potential between the polarization potential $E\,(=-\varepsilon_{F(SC)}/e)$ and the equilibrium redox potential

E_{eq} (= $-\varepsilon_{F(REDOX)}/e$) as defined in Eqn. 10–28:

$$\eta = -\frac{\varepsilon_{F(SC)} - \varepsilon_{F(REDOX)}}{e} \ . \tag{10–28}$$

10.4.2 Generation and transport of holes

The overvoltage for the *generation and transport of holes*, $\eta_{p,\,SC}$, is the difference between the quasi-Fermi level $_p\varepsilon_{F,s}^*$ of interfacial holes and the Fermi level $\varepsilon_{F(SC)}$ of electrons in the electrode interior as defined in Eqn. 10–29:

$$\eta_{p,\,SC} = -\frac{\varepsilon_{F(SC)} - {_p\varepsilon_{F,s}^*}}{e} \ . \tag{10–29}$$

For p-type electrodes in the dark and in the photoexcited state, the concentration of majority charge carriers (holes) is sufficiently great that the Fermi level $\varepsilon_{F(SC)}$ of the electrode interior nearly equals the quasi-Fermi level $_p\varepsilon_{F,s}^*$ of interfacial holes; hence, the overvoltage $\eta_{p,\,SC}$ for the generation and transport of holes in the space charge layer is zero even as the transfer of anodic holes progresses as expressed in Eqn. 10–30:

$$\eta_{p,\,SC} \doteq 0 \ ; \qquad \text{p-type electrodes in both dark and photoexcited states.} \tag{10–30}$$

For n-type electrodes in the dark, the transfer of anodic holes reduces the concentration of interfacial holes (minority carriers) so that the quasi-Fermi level $_p\varepsilon_{F,s}^*$ of interfacial holes increases beyond the Fermi level $\varepsilon_{F(SC)}$ ($_p\varepsilon_{F,s}^* > \varepsilon_{F(SC)}$) of the electrode interior; hence, a positive overvoltage $\eta_{p,\,SC}$ emerges due to the diffusion of holes in the electrode as shown in Eqn. 10–31:

$$\eta_{p,\,SC} > 0 \ ; \qquad \text{n-type electrodes in the dark state.} \tag{10–31}$$

On the other hand, for photoexcited n-type electrodes, the concentration of interfacial holes becomes greater than the concentration of holes in the electrode interior so that the quasi-Fermi level $_p\varepsilon_{F,s}^*$ of interfacial holes decreases below the Fermi level $\varepsilon_{F(SC)}$ ($_p\varepsilon_{F,s}^* \ll \varepsilon_{F(SC)}$) of the electrode interior; hence, a *negative overvoltage* arises which is caused by the back diffusion of photogenerated holes from the electrode interface toward the electrode interior. This negative overvoltage involving reactions of photoexcited anodic holes is not in the anodic direction but in the cathodic direction corresponding to the cathodic back diffusion of photoexcited holes in the electrode.

This negative overvoltage in the anodic hole transfer reaction may better be called the *undervoltage* or the "*inverse overvoltage*" (the negative overvoltage in the anodic reaction) rather than the usual overvoltage which is positive in the anodic reaction. Eqn. 10–32 expresses that the overvoltage for the generation and transport of holes associated with anodic hole transfer reactions is negative in photoexcited n-type semiconductor electrodes:

$$\eta_{p,SC} < 0 \ ; \qquad \text{n-type electrodes in the photoexcited state.} \qquad (10\text{–}32)$$

The energy equivalent to the inverse overvoltage corresponds to the *gain of energy* due to the absorption of photons in n-type semiconductor electrodes.

10.4.3 Interfacial overvoltage of hole transfer

We discuss the interfacial overvoltage, η_H, for the transfer of holes across the interface of semiconductor electrodes. The interfacial overvoltage denoted by η_H in this section includes not only the overvoltage due to the compact layer (Refer to Sec. 8.3.), which is zero in the state of band edge level pinning, but also the overvoltage due to the quasi-Fermi level that determines the concentration of interfacial holes. The interfacial overvoltage η_H is defined by the difference between the quasi-Fermi level $_p\varepsilon^*_{F,s}$ of interfacial holes and the Fermi level $\varepsilon_{F(REDOX)}$ of the redox reaction as in Eqn. 10–33:

$$\eta_H = -\frac{_p\varepsilon^*_{F,s} - \varepsilon_{F(REDOX)}}{e} \ , \qquad (10\text{–}33)$$

where the diffusion overvoltage η_{diff} in aqueous solution is disregarded. This interfacial overvoltage is of course the positive overvoltage in the anodic direction. For electrodes in the dark the quasi-Fermi level $_p\varepsilon^*_{F,s}$ in Eqn. 10–33 is replaced for the Fermi level ε^s_F of interfacial electrons in the electrode; $_p\varepsilon^*_{F,s}$ equals ε^s_F in the dark.

Fig. 10–22 shows the energy level diagrams for redox reactions involving transfer of anodic redox holes at photoexcited n-type and dark p-type electrodes of the same semiconductor. For p-type electrodes, of which the majority charge carriers are holes, the overvoltage for the generation and transport of holes is zero ($\eta_{p,SC} \doteq 0$); and the total overvoltage is composed of only the interfacial overvoltage ($\eta \doteq \eta_H$) in both the dark and the photoexcited states. For n-type electrodes in the photoexcited state, however, the total overvoltage includes the overvoltage $\eta_{p,SC}$ for the generation and transport of holes and the interfacial overvoltage η_H as defined in Eqn. 10–34:

$$\eta = \eta_{p,SC} + \eta_H \ , \qquad \text{n-type in the photoexcited state.} \qquad (10\text{–}34)$$

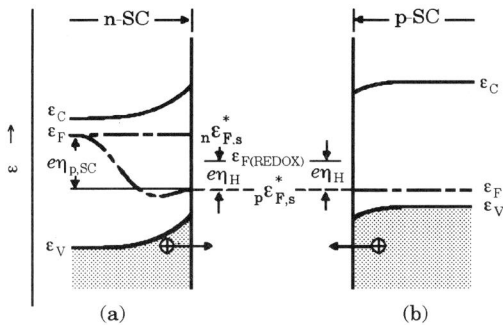

Fig. 10–22. Overvoltages in an anodic hole transfer (a) at a photoexcited n-type electrode and (b) at a p-type electrode of the same semiconductor: η_H = overvoltage for the hole transfer across the interface; $\eta_{p,SC}$ = inverse overvoltage due to the generation and transport of photoexcited holes in the n-type electrode.

Since the overvoltage $\eta_{p,SC}$ for the generation and transport of holes is negative, the total overvoltage becomes negative when the magnitude of $\eta_{p,SC}$ exceeds η_H; the condition usually occurs with photoexcited n-type electrodes. This is the basis for the potential for the onset of anodic hole transfer at photoexcited n-type electrodes being more cathodic (negative) than the potential for the onset of anodic hole transfer at p-type electrodes of the same semiconductor or at metal electrodes.

In general, the reaction current for hole transfer i across the electrode interface is given by Eqn. 10–35:

$$i = e\,(k^+ c_{RED} - k^- c_{OX})\,, \tag{10-35}$$

where c_{RED} and c_{OX} are the concentrations of reductants and oxidants in aqueous solution, k^+ and k^- are the rate constants of the anodic and cathodic reactions, respectively. The rate constants of k^+ and k^- depend on η_H as shown in Eqns. 10–36 and 10–37:

$$k^+ = k_0 \exp\left(\frac{e\,\alpha^+ \eta_H^0}{kT}\right)\,, \tag{10-36}$$

$$k^- = k_0 \exp\left(\frac{-e\,\alpha^- \eta_H^0}{kT}\right)\,, \tag{10-37}$$

where η_H^0 is the interfacial overvoltage referred to the standard redox potential

E^0_{REDOX} ($= -\varepsilon^0_{\text{REDOX}}/e$), k_0 is the exchange rate of reaction at the standard redox potential, and α is the Tafel constant (the transfer coefficient). Based on the discussion in Sec. 8.3, α^+ is one ($\alpha^+ = 1$) and α^- is zero ($\alpha^- = 0$) for hole transfer reactions. Accordingly, from Eqns. 10–35, 10–36 and 10–37, we obtain Eqns. 10–38 and 10–39:

$$\frac{k^+}{k^-} = \exp\left(\frac{e\,\eta^0_H}{kT}\right), \tag{10--38}$$

$$i = e(k^+ c_{\text{RED}} - k^- c_{\text{OX}}) = e\,k_0\left\{c_{\text{RED}}\exp\left(\frac{e\,\eta^0_H}{kT}\right) - c_{\text{OX}}\right\}. \tag{10--39}$$

The interfacial overvoltage η^0_H referred to the standard redox potential is defined, generally, by Eqn. 10–41:

$$\eta^0_H = -\frac{{}_p\varepsilon^*_{F,s} - \varepsilon^0_{F(\text{REDOX})}}{e}. \tag{10--41}$$

Eqn. 10–41 leads to $\eta^0_H = E - E^0_{\text{REDOX}}$ for p-type electrodes (${}_p\varepsilon^*_{F,s} \doteq \varepsilon_{F(\text{SC})}$) at which the total overvoltage η equals the interfacial overvoltage η^0_H; for photoexcited n-type electrodes (${}_p\varepsilon^*_{F,s} = \varepsilon_{F(\text{SC})} + e\,\eta_{p,\text{SC}} \ll \varepsilon_{F(\text{SC})}$), the total overvoltage η is the sum of the overvoltages due to both the hole generation plus transport $\eta_{p,\text{SC}}$ and the hole transfer across the electrode interface η^0_H. Hence, Eqn. 10–42 is obtained:

$$\eta^0_H = -\frac{\varepsilon_{F(\text{SC})} + e\,\eta_{p,\text{SC}} - \varepsilon^0_{F(\text{REDOX})}}{e} = E - E^0_{\text{REDOX}} - \eta_{p,\text{SC}}, \tag{10--42}$$

where E is the electrode potential and $\eta_{p,\text{SC}}$ is the inverse overvoltage (negative sign) in photoexcited n-type electrodes.

10.4.4 Recombination of photoexcited holes in anodic reactions

In photoexcited n-type semiconductor electrodes, the photoexcited electron–hole pairs recombine in the electrodes in addition to the transfer of holes or electrons across the electrode interface. The recombination of the photoexcited holes with electrons in the space charge layer requires a cathodic electron flow from the electrode interior towards the electrode interface. The current associated with the recombination of cathodic holes, i_{rec}, in n-type electrodes, at which the interfacial reaction is in equilibrium, has already been given by Eqn. 8–70. Assuming that Eqn. 8–70 applies not only to equilibrium but also to non-equilibrium transfer reactions involving interfacial holes, we obtain Eqn. 10–43:

$$i_{\text{rec}} = -i_{p,\lim}\exp\left(\frac{{}_n\varepsilon_F - {}_p\varepsilon^*_{F,s}}{\xi kT}\right), \tag{10--43}$$

where $i_{p,\,lim}$ is the limiting current of hole diffusion in the dark n-type electrode, ${}_n\varepsilon_F$ is the Fermi level of the n-type electrode, ${}_p\varepsilon_{F,s}^*$ is the quasi-Fermi level of interfacial holes in photoexcited n-type electrodes, and ξ is a constant. The term of $({}_n\varepsilon_F - {}_p\varepsilon_{F,s}^*)$ in the right hand side of Eqn. 10–43 is the affinity for the recombination of photoexcited electron–hole pairs.

The current i flowing in photoexcited n-type semiconductor electrodes equals the sum of the photoexcited hole current i_{ph}, the limiting current of hole diffusion $i_{p,\,lim}$, and the current of hole recombination i_{rec} as shown in Eqn. 10–44:

$$i = i_{ph} + i_{p,\,lim} + i_{rec} = i_{ph} + i_{p,\,lim}\left\{1 - \exp\left(\frac{{}_n\varepsilon_F - {}_p\varepsilon_{F,s}^*}{\xi k T}\right)\right\}. \tag{10-44}$$

In the photostationary state, Eqn. 10–44 equals the transfer current of anodic holes across the electrode interface shown in Eqn. 10–39.

As described in this section, the Fermi level ${}_n\varepsilon_F$ of a photoexcited n-type electrode associated with a transfer current, i, of anodic holes corresponds to a polarization potential ${}_nE(i)$ of the electrode; whereas, the quasi-Fermi level ${}_p\varepsilon_{F,s}^*$ associated with interfacial holes in the photoexcited n-type electrode corresponds to a polarization potential ${}_pE(i)\,(= -{}_n\varepsilon_F/e = -{}_p\varepsilon_{F,s}^*/e)$ depending on transfer current i of anodic holes for a p-type electrode of the same semiconductor. Accordingly, from Eqn. 10–44 we obtain an expression, Eqn. 10–45, for the difference in these polarization potentials:

$$_pE(i) - {}_nE(i) = \frac{\xi k T}{e}\ln\left(\frac{i_{ph} - i}{i_{p,\,lim}} + 1\right). \tag{10-45}$$

Fig. 10–23 shows the electron levels and the polarization curves for transfer reactions of anodic holes both at a photoexcited n-type electrode and at a dark p-type electrode of the same semiconductor. The range of potential where the anodic hole current occurs at the photoexcited n-type electrode is more cathodic (more negative) than the range of potential for the anodic hole current at the dark p-type electrode. The difference between the polarization potential ${}_nE(i)$ (point N in the figure) of the photoexcited n-type electrode and the polarization potential ${}_pE(i)$ (point P in the figure) of the dark p-type electrode at a constant anodic current i is equivalent to the difference between the quasi-Fermi level ${}_p\varepsilon_{F,s}^*$ of interfacial holes and the Fermi level ${}_n\varepsilon_F$ of interior electrons in the photoexcited n-type electrode; this difference of polarization potential, in turn, equals the inverse overvoltage $\eta_{p,\,SC}(i)$ defined in Eqn. 10–46:

$$_pE(i) - {}_nE(i) = \frac{{}_n\varepsilon_F(i) - {}_p\varepsilon_{F,s}^*(i)}{e} = -\eta_{p,\,SC}(i). \tag{10-46}$$

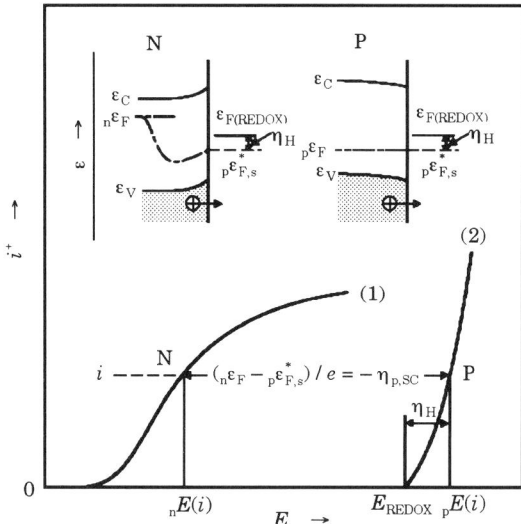

Fig. 10–23. Energy levels and polarization curves for a redox reaction associated with the transfer of anodic redox holes at a photoexcited n-type electrode and at a dark p-type electrode of the same semiconductor: curve (1) = polarization curve of the anodic transfer of photoexcited holes at an n-type electrode; curve (2) = polarization curve of the anodic transfer of holes at a p-type electrode in the dark (equivalent to the curve representing the anodic current as a function of the quasi-Fermi level of interfacial holes in the photoexcited n-type electrode); i^+ = anodic transfer current of holes; E_{REDOX} = equilibrium potential of redox hole transfer; N = anodic polarization at potential $_nE(i)$ of a photoexcited n-type electrode; P = anodic polarization at potential $_pE(i)$ of a dark p-type electrode.

Consequently, by measuring the polarization curves for the transfer reaction of anodic holes both at the photoexcited n-type and at the dark p-type electrodes of the same semiconductor, we obtain the *relationship* between the Fermi level $_n\varepsilon_F$ of the electrode (polarization potential E) and the quasi-Fermi level $_p\varepsilon_{F,s}^*$ of interfacial holes in the photoexcited n-type electrode as a function of the transfer current i of anodic holes.

10.4.5 Cathodic hole injection reactions

The generation of holes in n-type electrodes may result not only from photoexcitation but also from the *injection* of *cathodic hole* according to Eqn. 10–47:

$$RED \rightarrow OX + h \ . \tag{10–47}$$

The current of cathodic hole injection, i, can be obtained from Eqn. 10–37.

The holes injected by a cathodic redox reaction (Eqn. 10–47) diffuse toward the electrode interior and recombine with electrons of the majority charge carriers in the same way as the photogenerated holes, producing a cathodic current i_{rec}, which is equivalent to the rate of recombination of holes. The cathodic current i actually observed is the sum of the current of recombination i_{rec} and the limiting diffusion current of holes $i_{\text{p, lim}}$ as shown in Eqn. 10–48:

$$i = i_{\text{p, lim}} + i_{\text{rec}} = i_{\text{p, lim}} \left\{ 1 - \exp\left(\frac{{}_n\varepsilon_F - {}_p\varepsilon_{F,s}^*}{\xi k T} \right) \right\} . \tag{10-48}$$

Eqn. 10–48 is obtained by excluding the photocurrent i_{ph} from the reaction current of Eqn. 10–44. In the stationary state, the total current i in Eqn. 10–48 equals the transfer current of cathodic redox holes across the electrode interface.

The Fermi level ${}_n\varepsilon_F$ of the electrode corresponds to the polarization potential ${}_nE(i)$ of the photoexcited n-type electrode, and the quasi-Fermi level ${}_p\varepsilon_{F,s}^*$ of interfacial holes corresponds to the polarization potential ${}_pE(i)$ of the dark p-type electrode of the same semiconductor. Then, Eqn. 10–48 becomes Eqn. 10–49:

$$_pE(i) - {}_nE(i) = \frac{\xi k T}{e} \ln\left(1 - \frac{i}{i_{\text{p, lim}}} \right) . \tag{10-49}$$

Fig. 10–24 shows schematically the electron levels and the polarization curves for a cathodic hole injection in n-type and p-type electrodes of the same semiconductor in the dark. The range of potential where the cathodic reaction occurs on the n-type electrode is more cathodic (more negative) than the range of potential for the cathodic reaction on the p-type electrode. The difference between the polarization potential ${}_nE(i)$ (point N in the figure) of the n-type electrode and the polarization potential ${}_pE(i)$ (point P in the figure) of the p-type electrode at a constant cathodic current i is equivalent to the difference between the Fermi level ${}_n\varepsilon_F$ of interior electrons and the quasi-Fermi level ${}_p\varepsilon_{F,s}^*$ of interfacial holes in the hole-injected n-type electrode; this difference of polarization potential, in turn, equals the overvoltage $\eta_{\text{p, SC}}(i)$ for the recombination of cathodically injected holes with electrons of the majority carriers.

For cathodic hole injection, the overvoltage $\eta_{\text{p, SC}}(i)$ includes both diffusion and recombination of holes in the electrode; this overvoltage occurs in the same cathodic direction as the cathodic hole injection so that $\eta_{\text{p, SC}}$ is the usual overvoltage (a negative overvoltage in the cathodic reaction) rather than the inverse overvoltage. Then, we obtain Eqn. 10–50:

$$_pE(i) - {}_nE(i) = \frac{{}_n\varepsilon_F(i) - {}_p\varepsilon_{F,s}^*(i)}{e} = -\eta_{\text{p, SC}}(i) . \tag{10-50}$$

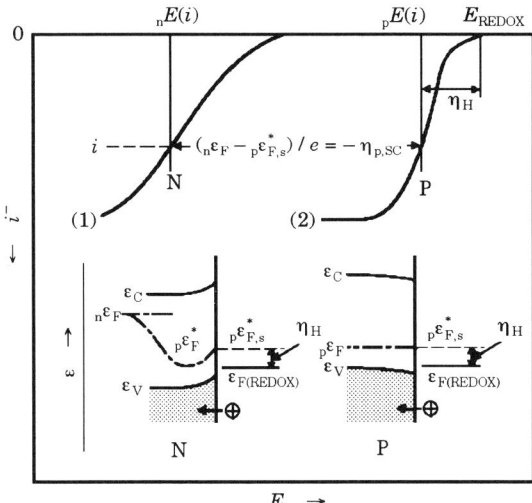

Fig. 10–24. Electron levels and polarization curves for a redox reaction associated with the transfer of cathodic holes both at an n-type and at a p-type electrode of the same semiconductor in the dark: curve (1) = polarization curve of the cathodic hole injection in the dark n-type electrode; curve (2) = polarization curve of the cathodic hole injection in the dark p-type electrode (equivalent to the curve representing the cathodic hole injection current as a function of the quasi-Fermi level of interfacial holes in the dark n-type electrode); i^- = cathodic hole injection current; N = polarization of the cathodic hole injection at potential $_nE(i)$ of the dark n-type electrode, P = polarization of the cathodic hole injection at potential $_pE(i)$ of the dark p-type electrode.

Based on Eqn. 10–50, the *relationship* between the Fermi level $_n\varepsilon_F$ (polarization potential E) and the quasi-Fermi level $_p\varepsilon_{F,s}^*$ of interfacial holes in the hole-injected n-type electrode as a function of the transfer current i of cathodic holes can be obtained by measuring the polarization curves for the redox reaction of cathodic hole injection on both the n-type and the p-type electrodes of the same semiconductor.

10.5 Photoelectrochemical Cells

Photoelectrochemical semiconductor cells are used to convert photon energy into chemical substances or into electricity; the former is a *photoelectrolytic cell* and the latter is a *photovoltaic cell*. A photoelectrochemical semiconductor cell consists of either a pair of metal and semiconductor electrodes or a pair of two semiconductor electrodes.

A typical photoelectrolytic cell decomposes water to produce gaseous hydrogen and oxygen molecules, the overall reaction of which is expressed in Eqn. 10–51:

$$H_2O + h\nu \rightarrow H_2 + \frac{1}{2} O_2 \ . \tag{10-51}$$

This cell reaction is composed of an anodic reaction and a cathodic reaction shown, respectively, in Eqns. 10–51a and. 10–51b:

$$H_2O + 2 h \rightarrow 2 H^+_{aq} + \frac{1}{2} O_2 \ , \qquad \text{anodic reaction,} \tag{10-51a}$$

$$2 H^+_{aq} + 2 e \rightarrow H_2 \ , \qquad \text{cathodic reaction.} \tag{10-51b}$$

In the photoelectrolytic reaction, the two electrodes are short-circuited and the cell voltage is small so that no significant electric energy may be produced. The photopotential provides the overvoltages for both anodic and cathodic reactions.

The photovoltaic cell usually involves a single redox reaction consisting of an anodic reaction at the anode and a cathodic reaction at the cathode as shown in Eqns. 10–52a and 10–52b:

$$\text{RED} + h \rightarrow \text{OX} \ , \qquad \text{anodic reaction,} \tag{10-52a}$$

$$\text{OX} + e \rightarrow \text{RED} \ , \qquad \text{cathodic reaction.} \tag{10-52b}$$

Here, no net chemical change occurs. In the photovoltaic cell, almost all the photopotential generated exists between the two electrodes and can be used to produce electric energy.

10.6 Photoelectrolytic Cells

10.6.1 Photoelectrolytic cells of metal and semiconductor electrodes

Fig. 10–25 shows an energy diagram of a photoelectrolytic cell for decomposing water; this cell is composed of a metallic cathode and an n-type semiconductor anode, on which the following anodic and cathodic reactions, Eqns. 10–53(a) and 10–53(b), proceed:

$$H_2O + 2\,h^+ \rightarrow 2\,H^+_{aq} + \frac{1}{2}O_2\;, \qquad \varepsilon_{F(H^+/H_2)}\,, \tag{10-53a}$$

$$2\,H^+_{aq} + 2\,e^- \rightarrow H_2\;, \qquad \varepsilon_{F(O_2/H_2O)}\,, \tag{10-53b}$$

where $\varepsilon_{F(H^+/H_2)}$ and $\varepsilon_{F(O_2/H_2O)}$ are the Fermi levels of the hydrogen and oxygen redox reactions, respectively.

In order for the photoelectrolytic decomposition of liquid water to proceed, the Fermi levels of the redox reactions in Eqns. 10–53a and 10–53b need to be located within the band gap of the n-type semiconductor anode. In Fig. 10–25(a), we have assumed that the Fermi level $\varepsilon_{F(SC)}$ of the n-type semiconductor anode at the flat band potential E_{fb} is higher than the Fermi level $\varepsilon_{F(H^+/H_2)}$ of the hydrogen redox reaction; we have also assumed that the Fermi level $\varepsilon_{F(M)}$ of the metallic cathode is lower than $\varepsilon_{F(H^+/H_2)}$. Further, we have assumed that the edge level of

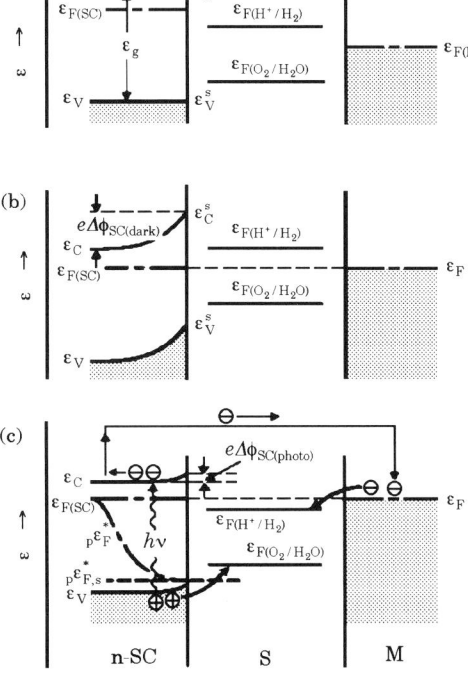

Fig. 10–25. Energy diagrams of a cell for photoelectrolytic decomposition of water consisting of a metal cathode (M) and an n-type semiconductor anode (n-SC) of which the Fermi level is higher than the Fermi level of the hydrogen redox reaction ($\varepsilon_{F(SC)} > \varepsilon_{F(H^+/H_2)}$): (a) the cell circuit is open in the dark, (b) the cell circuit is closed in the dark, (c) the cell circuit is closed in the photoexcited state (the cell reaction proceeds.). $\Delta\phi_{SC}$ = potential barrier of the space charge layer.

the conduction band ε_C^s is higher than the Fermi level $\varepsilon_{F(H^+/H_2)}$ of the hydrogen redox reaction and the edge level of the valence band ε_V^s is lower than the Fermi level $\varepsilon_{F(O_2/H_2O)}$ of the oxygen redox reaction.

When the cell circuit is closed in the dark, as shown in Fig. 10–25(b), the Fermi level is equilibrated between the metallic cathode and the n-type semiconductor anode. As a result, a space charge depletion layer (potential barrier, $\Delta\phi_{SC(dark)}$) is formed in the semiconductor anode, thereby shifting the potential of the anode from the flat band potential E_{fb} to the more anodic (more positive) potential E (= E_{fb} + $|\Delta\phi_{SC(dark)}|$). In the dark, however, the anodic hole transfer associated with the oxygen evolution reaction cannot occur at the n-type semiconductor anode because the Fermi level $\varepsilon_{F(SC)}$ of the anode is higher than the Fermi level $\varepsilon_{F(O_2/H_2O)}$ of the oxygen redox reaction and the concentration of holes is small; no cathodic electron transfer associated with the hydrogen evolution is expected thermodynamically at the metallic cathode because the Fermi level $\varepsilon_{F(M)}$ is lower than the Fermi level $\varepsilon_{F(H^+/H_2)}$ of the hydrogen redox reaction.

When the n-type semiconductor anode is photoexcited, as shown in Fig. 10–25(c), the Fermi level of the anode is raised (the potential of the anode is lowered) by an energy equivalent to the photopotential ΔE_{ph}; at the same time, the Fermi level of the metallic cathode short-circuited with the semiconductor anode is also raised so as to be higher than the Fermi level $\varepsilon_{F(H^+/H_2)}$ of the hydrogen redox reaction; this shift of the Fermi level makes it thermodynamically possible for hydrogen evolution to proceed via the cathodic transfer of electrons at the metallic cathode. Further, photoexcitation increases the concentration of holes (minority charge carriers) at the anode interface and shifts the quasi-Fermi level of interfacial holes to a level which is lower than the Fermi level $\varepsilon_{F(O_2/H_2O)}$ of the oxygen redox reaction; this quasi-Fermi level provides the thermodynamic basis for the oxygen evolution via the anodic transfer of holes at the photoexcited n-type semiconductor anode. As a result, the photoelectrolytic decomposition of water occurs with both oxygen evolution via the anodic hole transfer at the n-type semiconductor anode and hydrogen evolution via the cathodic electron transfer at the metallic cathode.

The energy balance in the photoelectrolytic decomposition of water can be obtained from Fig. 10–25(c) as shown in Eqn. 10–54:

$$\varepsilon_g - (\varepsilon_C^s - \varepsilon_{F(SC)}) - (_p\varepsilon_{F,s}^* - \varepsilon_V^s) = -e\,\eta_{p,SC}$$

$$= (\varepsilon_{F(H^+/H_2)} - \varepsilon_{F(O_2/H_2O)}) + e\,\eta_H^+ - e\,\eta_H^- + e\,i\,R \quad , \qquad (10\text{–}54)$$

where: ε_g is the band gap of the n-type anode; $\eta_{p,SC}$ is the inverse overvoltage (negative) for the generation of photoexcited holes in the n-type anode (equivalent to the energy gain due to photoexcitation $-e\,\eta_{p,SC}$); $\eta_H^- = (\varepsilon_{F(M)} - \varepsilon_{F(H^+/H_2)})/e$ is the cathodic overvoltage (negative) of the hydrogen redox reaction at the metallic cathode; $\eta_H^+ = -(\varepsilon_{F(O_2/H_2O)} - {}_p\varepsilon_{F,s}^*)/e$ is the anodic interfacial overvoltage (positive) of the oxygen redox reaction at the photoexcited n-type electrode; and iR is the resistance overvoltage (positive) for ion transport in the cell electrolyte. Such a photoelectrolytic cell for decomposing water has been observed with an n-type semiconductor anode of strontium titanate, $SrTiO_3$ [Okuda-Yoshida-Tanaka, 1971].

Fig. 10–26 is an energy diagram for the photoelectrolytic decomposition of water in a cell consisting of a metallic cathode and an n-type anode of strontium titanate, of which the band gap is 3.2 eV and the flat band potential at pH 0 is $-0.1\ V_{NHE}$.

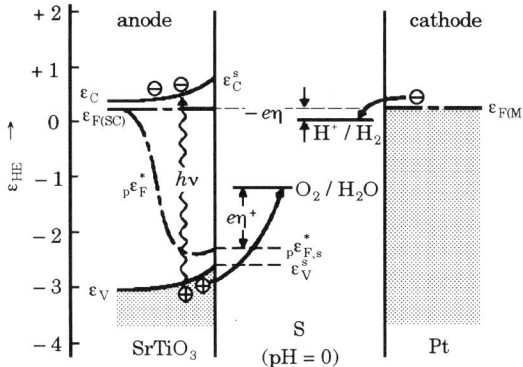

Fig. 10–26. Energy diagram for a cell for photoelectrolytic decomposition of water consisting of a platinum cathode and an n-type semiconductor anode of strontium titanate of which the Fermi level at the flat band potential is higher than the Fermi level of the hydrogen redox reaction ($\varepsilon_{F(SC)} > \varepsilon_{F(H^+/H_2)}$): ε_{HE} = electron energy level referred to the normal hydrogen electrode; η^+ = anodic overvoltage (positive) of the hole transfer across the n-type anode interface; η^- = cathodic overvoltage (negative) of the electron transfer across the metallic cathode interface.

Since the highest possible Fermi level of the photoexcited n-type anode corresponds to the flat band potential of a semiconductor anode, the Fermi level of a metallic cathode short-circuited with the photoexcited n-type anode can also be raised up to the level equivalent to the flat band potential of the semiconductor anode. In order for the cathodic electron transfer of the hydrogen redox reaction to proceed at the metallic cathode, the Fermi level $\varepsilon_{F(M)}$ of the cathode needs to be higher than the Fermi level $\varepsilon_{F(H^+/H_2)}$ of the hydrogen redox reaction. Consequently, in order for the photoelectrolytic decomposition of water to occur, the Fermi level at the flat band potential of the photoexcited anode needs to be higher than the Fermi level $\varepsilon_{F(H^+/H_2)}$ of the hydrogen redox reaction as shown in Fig. 10–26.

For n-type anodes that do no meet this requirement, therefore, an external voltage, V_{ex}, must be applied to the photoelectrolytic cell until the Fermi level $\varepsilon_{F(M)}$ of the metallic cathode exceeds the Fermi level $\varepsilon_{F(H^+/H_2)}$ of the hydrogen redox reaction ($\varepsilon_{F(M)} > \varepsilon_{F(H^+/H_2)}$) for the photoelectrolytic decomposition of water to proceed as shown in Fig. 10–27.

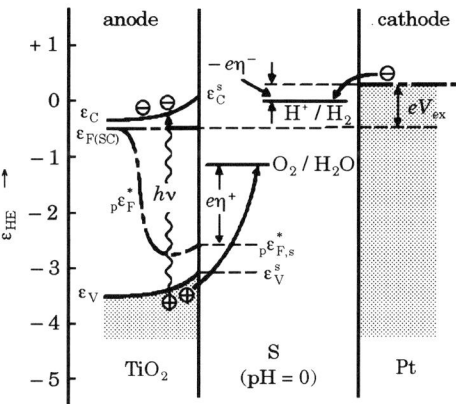

Fig. 10–27. Energy diagram for a cell for photoelectrolytic decomposition of water consisting of a platinum cathode and an n-type anode of titanium oxide of which the Fermi level at the flat band potential is lower than the Fermi level of the hydrogen redox reaction ($\varepsilon_{F(SC)} < \varepsilon_{F(H^+/H_2)}$): V_{ex} = external voltage required for the cell reaction to occur; S = aqueous solution.

Fig. 10–28 illustrates the polarization curves of both a photoexcited n-type anode and a metallic cathode in a photoelectrolytic cell shown in Fig. 10–26. The cathodic current of the hydrogen evolution at the metallic cathode occurs in the range of potential which is more cathodic (more negative) than the equilibrium potential E_{H^+/H_2} of the hydrogen redox reaction as expected from electrochemical thermodynamics; whereas, the anodic current (solid curve, n-SC⁺) of the oxygen evolution at the photoexcited n-type semiconductor anode occurs in the range of potential which is more cathodic (more negative) than the equilibrium potential E_{O_2/H_2O} of the oxygen redox reaction but more anodic (more positive) than the flat band potential of the n-type semiconductor anode E_{fb} (Refer to Fig. 10–14 and Fig. 10–23.). Note that the anodic current (dotted curve, p-SC⁺) of the oxygen evolution at p-type anodes of the same semiconductor as the photoexcited n-type

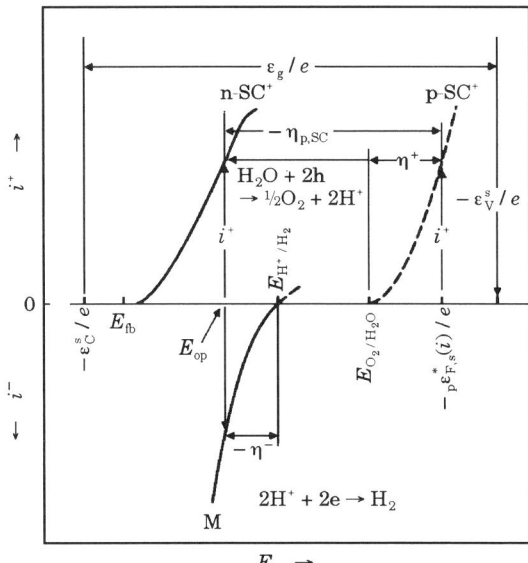

Fig. 10–28. Polarization curves for the cell reactions for photoelectrolytic decomposition of water at a photoexcited n-type anode and at a metal cathode: solid curve M = cathodic polarization curve of hydrogen evolution at the metal cathode; solid curve n-SC* = anodic polarization curve of the oxygen evolution at the photoexcited n-type anode (Fermi level versus current curve); dotted curve p-SC* = quasi-Fermi level of interfacial holes as a function of the anodic reaction current at the photoexcited n-type anode (anodic polarization curve represented by the interfacial hole level); E_{op} = electrode potential of the two operating electrodes in the photoelectrolytic cell; $\eta_{p,SC}$ = inverse overvoltage of the generation and transport of photoexcited holes in the n-type anode.

anode occurs in the range of potential which is more anodic (more positive) than the equilibrium potential E_{O_2/H_2O} of the oxygen redox reaction.

As shown in Eqn. 10–46, the difference of the polarized potential at constant anodic current between the photoexcited n-type and the dark p-type anodes of the same semiconductor represents the *inverse* overvoltage $\eta_{p,SC}$ for the generation and transport of photo-excited holes.

The cathodic and anodic polarization potentials, E_C and E_A, in the stationary state of the cell for photoelectrolytic decomposition of water in which the metallic cathode and the n-type semiconductor anode are short-circuited are given, respectively, in Eqns. 10–55 and 10–56:

$$E_\text{C} = E_{\text{H}^+/\text{H}_2} + \eta_\text{H}^- , \qquad \text{cathode} , \qquad (10\text{--}55)$$

$$E_\text{A} - \eta_{\text{p, SC}} = E_{\text{O}_2/\text{H}_2\text{O}} + \eta_\text{H}^+ , \qquad \text{anode} . \qquad (10\text{--}56)$$

where: $\eta_{\text{p, SC}}$ is the *inverse* overvoltage (negative) for the generation and transport of photoexcited holes; η_H^- is the cathodic overvoltage (negative) for hydrogen evolution at the metallic cathode; and η_H^+ is the anodic interfacial overvoltage (positive) for oxygen evolution which equals the difference between the Fermi level $\varepsilon_{\text{F}(\text{O}_2/\text{H}_2\text{O})}$ of the oxygen redox reaction and the quasi-Fermi level $_\text{p}\varepsilon_{\text{F,s}}^*$ of interfacial holes in the photoexcited n-type electrode, $\eta_\text{H}^+ = -(\varepsilon_{\text{F}(\text{O}_2/\text{H}_2\text{O})} - {_\text{p}\varepsilon_{\text{F,s}}^*})/e$. When the *iR* drop in the electrolyte is negligible, the cathodic and anodic polarization potentials E_C and E_A equal each other thereby representing the *operational* potential E_op ($= E_\text{C} = E_\text{A}$) as shown in Fig. 10–28.

The photoelectrolytic cell may also consist of a cathode of a photoexcited p-type semiconductor (photoexcited hydrogen evolution) and a metallic anode (oxygen evolution); a p-type semiconductor of gallium phosphide GaP with the band gap of 2.25 eV and a flat band potential of -1.2 V$_\text{NHE}$ at pH = 0 is a good example of the p-type semiconductor cathode. [Yoneyama-Sakamoto-Tamura, 1975]. Fig. 10–29 shows an energy diagram of a photoelectrolytic cell composed of both a photoexcited p-type cathode of gallium phosphide and a metallic anode; this cell can be compared with the cell composed of both a photoexcited n-type anode of titanium oxide and a metallic platinum cathode shown in Fig. 10–27.

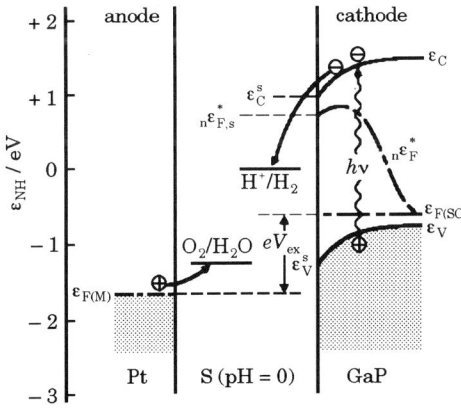

Fig. 10–29. Energy diagram for a cell for photoelectrolytic decomposition of water composed of a platinum anode and a photoexcited p-type cathode of gallium phosphide: V_ex = external voltage required for the cell reaction to proceed.

10.6.2 Photoelectrolytic cells of two semiconductor electrodes

The photoelectrolytic cell can be composed of both a photoexcited n-type anode and a photoexcited p-type cathode; here, the anodes and cathodes can be the same semiconductor or different semiconductors. The energy diagrams for a photoelectrolytic cell of an n-type anode and a p-type cathode of different semiconductors is illustrated in Fig. 10–30.

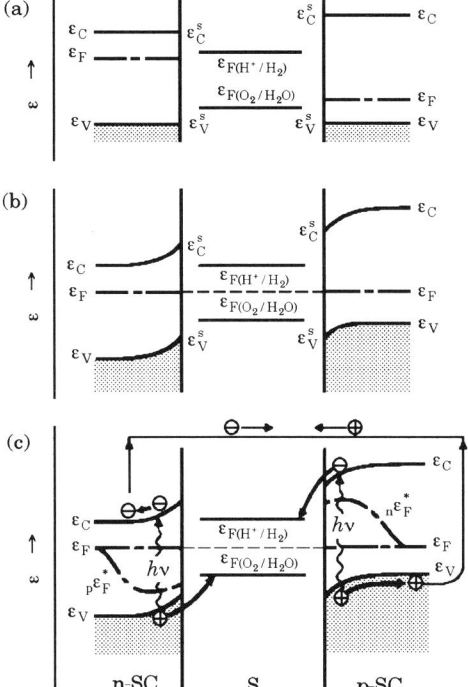

Fig. 10–30. Energy diagrams for a cell for photoelectrolytic decomposition of water consisting of a p-type semiconductor cathode and an n-type semiconductor anode: (a) the cell circuit is open in the dark, (b) the cell circuit is closed in the dark, (c) the cell circuit is closed in the photoexcited state.

Fig. 10–30(a) applies to an open circuit cell in the dark; Fig. 10–30(b) applies to a short circuit cell in the dark. After the cell circuit is closed in the dark, the Fermi level is equilibrated between the two electrodes thereby forming a space charge layer both in the n-type anode and in the p-type cathode. The overall potential difference, ΔE, generated in the two space charge layers nearly equals the difference of the flat band potential between the n-type anode and the p-type cathode as expressed in Eqn. 10–57:

$$\Delta E = {}_p E_{fb} - {}_n E_{fb} \tag{10-57}$$

When the n-type anode and p-type cathode are in the state of band edge level pinning, the overall difference of potential due to the two space charge layers remains constant even in the state of photoexcitation.

Fig. 10–30(c) applies to the photoexcited cell, where oxygen evolution proceeds via the anodic transfer of holes at the n-type anode and hydrogen evolution proceeds via the cathodic transfer of electrons at the p-type cathode. In order for the photoelectrolytic decomposition of water to proceed in such a cell, the edge level of the valence band ${}_n\varepsilon_V^s$ of the n-type anode needs to be lower than the Fermi level $\varepsilon_{F(O_2/H_2O)}$ of the oxygen redox reaction; and the edge level of the conduction band ${}_p\varepsilon_C^s$ of the p-type cathode needs to be higher than the Fermi level $\varepsilon_{F(H^+/H_2)}$ of the hydrogen redox reaction as shown in Fig. 10–30.

Fig. 10–31 shows the energy diagram of a cell for photoelectrolytic decomposition of water consisting of an n-type anode of titanium oxide and a p-type cathode of gallium phosphide in which the photoelectrolytic reaction proceeds without applied external voltages. In this diagram the *iR drop* in the electrolyte is assumed to be zero. The overall energy balance in the reaction of the photoelectrolytic cell can be obtained in Eqn. 10–58 in the same way as Eqn. 10–54:

$$\left\{ \varepsilon_g - (\varepsilon_C^s - \varepsilon_{F(SC)}) - ({}_p\varepsilon_{F,s}^* - \varepsilon_V^s) \right\}_n + \left\{ \varepsilon_g - (\varepsilon_{F(SC)} - \varepsilon_V^s) - (\varepsilon_C^s - {}_n\varepsilon_{F,s}^*) \right\}_p$$

$$= ({}_n\varepsilon_{F,s}^* - \varepsilon_F)_p + (\varepsilon_F - {}_p\varepsilon_{F,s}^*)_n = (\eta_{n,SC})_p + (-\eta_{p,SC})_n = {}_n\varepsilon_{F,s}^* - {}_p\varepsilon_{F,s}^*$$

$$= (\varepsilon_{F(H^+/H_2)} - \varepsilon_{F(O_2/H_2O)}) + e\,\eta_H^+ - e\,\eta_H^- + e\,i\,R \quad, \tag{10-58}$$

where $(\eta_{n,SC})_p$ is the *inverse* overvoltage (positive) for the generation of photoexcited electrons in the p-type cathode; $(\eta_{p,SC})_n$ is the *inverse* overvoltage (negative) for the generation of photoexcited holes in the n-type anode. The left hand side of Eqn. 10–58 represents the affinity for the overall photoelectrolytic reaction; the first term represents the affinity for the anodic reaction and the second term represents the affinity for the cathodic reaction.

The photoelectrolytic cell consisting of the n-type and p-type semiconductor electrodes provides an advantage over the cell consisting of the semiconductor and metal electrodes: a cell consisting of two semiconductor electrodes with their small band gaps adsorb the energy of solar photons more efficiently than the cell consisting of semiconductor and metal electrodes in which the single semiconductor electrode would require a relatively wide band gap for the decomposition of water.

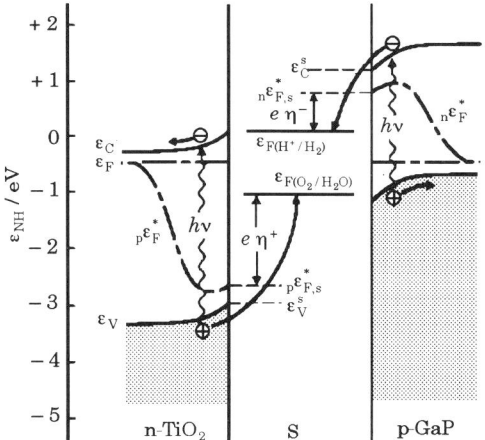

Fig. 10-31. Energy diagram for a photoelectrolytic cell for decomposition of water consisting of a p-type cathode of gallium phosphide and an n-type anode of titanium oxide.

Fig. 10–32 shows the polarization curves for both the anodic oxygen evolution at an n-type anode and the cathodic hydrogen evolution at a p-type cathode. The anodic current (solid curve, n-SC$^+$) of the photoexcited n-type anode occurs in the range of potential which is more cathodic (more negative) than the range of potential for the anodic current (dotted curve n-SC$^+$) of a p-type anode of the same semiconductor as the photoexcited n-type anode; and the cathodic current (solid curve, p-SC$^-$) of the photoexcited p-type cathode occurs in the range of potential which is more anodic (more positive) than the range of potential for the cathodic current (dotted curve, n-SC$^-$) of an n-type cathode of the same semiconductor as the photoexcited p-type cathode.

The cathodic and anodic polarization potentials would equal each other representing the operational potential E_{op} if the *iR drop* in the electrolyte is negligible as shown in Fig. 10–32.

It follows from Eqn. 10–58 that the affinity of the reaction of the photoelectrolytic cell for which the polarization curves are shown in Fig. 10–32 is represented by the difference of the quasi-Fermi levels of interfacial minority charge carriers, $({}_n\varepsilon_{F,s}^*)_p - ({}_p\varepsilon_{F,s}^*)_n$, between the anode and the cathode. Eqn. 10–58 shows that the dissipation of energy in the reaction of the photoelectrolytic cell corresponds to the gain of energy due to the formation of photoexcited electrons and holes, $e\,(\eta_{n,SC})_p + e\,(-\eta_{p,SC})_n$; this affinity is equivalent to the sum of both the interfacial overvoltages and the electromotive force (thermodynamic) of the cell for the decomposition of water, $e\,\eta_H^+ + e\,(-\eta_H^-) + e\,(E_{O_2/H_2O} - E_{H^+/H_2})$.

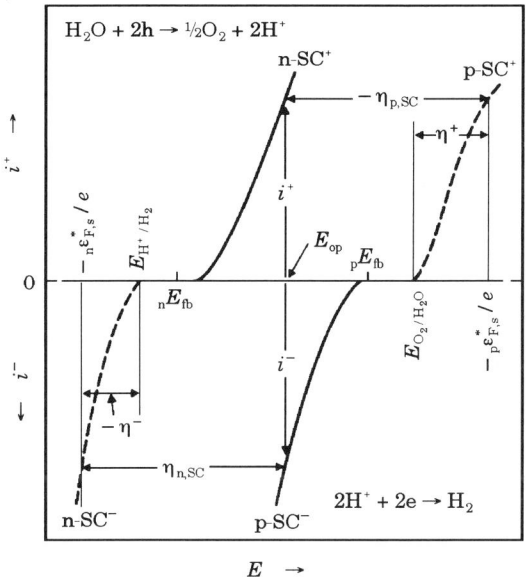

Fig. 10–32. Polarization curves of the cell reaction for photoelectrolytic decomposition of water at a photoexcited n-type anode and at a photoexcited p-type cathode: solid curve n-SC$^+$ = anodic polarization curve of the oxygen evolution at the photoexcited n-type anode (Fermi level versus current curve); dotted curve n-SC$^+$ = anodic polarization curve of the oxygen evolution at dark p-type anodes of the same semiconductor as the photoexcited n-type anode (equivalent to the curve of current versus. quasi-Fermi level of interfacial holes in the photoexcited n-type anode); solid curve p-SC$^-$ = cathodic polarization curve of the hydrogen evolution at the photoexcited p-type cathode (Fermi level versus current curve); dotted curve n-SC$^-$ = cathodic polarization curve of the hydrogen evolution at dark n-type electrodes of the same semiconductor as the photoexcited p-type cathode (equivalent to the curve of current versus quasi-Fermi level of interfacial electrons in the photoexcited p-type cathode); $_nE_{fb}$ ($_pE_{fb}$) = flat band potential of the n-type (p-type) electrode; $\eta_{n,SC}$ ($\eta_{p,SC}$) = inverse overvoltage for the generation of photoexcited electrons (holes) in the p-type (n-type) electrode.

10.7 Photovoltaic Cells

In photovoltaic cells, the same redox reaction, OX + e = RED, may be used for the anode and cathode. Fig. 10–33 shows an energy diagram of an operating photovoltaic cell; this cell consists of a metallic cathode and a photoexcited n-type semiconductor anode. The electromotive force (the open cell voltage), ΔE_{ph}, equals approximately the difference in potential between the flat band potential $_nE_{fb}$ of the n-type anode and the equilibrium electrode potential E_{REDOX} of the redox reaction as shown in Eqn. 10–59:

$$\Delta E_{\text{ph}} = E_{\text{REDOX}} - {}_p E_{\text{fb}} \;. \tag{10–59}$$

From the energy diagram shown in Fig. 10–33, the operating cell voltage, V_{ph}, is obtained, as expressed in Eqn. 10–60, by subtracting from the electromotive force ΔE_{ph} the potential barrier $\Delta \phi_{\text{SC}}$ of the space charge layer, the cathodic overvoltage η_{H}^-, and the *iR drop* in the electrolyte:

$$V_{\text{ph}} = |\Delta E_{\text{ph}}| - |\Delta \phi_{\text{SC}}| - |\eta_{\text{H}}^-| - |i\,R| \;. \tag{10–60}$$

Introducing the gain of photoenergy due to the generation of photoexcited holes (*inverse* overvoltage), $|e\,\eta_{\text{p, SC}}| = |\varepsilon_{\text{F(SC)}} - {}_p\varepsilon_{\text{F,s}}^*|$, we obtain Eqn. 10–61 from the energy diagram shown in Fig. 10–33:

$$V_{\text{ph}} = |\eta_{\text{p, SC}}| - |\eta_{\text{H}}^+| - |\eta_{\text{H}}^-| - |i\,R| \;. \tag{10–61}$$

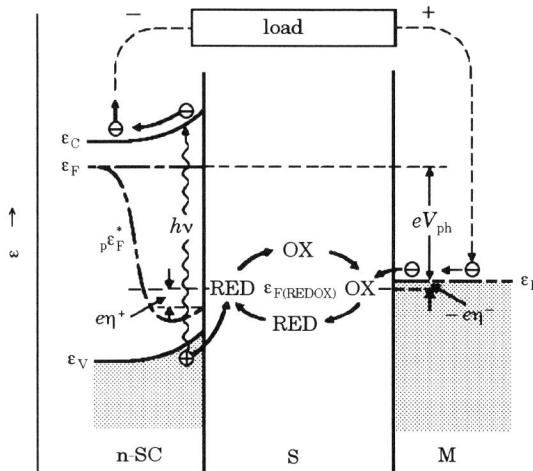

Fig. 10–33. Energy diagram for a photovoltaic cell composed of a metal cathode and an n-type semiconductor anode: V_{ph} = cell voltage in operation at current i.

Fig. 10–34 shows the energy diagram of an operating photovoltaic cell which consists of a photoexcited anode of n-type semiconductor and a photoexcited cathode of p-type semiconductor. The electromotive force, ΔE_{ph}, of this type of photovoltaic cell equals approximately the difference in potential between the flat band potential $_nE_{fb}$ of the n-type anode and the flat band potential $_pE_{fb}$ of the p-type cathode as shown in Eqn. 10–62:

$$\Delta E_{ph} = {_pE_{fb}} - {_nE_{fb}} \quad . \tag{10-62}$$

The operating cell voltage is obtained, as shown in Eqn. 10–63, by subtracting from the electromotive force ΔE_{ph} the potential barriers of the space charge layers $\Delta \phi_{n-SC}$ and $\Delta \phi_{p-SC}$ in the two electrodes and the iR drop in the electrolyte:

$$V_{ph} = |\Delta E_{ph}| - |\Delta \phi_{n-SC}| - |\Delta \phi_{p-SC}| - |iR| \quad . \tag{10-63}$$

Introducing, again, the gains of photoenergy (inverse overvoltage), $|e\,\eta_{p,SC}| = |_n\varepsilon_{F(SC)} - {_p\varepsilon^*_{F,s}}|$ and $|e\,\eta_{n,SC}| = |_p\varepsilon_{F(SC)} - {_n\varepsilon^*_{F,s}}|$, due to the generation of the photoexcited holes and electrons in the two electrodes, we also obtain Eqn. 10–64 from the energy diagram shown in Fig. 10–34:

$$V_{ph} = |\eta_{p,SC}| + |\eta_{n,SC}| - |\eta^+_H| - |\eta^-_H| - |iR| \quad . \tag{10-64}$$

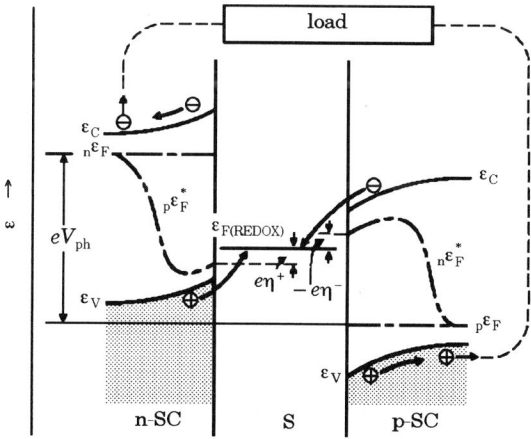

Fig. 10–34. Energy diagram for a photovoltaic cell composed of an n-type semiconductor anode and a p-type semiconductor cathode.

Fig. 10–35 shows the anodic and cathodic polarization curves for the operating photovoltaic cell in Fig. 10–34. The anodic current (solid curve, n-SC$^+$) of the photoexcited n-type anode occurs in the range of potential which is more cathodic (more negative) by a magnitude of the inverse overvoltage (the gain of photoenergy) than the range of potential for the anodic current (dotted curve, p-SC$^+$) of p-type anodes of the same semiconductor as the photoexcited n-type anode. Similarly, the cathodic current (solid curve, p-SC$^-$) of the photoexcited p-type cathode occurs in the range of potential which is more anodic (more positive) by a magnitude of the inverse overvoltage (the gain of photoenergy) than the range of potential for the cathodic current (dotted curve, n-SC$^-$) of n-type cathodes of the same semiconductor as the photoexcited p-type cathode. The photovoltaic reaction current (dotted curves) depends on the quasi-Fermi levels of minority charge carriers at the interface, but the observed anodic and cathodic polarization curves (solid curves) provide an operating photovoltaic cell voltage V_{ph} as shown in Fig. 10–35.

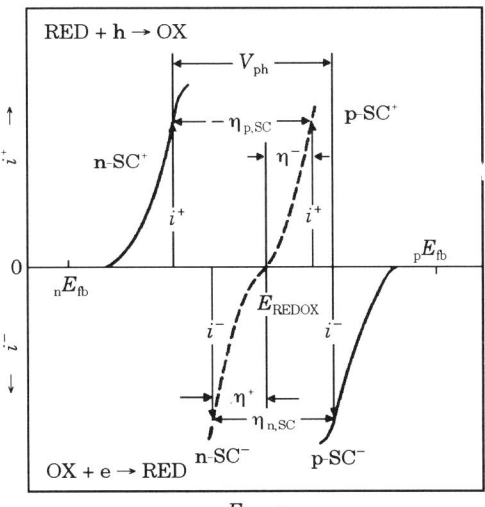

Fig. 10–35. Polarization curves for the redox reactions at an n-type anode and at a p-type cathode in a photovoltaic cell: solid curve n-SC$^+$ = anodic current at the photoexcited n-type anode (Fermi level versus current curve); dotted curve p-SC$^+$ = anodic current at the dark p-type anode (the current versus quasi-Fermi level of interfacial holes in the photoexcited n-type anode); solid curve p-SC$^-$ = cathodic current at the photoexcited p-type cathode (Fermi level versus current curve); dotted curve n-SC$^-$ = cathodic current at the dark n-type cathode (the current versus quasi-Fermi level of interfacial electrons in the photoexcited p-type cathode).

In practical application of both photoelectrolytic and photovoltaic cells, it is a serious matter of concern to prevent the corrosion of the electrodes at which not only the cell reactions but also the corrosion reactions frequently proceed.

References

[Butler, 1977]: M. A. Butler, *J. Appl. Phys.*, **48**, 1914(1977).
[David-Butler, 1988]: S. David and M. A. Butler, *Semiconductor Electrodes*, (Edited by H. O. Flinklea), p. 329, Elsevier, Amsterdam, (1988).
[Fujishima-Honda-Kikuchi, 1969]: A. Fujishima, K. Honda and S. Kikuchi, *J. Chem. Soc. Japan* , **72**, 108(1969).
[Gerischer, 1990]: H. Gerischer, *Electrochimica Acta*, **35**, 1677(1990).
[Harten, 1960]: H. U. Harten, *J. Phys. Chem. Solids,* **14,** 220(1960).
[McEvoy-Etman-Memming, 1985]: A. J. McEvoy, M. Etman and R. Memming, *J. Electroanal. Interface. Electrochem.*, **190**, 225(1985).
[Memming, 1983]: R. Memming, *Comprehensive Treatise of Electrochemistry*, **Vol. 7**, (Edited by E. Conway et al.) , p. 529, Plenum Press, New York, (1983).
[Memming, 1983]: R. Memming, *Ber. Bunsenges. Phys. Chem.*, **91**, 353(1987).
[Memming-Kelly, 1981]: R. Memming and J. Kelly, *Proc. 3rd. Int. Conf. Photochem. Energy Conversion and Storage,* p. 243, Acad. Press, New York, (1981).
[Myamlin-Pleskov, 1967]: V. A. Myamlin and Y. V. Pleskov, *Electrochemistry of Semiconductors,* Plenum press, New York, (1967).
[Okuda-Yoshida-Tanaka, 1971]: M. Okuda, K. Yoshida and N. Tanaka, *Japan. J. Appl. Phys.*, **15**, 1599(1971).
[Sato, 1994]: N. Sato, *Electrode Chemistry* **2** , p. 174, Japan Technical Information Service, Tokyo, (1994).
[Sinn-Meissner-Memming, 1990]: C. Sinn, D. Meissner and R. Memming, *J. Electrochem. Soc.*, **137**, 168(1990).
[Schockley, 1950]: W. Schockley, *Electrons and Holes in Semiconductors*, Van Nostrand, New York, (1950).
[Williams, 1960]: R. Williams, *J. Chem, Phys.*, **32**, 1505(1960).
[Yoneyama-Sakamoto-Tamura, 1975]: H. Yoneyama, A. Sakamoto and H. Tamura, *Electrochimica Acta*, **20**, 341(1975).

CHAPTER 11

MIXED ELECTRODES

11.1 The Single Electrode and The Mixed Electrode

The following reaction, involving the oxidation of iron by acid, illustrates a type of reaction that is often of interest to the corrosion of materials and to various other applications such as catalyses: $Fe + 2 H_{aq}^+ \rightarrow Fe_{aq}^{2+} + H_2$. As described in Sec. 6.3.2, this reaction constitutes an electrochemical cell which is composed of the anodic dissolution of iron, $Fe \rightarrow Fe_{aq}^{2+} + 2\,e$, at the anode and the cathodic reduction of hydrogen ions, $2\,H_{aq}^+ + 2\,e \rightarrow H_2$, at the cathode. These separated electrode reactions are *"coupled"* and proceed as an integrated reaction at two separated electrodes called *"half cells"*. Such a half cell at which only a single electrode reaction proceeds is also called the *"single electrode"*. If the single electrode is isolated from the counter electrode (if the cell is in the open circuit state), the electrode reaction would be in the state of equilibrium and the potential of the single electrode (the open circuit potential, rest potential or spontaneous potential) would be the equilibrium potential of the electrode reaction.

Further, these anodic and cathodic reactions can occur spatially at adjacent locations on the surface of a metal electrode rather than on two separated metal electrodes as shown in Fig. 11–1, where the anodic dissolution of iron and the cathodic reduction of hydrogen ions proceed simultaneously on an iron electrode in aqueous solutions. The electrons produced in the anodic dissolution of iron are the same electrons involved in the cathodic reduction of hydrogen ions; hence, the anodic reaction cannot proceed more rapidly than that the electrons can be accepted by the cathodic reaction and vice versa. Such an electrode at which a pair of anodic and cathodic reactions proceeds is called the *"mixed electrode"*. For the mixed electrodes, the anode (current entrance) and the cathode (current exit) coexist on the same electrode interface. The concept of the mixed electrode was first introduced in the field of corrosion science of metals [Evans, 1946; Wagner-Traud, 1938].

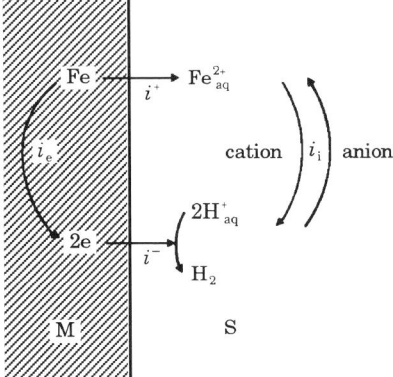

Fig. 11-1. Mixed electrode model (the local cell model) for the corrosion of metals: i^+ = anodic current for the transfer of iron ions; i^- = cathodic current of electron transfer for the reduction of hydrogen ions.

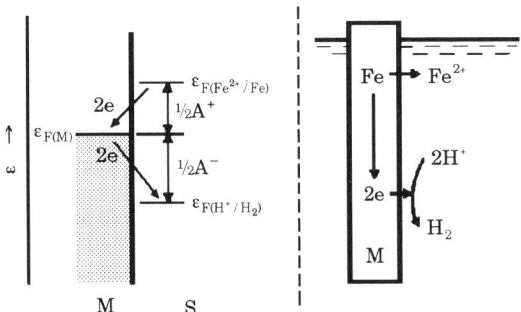

Fig. 11-2. Electron energy levels for the mixed electrode reaction of iron corrosion in acidic solutions: $\varepsilon_{F(M)}$ = Fermi level of an iron electrode; $\varepsilon_{F(H^+/H_2)}$ = Fermi level of the hydrogen redox reaction; $\varepsilon_{F(Fe^{2+}/Fe)}$ = Fermi level of the transfer reaction of iron ions; A^+ = affinity of the anodic transfer of iron ions; A^- = affinity of the cathodic reduction of hydrogen ions.

Thus, the corrosion reaction of metallic iron is a coupled reaction of anodic oxidation of iron and cathodic reduction of hydrogen ions as shown in Eqn. 11–1:

$Fe \rightarrow Fe^{2+}_{aq} + 2\,e$ anodic reaction
$2\,H^+_{aq} + 2\,e \rightarrow H_2$ cathodic reaction

$Fe + 2\,H^+_{aq} \rightarrow Fe^{2+}_{aq} + H_2$ overall reaction .   (11–1)

In order for this mixed electrode reaction shown in Fig. 11–2 to proceed, the Fermi level $\varepsilon_{F(M)}$ of the iron electrode must be higher than the Fermi level $\varepsilon_{F(H^+/H_2)}$ of the hydrogen redox reaction and also must be lower than the Fermi level $\varepsilon_{F(Fe^{2+}/Fe)}$ for the transfer reaction of iron ions. In other words, the potential E of the iron electrode must be lower than the equilibrium potential $E_{(H^+/H_2)}$ of the hydrogen redox reaction and also must be higher than the equilibrium potential $E_{(Fe^{3+}/Fe^{2+})}$ of the transfer of iron ions (the dissolution–precipitation of iron ions).

On the mixed electrode of metallic iron immersed in acidic solutions the anodic and cathodic charge transfer reactions (the anodic transfer of iron ions and the cathodic transfer of electrons) proceed across the electrode interface at which the anodic current (the positive charge current) is exactly balanced with the cathodic current (the negative charge current) producing thereby zero net current.

11.2 Catalytic Reactions on Mixed Electrodes

The presence of a metal surface can *catalyze* redox reactions which also constitute mixed electrode reactions. For example, the oxidation of hydrogen molecules ($2 H_{2,\,aq} + O_{2,\,aq} \rightarrow 2 H_2O$) does not occur in aqueous solutions but this oxidation *catalytically* proceeds on platinum electrodes as a coupled process of the anodic and the cathodic reactions shown in Eqn. 11–2 and in Fig. 11–3:

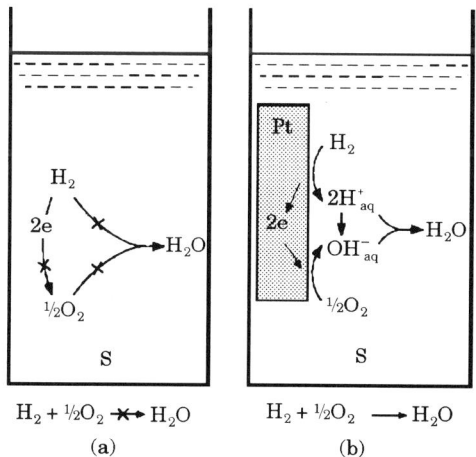

Fig. 11–3. Catalytic oxidation of hydrogen molecules on metal electrodes: (a) the oxidation of hydrogen does not occur in aqueous solutions; (b) the oxidation of hydrogen occurs catalytically on metal electrodes as a mixed electrode reaction.

$H_{2,\,aq} \rightarrow 4\,H^+_{aq} + 4\,e$	anodic reaction
$O_{2,\,aq} + 4\,H^+_{aq} + 4\,e \rightarrow 2\,H_2O$	cathodic reaction

$$H_{2,\,aq} + O_{2,\,aq} \rightarrow 2\,H_2O \qquad \text{overall reaction} \qquad (11\text{–}2)$$

The direct transfer of electrons from the frontier orbital of hydrated hydrogen molecules to the frontier orbital of hydrated oxygen molecules does not take place because its activation energy is high; but the indirect transfer of electrons via both the electron level of metallic electrodes and the redox electron level of adsorbed reaction intermediates proceeds at an appreciable rate on metal electrodes.

For further example, the decomposition of hydrogen peroxide, $2\,H_2O_{2,\,aq} \rightarrow O_{2,\,aq} + 2\,H_2O$, which hardly occurs in aqueous basic solutions, catalytically proceeds on mercury electrodes as a mixed electrode process of anodic and cathodic reactions shown in Eqn. 11–3:

$H_2O_{2,\,aq} + 2\,OH^-_{aq} \rightarrow 2\,H_2O + O_{2,\,aq} + 2\,e$	anodic reaction
$H_2O_{2,\,aq} + 2\,e \rightarrow 2\,OH^-_{aq}$	cathodic reaction

$$2\,H_2O_{2,\,aq} \rightarrow O_{2,\,aq} + 2\,H_2O \qquad \text{overall reaction} \qquad (11\text{–}3)$$

The catalytic effect of the mixed electrodes occurs also in a number of bio-redox reactions on bio-membranes and also on enzymes as shown in Fig. 11–4:

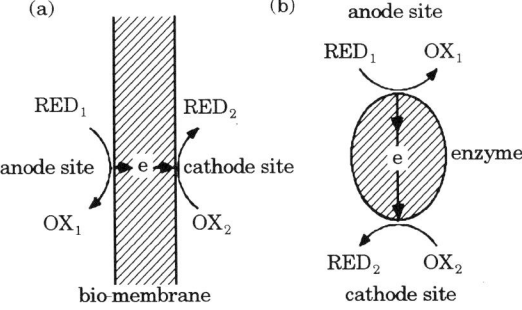

Fig. 11–4. Catalytic redox reactions in the mechanism of mixed electrode reactions (a) on bio-membranes and (b) on enzymes.

| RED$_1$ → OX$_1$ + e | anodic reaction |
| OX$_2$ + e → RED$_2$ | cathodic reaction |

RED$_1$ + OX$_2$ → OX$_1$ + RED$_2$. \qquad overall reaction

Such bio-redox reactions are catalyzed by the transfer of electrons or holes via electron levels in membranes and enzymes; reactions proceed by the same mechanisms as those of mixed electrode reactions.

11.3 Mixed Electrode Potential

The electrode potential of a mixed electrode at which a coupled reaction of charge transfer proceeds is called the *mixed electrode potential*; this mixed electrode potential is obviously different from the single electrode potential at which a single reaction of charge transfer is at equilibrium. For corroding metal electrodes, as shown in Fig. 11–2, the mixed potential is often called the *corrosion potential*, E_{corr}. At this corrosion potential E_{corr} the anodic transfer current of metallic ions i^+, which corresponds to the corrosion rate (the corrosion current i_{corr}), is exactly balanced with the cathodic transfer current of electrons for reduction of the oxidants (e.g. hydrogen ions) i^- as shown in Eqn. 11–4:

$$i_{corr} = i^+ = i^-, \qquad \text{at the corrosion potential.} \qquad (11\text{–}4)$$

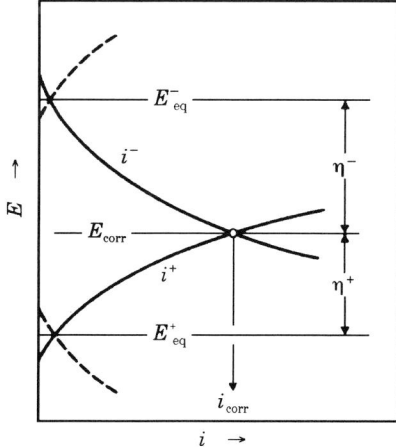

Fig. 11–5. Polarization curves of anodic metal dissolution and of cathodic oxidant reduction at a corroding metallic electrode (mixed electrode): E^+_{eq} (E^-_{eq}) = equilibrium potential of the metal dissolution (the oxidant reduction); i^+ (i^-) = anodic (cathodic) current of metal dissolution (oxidant reduction); η^+ (η^-) = anodic (cathodic) overvoltage; dotted curve = reaction current in the reverse direction.

Therefore, it follows that the anodic polarization curve intersects the cathodic polarization curve ($i^+ = i^-$) at the corrosion potential E_{corr} as illustrated in Fig. 11–5.

The overvoltages η^+ and η^- for the anodic and cathodic reactions at the corrosion potential are, respectively, given in Eqn. 11–5:

$$\eta^+ = E_{corr} - E^+_{eq}, \qquad \eta^- = E_{corr} - E^-_{eq}, \qquad (11\text{--}5)$$

where E^+_{eq} and E^-_{eq} are, respectively, the equilibrium potentials for the anodic metal dissolution and cathodic oxidant reduction. Eqn. 11–5 gives Eqn. 11–6:

$$E^-_{eq} - E^+_{eq} = |\eta^+| + |\eta^-|. \qquad (11\text{--}6)$$

Eqn. 11–6 corresponds to the affinity for the reaction of metallic corrosion. As described in Chaps. 8 and 9, the anodic transfer current i^+ of metal ions and the cathodic transfer current i^- of electrons across the interface of corroding metallic electrodes are, respectively, given in Eqns. 11–7 and 11–8:

$$i^+ = i^+_0 \exp\left\{\frac{\alpha^+ e (E - E^+_{eq})}{kT}\right\}, \qquad (11\text{--}7)$$

$$i^- = i^-_0 \exp\left\{\frac{-\alpha^- e (E - E^-_{eq})}{kT}\right\}, \qquad (11\text{--}8)$$

where i_0 is the exchange current at equilibrium, and α is the Tafel constant. Introducing the corrosion potential E_{corr} and the corrosion current i_{corr} into Eqns. 11–7 and 11–8, we obtain Eqns. 11–9 and 11–10:

$$i^+ = i_{corr} \exp\left\{\frac{\alpha^+ e (E - E_{corr})}{kT}\right\}, \qquad (11\text{--}9)$$

$$i^- = i_{corr} \exp\left\{\frac{-\alpha^- e (E - E_{corr})}{kT}\right\}. \qquad (11\text{--}10)$$

The polarization curve (polarization current i_p versus polarization potential E) of a corroding metallic electrode can be measured by polarizing the electrode in the anodic and cathodic directions. In the range of electrode potential a short distance from the corrosion potential, the polarization curve follows the Tafel relation as shown in Fig. 11–6. Here, the polarization current, i_p, in the anodic direction equals the dissolution current of the metal i^+; and the polarization current, i_p, in the cathodic direction equals the reduction current of the oxidant i^-. However, in the range of potential near the corrosion potential, the polarization current, i_p, is the difference between the anodic dissolution current of the metal i^+ and the cathodic reduction current of the oxidant i^- as shown in Eqn. 11–11:

$$i_p = i^+ - i^- = i_{corr}\left[\exp\left\{\frac{\alpha^+ e(E-E_{corr})}{kT}\right\} - \exp\left\{\frac{-\alpha^- e(E-E_{corr})}{kT}\right\}\right]. \quad (11\text{-}11)$$

For a narrow range of polarization near the corrosion potential ($E - E_{corr}$ < 0.01 V), Eqn. 11–11 gives, to a first approximation, Eqns. 11–12 and 11–13. The polarized current, i_p, is related to the corrosion current, i_{corr}, by Eqns. 11–12 and 11–13:

$$i_p = i_{corr}\frac{e(\alpha^+ + \alpha^-)(E-E_{corr})}{kT} = \frac{E-E_{corr}}{R_p}, \quad (11\text{-}12)$$

$$R_p = \frac{kT}{e}\frac{1}{i_{corr}(\alpha^+ + \alpha^-)}, \quad (11\text{-}13)$$

where R_p is called the *polarization resistance* of the corroding metal electrode at the corrosion potential.

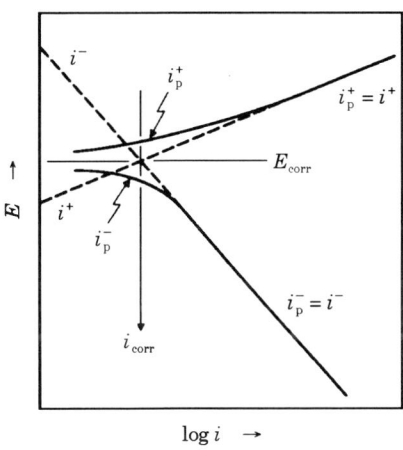

Fig. 11–6. Polarization curves that can be observed with a corroding metallic electrode (solid curve) compared with the anodic and cathodic reaction currents (dotted curve) as functions of electrode potential: i_p^+ (i_p^-) = anodic (cathodic) polarization current; i^+ (i^-) = anodic (cathodic) reaction current.

Fig. 11–7 shows the polarization curve of an iron electrode in an acidic solution, in which the anodic reaction is the anodic transfer of iron ions for metal dissolution (Tafel slope 40 mV/decade); the cathodic reaction is the cathodic transfer of electrons for reduction of hydrogen ions (Tafel slope 120 mV/decade) across the interface of iron electrodes.

When the cathodic reaction is the reduction of oxygen molecules for which the equilibrium potential is relatively high (much more anodic than the corrosion

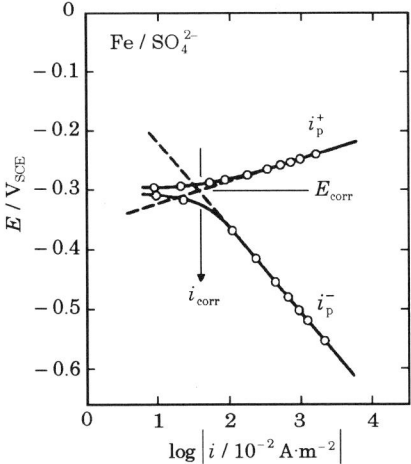

Fig. 11–7. Polarization curves for a corroding iron electrode in a solution of sulfuric acid (0.45 M Na_2SO_4 + 0.05 M H_2SO_4, 25°C, pH 1.7). [From Kaesche, 1979.]

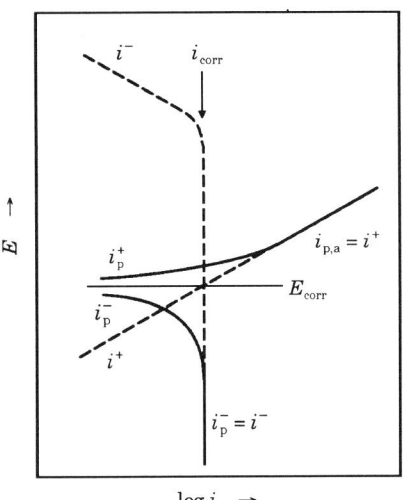

Fig. 11–8. Polarization curves for a corroding metallic electrode of which the corrosion rate is controlled by the diffusion of oxidants in aqueous solution: solid curve = observable polarization curve.

potential), the corrosion current is frequently controlled by the diffusion of hydrated oxygen molecules towards the corroding metal electrode; thus, the corrosion current equals the diffusion current of oxygen molecules as shown in Fig. 11–8. For this mode of diffusion-controlled corrosion of metals the cathodic Tafel constant is zero ($\alpha^- = 0$), since the diffusion current is independent of electrode potentials.

Eqn. 11–13 also applies to the diffusion-controlled corrosion yielding Eqn. 11–14:

$$R_p = \frac{kT}{e} \frac{1}{i_{corr} \alpha^+} \quad . \tag{11-14}$$

In both cases of Eqns. 11–13 and 11–14, in which the corrosion reaction is controlled by the transfer of interfacial charge or by the diffusion of oxidants, the polarization resistance, R_p, can be used to estimate the corrosion rate of the metal (corrosion current, i_{corr}).

11.4 Passivation of Metal Electrodes

11.4.1 Polarization curves of anodic metal dissolution

For some metallic electrodes, such as transition metals, metal ions dissolve directly from the metallic phase into acidic solutions; this direct dissolution of metal ions proceeds at relatively low (less anodic) electrode potentials. The direct dissolution of metal ions is inhibited by the formation of a thin oxide film on metallic electrodes at higher (more anodic) electrode potentials. At still higher electrode potentials this inhibitive film becomes electrochemically soluble (or apparently broken down) and the dissolution rate of the metal increases substantially. These three states of direct dissolution, inhibition, and indirect dissolution via a film (or a broken film) are illustrated in Fig. 11–9.

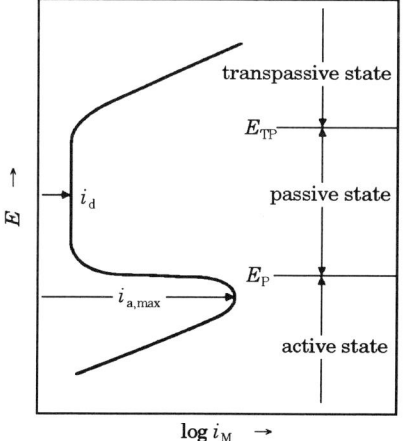

Fig. 11-9. Anodic polarization curve of a metallic electrode for the active dissolution, passivation, and transpassivation in aqueous acidic solutions: i_M = anodic current of metal dissolution; E_P = passivation potential; E_{TP} = transpassivation potential; $i_{a,max}$ = maximum metal dissolution current in the active state (critical current required for passivation); i_d = passive dissolution current (minimum current required for maintaining the passive state).

The state in which the anodic dissolution of metals proceeds from the bare metal surface at relatively low electrode potentials is called the *active state*; the state in which metal dissolution is inhibited substantially by an superficial oxide film at higher electrode potentials is called the *passive state*; the state in which the anodic dissolution of metals increases again at still higher (more anodic) potentials is called the *transpassive state*.

The transition from the active state to the passive state is the *passivation*, and the transition in the reverse direction is the *activation* or *depassivation*. The threshold of potential between the active and the passive states is called the *passivation potential* or the passivation–depassivation potential. Similarly, the transition from the passive state to the tanspassive state is the *transpassivation*, and the critical potential for the transpassivation is called the *transpassivation potential*. Further, the superficial thin film formed on metals in the passive state is often called the "*passive film*" (or *passivation film*), the thickness of which is in the order of 1 to 5 nm on transition metals such as iron and nickel.

In the active state, the dissolution of metals proceeds through the anodic transfer of metal ions across the compact electric double layer at the interface between the bare metal and the aqueous solution. In the passive state, the formation of a thin passive oxide film causes the interfacial structure to change from a simple metal/solution interface to a three-phase structure composed of the metal/film interface, the thin film layer, and the film/solution interface [Sato, 1976, 1990]. Therefore, the rate of metal dissolution in the passive state is controlled by the transfer rate of metal ions across the film/solution interface (the dissolution rate of the passive semiconductor oxide film); this rate is a function of the potential difference across the film/solution interface. Since the potential difference across the film/solution interface is constant in the stationary state of the passive oxide film (in the state of *band edge level pinning*), the rate of the film dissolution is independent of the electrode potential in the range of potential of the passive state. However, in the transpassive state the potential difference across the film/solution interface becomes dependent on the electrode potential (in the state of *Fermi level pinning*), and the dissolution of the thin transpassive film depends on the electrode potential as described in Sec. 11.4.2.

Anodic passivation can be observed easily and clearly with iron group metals and alloys as shown in Fig. 11–10. In principal, anodic passivation occurs with most metals. For instance, even with noble metals such as platinum, which is resistant to anodic dissolution in sulfuric acid solutions, a bare metal surface is realized in the active state and a superficial thin oxide film is formed in the passive state. For less noble metals of which the affinity for the oxide formation is high, the active state is not observed because the metal surface is always covered with an oxide film.

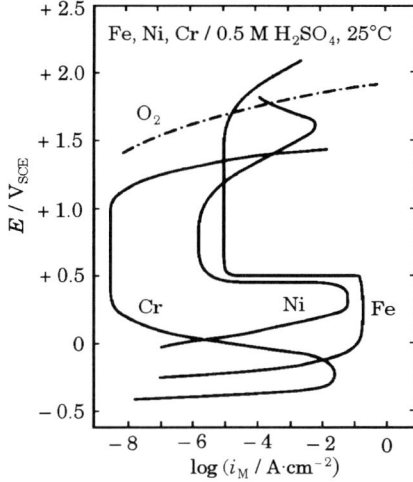

Fig. 11-10. Anodic polarization curves observed for metallic iron, nickel, and chromium electrodes in a sulfuric acid solution (0.5 M H_2SO_4) at 25°C: solid curve = anodic metal dissolution current; dot-dash curve = anodic oxygen evolution current. [Sato-Okamoto, 1981.]

11.4.2 Metal dissolution in the passive and transpassive states

In the stationary state of anodic dissolution of metals in the passive and transpassive states, the anodic transfer of metallic ions (*metal ion dissolution*) across the film/solution interface takes place, but the anodic transfer of oxygen ions across the film/solution interface is in the equilibrium state. In other words, the rate of film formation (the anodic transfer of metal ions across the metal/film interface combined with anodic transfer of oxygen ions across the film/solution interface) equals the rate of film dissolution (the anodic transfer of metal ions across the film/solution interface combined with cathodic transfer of oxygen ions across the film/solution interface).

Therefore, in the stationary state, the rate of anodic transfer of metal ions across the metal/film interface equals the rate of anodic transfer of metal ions across the film/solution interface; this rate of metal ion transfer represents the dissolution rate of the passive film. The thickness of the passive film at constant potential remains generally constant with time in the stationary state of dissolution, although the thickness of the film depends on the electrode potential and also on the dissolution current of the passive film.

The passive film is composed of metal oxides which can be semiconductors or insulators. Then, the electron levels in the passive film are characterized by the conduction and valence bands. However, we need to examine whether the band model can apply to the thin passive oxide film the thickness of which is in the range of nanometers. The passive film has a two-dimensional periodic lattice structure on the plane parallel to the metal/film interface so that the band model can apply in the two-dimensional sense. If the thickness of the passive film is less than 1 nm, the electron level would be localized in individual ions rather than forming the delocalized energy band in the direction vertical to the interface. As the thickness of the film increases, the macroscopic lattice periodicity gradually develops in the thickness direction where the band model can apply.

Observations with microcrystals of semiconductor silicon have shown that the transition from the model of localized electron levels (*quantum size*) to the band model of delocalized electron levels (microscopic or macroscopic size) occurs at about 2 nm [Kanemitsu-Uto-Masumoto, 1993]. It appears, therefore, that the band model can apply to passive films thicker than 2 nm. Further, accounting that the film interacts with the substrate metal, the band model may apply even to the range of thickness less than 2 nm.

In the range of potential of the passive state the passive oxide film is in the state of *band edge level pinning* at the film/solution interface; hence, the potential difference $\Delta\phi_H$ across the film/solution interface remains constant irrespective of the electrode potential of the passive metal. Therefore, with increasing anodic

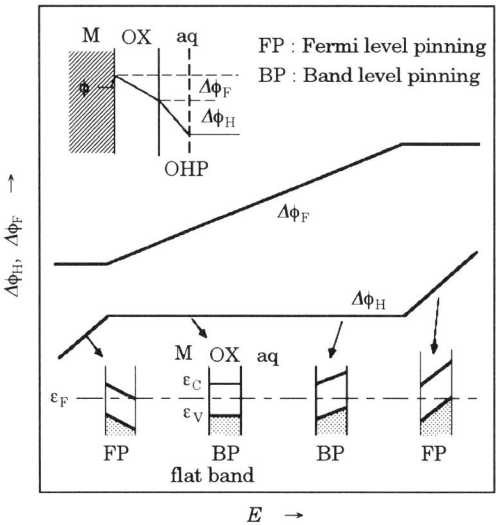

Fig. 11–11. Potential drops $\Delta\phi_H$ at the film/solution interface and $\Delta\phi_F$ in the passive film as a function of anodic potential E for passive metal electrodes in the stationary state: the interface is in the state of band edge level pinning to the extent that the Fermi level ε_F is within the band gap, but the interface changes to the state of Fermi level pinning as ε_F coincides with the valence band edge ε_V.

polarization and in the stationary state of polarization, the potential drop in the film $\Delta\phi_F$, instead of the potential difference $\Delta\phi_H$ at the film/solution interface, increases thereby producing both a potential-dependent thickness of the film and a potential-independent dissolution current of the film. Fig. 11–11 shows that band edge level pinning is maintained to the extend that the Fermi level is within the band gap at the film/solution interface (Refer to Sec. 5.9.).

Fig. 11–12 shows the anodic polarization curve of the iron electrode in the stationary state in a solution of sulfuric acid. The passive film on the iron electrode is an n-type semiconductor of iron oxide (γ-Fe_2O_3) with the band gap about 1.6 eV [Ohtsuka-Azumi-Sato, 1983]. There is a relatively wide energy gap ($\sim$ 1.3 eV) between the valence band edge and the Fermi level at the flat band potential ($E_{fb} \doteq 0.43\ V_{NHE}$) [Stimming-Schultze, 1976]. As shown in Fig. 11–12, the transpassivation potential E_{TP} is 1.3 V more anodic (more positive) than the flat band potential E_{fb}. With increasing anodic polarization from the range of potential of the passive state to the range of potential of the transpassive state, the film/solution interface changes from the state of *band edge level pinning* to the state of *Fermi level pinning*. Therefore, in the range of potential of the transpassive state, the potential difference $\Delta\phi_H$ across the film/solution interface increases with increasing anodic polarization producing thereby a film dissolution current dependent on the electrode potential (the *transpassive dissolution*).

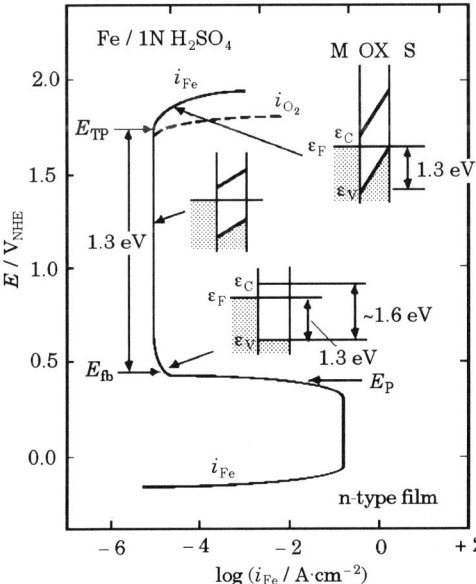

Fig. 11–12. Anodic polarization curve for an iron electrode in a sulfuric acid solution: the transpassivation starts at a potential at which the Fermi level ε_F is pinned at the valence band edge ε_V. [From Sato, 1982.]

In the case of the nickel electrodes on which the passive film is a p-type nickel oxide (NiO), the energy gap (~ 0.2 eV) between the valence band edge and the Fermi level at the flat band potential is small so that the transpassivation potential E_{TP} is relatively close to the flat band potential E_{fb} as in Fig. 11–13.

With respect to anodic dissolution, it appears, from comparing Fig. 11–12 with Fig. 11–13, that the passive films of *n-type* oxides are *more stable* against the anodic dissolution than the passive films of p-type oxides [Sato, 1982].

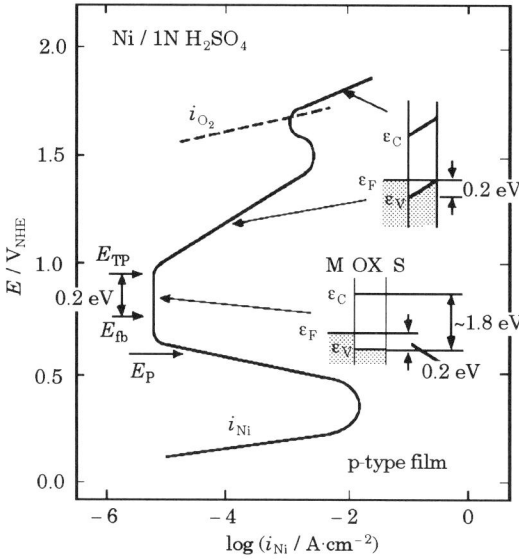

Fig. 11–13. Anodic polarization curve for a metallic nickel electrode in a sulfuric acid solution: the transpassivation arises at a potential relatively close to the flat band potential because of the p-type nature of the passive oxide film. [From Sato, 1982.]

For metallic iron and nickel electrodes, the transpassive dissolution causes no change of valence of metal ions during anodic transfer of metal ions across the film/solution interface (non-oxidative dissolution). However, there are some metals in which transpassive dissolution proceeds by the oxidative mode of film dissolution (Refer to Sec. 9.2.). For example, in the case of the chromium electrodes, on which the passive film is trivalent chromium oxide (Cr_2O_3), the transpassive dissolution proceeds via soluble hexavalent chromate ions. This process can be expressed by the oxidative dissolution: $Cr_2O_3 + 4\,H_2O_{aq} \rightarrow Cr_2O_{7,aq}^{2-} + 8\,H_{aq}^+ + 6\,e$.

11.4.3 Spontaneous passivation of metal electrodes

As described in Sec. 11.3, the spontaneous corrosion potential of a corroding metal is represented by the intersection of the anodic polarization curve of metal dissolution with the cathodic polarization curve of oxidant reduction (Figs. 11–5 and 11–6). Therefore, whether a metal electrode is in the active or passive states is determined by the intersection of the anodic and cathodic polarization curves.

It is well known that metallic iron corrodes violently in dilute nitric acid solutions, but metallic iron is passivated in concentrated nitric acid solutions as shown in Fig. 11–14(a). This passivation of metallic iron results from the strong oxidizing action of concentrated nitric acid that changes the iron electrode from the active state to the passive state.

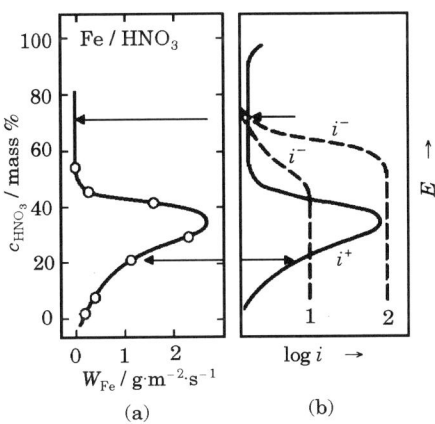

Fig. 11–14. (a) Corrosion rate of metallic iron in nitric acid solution as a function of the concentration of nitric acid and (b) schematic polarization curves for mixed electrode reactions of a corroding iron in nitric acid: W_{Fe} = iron corrosion rate; c_{HNO_3} = concentration of nitric acid; i^+ (i^-) = current of anodic iron dissolution (cathodic nitric acid reduction); dotted curve 1 = cathodic current of reduction of nitric acid in dilute solutions; dotted curve 2 = cathodic current of reduction of nitric acid in concentrated solutions. [From Tomashov, 1966 for (a).]

The cathodic reaction of nitric acid reduction is represented by Eqn. 11–15:

$$NO_{3,\,aq}^- + 3\,H_{aq}^+ + 2\,e \rightarrow HNO_{2,\,aq} + H_2O \,. \tag{11-15}$$

The cathodic current of this reaction increases with increasing concentration of nitric acid as shown by cathodic polarization curves (dotted curve) in Fig. 11–14(b).

The intersection of the anodic polarization curve of iron dissolution with the cathodic polarization curve of nitric acid reduction occurs in the range of potential of the active state in dilute nitric acid but is in the range of potential of the passive state in concentrated nitric acid as shown in Fig. 11–14(b). Consequently, the iron electrode hardly corrodes in concentrated nitric acid, although nitric acid is a strong corrosive medium. Such a passivation of metals in oxidizing environments is called *spontaneous passivation* (self-passivation) and occurs without applying external anodic polarization.

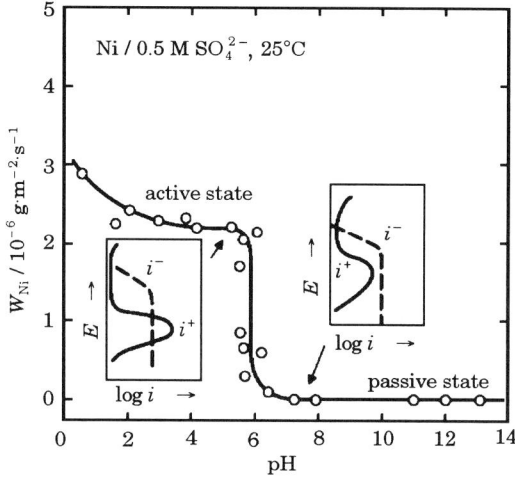

Fig. 11-15. Corrosion rate of metallic nickel in sulfate solutions (0.5 M Na_2SO_4) as a function of pH at 25°C: inserted sub-figures are polarization curves of nickel electrodes in acidic solutions and in basic solutions. [From Okamoto-Sato, 1959.]

Fig. 11–15 shows the corrosion rate observed for a metallic nickel electrode in aerated aqueous sulfate solutions as a function of pH. In acidic solutions, nickel corrodes in the active state at a rate which is controlled by the diffusion of hydrated oxygen molecules (oxidants). However, in solutions more basic than pH 6 nickel spontaneously passivates by hydrated oxygen molecules and corrosion is negligible. As shown in the inserted sub-figures in Fig. 11–15, the maximum current of anodic dissolution of nickel in the active state is greater in the range of acidic pH; however, the maximum current of anodic dissolution is smaller in the range of basic pH than the current of cathodic reduction of oxygen molecules (dotted curve) which is controlled by the diffusion of hydrated oxygen molecules. Consequently, metallic nickel remains in the active state in acidic solutions but is spontaneously passivated by oxygen molecules in basic solutions. It is obvious from the result of Fig. 11–15 that the transition from the active state to the passive state is not continuous, but occurs discontinuously at a threshold of pH.

Almost all metallic materials in practical environments perform their service in the state of spontaneous passivation, in which hydrated oxygen molecules or hydrogen ions act as oxidants to passivate the surfaces. Stainless steel is a good and widely known example of corrosion resistant metals; it is spontaneously passivated and remains in the passive state with a thin passive oxide film even in fairly corrosive environments.

References

[Evans, 1946]: U. R Evans, *Metallic Corrosion, Passivation and Protection*, Edward Arnold & Co., London, (1946).

[Kanemitsu-Uto-Masumoto]: Y. Kanemitsu, U. Uto and Y. Masumoto, *Phys. Rev.*, **B48**, 2827(1993).

[Kaesche, 1979]: H. Kaesche, *Die Korrosion der Metalle*, **2** Auflage, p. 122, Springer-Verlag, New York, (1979).

[Ohtsuka-Azumi-Sato, 1983]: T. Ohtsuka, K. Azumi and N. Sato, *J. Physique*, Colloque, **C10**, 191(1983).

[Okamoto-Sato, 1959]: G. Okamoto and N. Sato, *J. Japan Institute of Metals*, **23**, 725(1959).

[Sato, 1976]: N. Sato, *Passivity and Its Breakdown on Iron and Iron Base Alloys*, (Edited by R. W. Staehle and H. Okada), p. 1, National Association of Corrosion Engineers, Houston, (1976).

[Sato-Okamoto, 1981]: N. Sato and G. Okamoto, *Comprehensive Treatise of Electrochemistry*, **Vol. 4**, p. 193, (Edited by J. O' M. Bockris et al.), Plenum Publishing Corp., New York, (1981).

[Sato, 1982]: N. Sato, *J. Electrochem. Soc.*, **129**, 255(1982).

[Sato, 1990]: N. Sato, *Corrosion Science*, **31**, 1(1990).

[Stimming-Schultze, 1976]: U. Stimming and J. W. Schultze, *Ber. Bunsenges. Phy. Chem.*, **80**, 1297(1976).

[Stimming-Schultze, 1979]: U. Stimming and J. W. Schultze, *Electrochimica Acta*, **24**, 859(1979).

[Tomashov, 1966]: N. D. Tomashov, *Theory of Corrosion and Protection of Metals*, p. 325, Macmillan Co., New York, (1966).

[Wagner-Traud, 1938]: C. Wagner and W. Traud, *Z. Elektrochem.*, **44**, 391(1938).

LIST OF SYMBOLS

A	affinity of a reaction
A^+, A^-	forward or backward affinity of a reaction
A_{eq}^+, A_{eq}^-	forward or backward affinity of a reaction in equilibrium
A_i	electron affinity of a particle i
a_{eq}	activity of a particle in reaction equilibrium
a_i	activity of a particle i
C	electric capacity
C_d	electric capacity of a diffuse layer in aqueous solution
C_H	electric capacity of a compact layer (Helmholtz layer) at electrode interfaces
C_{SC}	electric capacity of a space charge layer in semiconductors
C_{ss}	electric capacity of the surface state in semiconductors
c_i	concentration of a particle i
c_i^0	concentration of a particle i in reaction equilibrium
$D(\varepsilon)$	state density of electrons
$D_C(\varepsilon)$	state density of electrons in the conduction band of semiconductors
D_{H_2O}	ionic dissociation energy of H_2O in aqueous solution
D_i	state density of electrons in a redox particle i
$D_M(\varepsilon)$	state density of electrons in metals
$D_{SC}(\varepsilon)$	state density of electrons in semiconductors
$D_{OX}(\varepsilon)$	state density of redox electrons in oxidant particles
$D_{RED}(\varepsilon)$	state density of redox electrons in reductant particles
$D_{REDOX}(\varepsilon)$	total state density of redox electrons in redox particles
$D_V(\varepsilon)$	state density of electrons in the valence band of semiconductors
D_{XY}	ionic dissociation energy of a compound XY
d_{SC}	thickness of a space charge layer in semiconductors
E_{av}	average electric field strength in a compact layer on electrodes
E_H	electric field strength in a compact layer (Helmholtz layer)
E_{SC}	electric field strength in a space charge layer
E	electrode potential (absolute electrode potential)
E_{eq}	electrode potential (absolute) in reaction equilibrium
E_0	standard electrode potential (absolute) in reaction equilibrium

$\boldsymbol{E}_{em}$	potential of "emersed" electrodes (absolute)
$\boldsymbol{E}_{fb}$	flat band potential (absolute) of semiconductor electrodes
$\boldsymbol{E}_{pzc}$	potential (absolute) of zero charge
E	electrode potential (relative)
E_C^s, E_V^s	electrode potential (relative) of the conduction or valence band edge at electrode surfaces (interfaces) of semiconductors
E_{cell}	electromotive force of electrochemical cells
E_{corr}	corrosion potential (relative) of metal electrodes
E_{fb}	flat band potential (relative) of semiconductor electrodes
$_nE_{fb}, E_{fb}^n$	flat band potential (relative) of n-type semiconductor electrodes
$_pE_{fb}, E_{fb}^p$	flat band potential (relative) of p-type semiconductor electrodes
E_{NHE}	electrode potential referred to the normal hydrogen electrode
ΔE_{ph}	photopotential
e_M	electron in metals
$e_{(REDOX)}$	redox electron in equilibrium with redox reactions
$e_{(NHE)}$	redox electron in equilibrium with the normal hydrogen electrode
$e_{(STD)}$	gaseous electron in the standard state
$e_{(vac)}$	isolated electron in vacuum
$f(\varepsilon)$	Fermi function
ΔG	change in the free enthalpy (Gibbs energy)
$g_{A/B}$	electrostatic potential difference across an interface A/B
g_{dip}	potential difference due to an electric dipole across an interface
g_{ion}	potential difference due to an excess ionic charge across an interface
Δg_r	change in the free enthalpy of an elemental step, r, in a reaction
Δg^*	activation energy of an elemental step in a reaction
$\Delta g^*_{(eq)}$	activation energy of an elemental step in reaction equilibrium
I_i	ionization energy of a particle i
i	reaction current
i_0	exchange reaction current in equilibrium
i_a, i^+	anodic reaction current
i_c, i^-	cathodic reaction current
i_{corr}	corrosion current of metals
i_{diff}	diffusion current of a particle
i_{lim}	limiting diffusion current of a particle
i_n, i_p	reaction current of electrons or holes at semiconductor electrodes
$i_{n,0}, i_{p,0}$	exchange reaction current of electrons or holes in equilibrium
i_p	polarization current at electrodes
i_{ph}	photocurrent (photoexcited reaction current)
i_{ph}^+, i_{ph}^-	anodic or cathodic photocurrent (photoexcited reaction current)
i_{rec}	recombination current of electron-hole pairs in semiconductors

LIST OF SYMBOLS

J_i	hydration energy of an ion i
$k_t(\varepsilon)$	rate constant of tunneling transfer of electrons
k^+, k^-	rate constant of forward or backward reactions
L_D	Debye length
L_i	sublimation energy of a constituent particle i in solid compounds
m	distribution ratio of the affinity to an elemental step in a reaction
m	Frumkin parameter (an interaction parameter) in adsorption
m_e	effective mass of an electron
m_i	concentration in molality of a particle i
m_{OX}	transport coefficient of an oxidant particle
m_{RED}	transport coefficient of a reductant particle
N_C	effective state density of electrons in the conduction band of semiconductors
N_{OX}	concentration of oxidants
N_{RED}	concentration of reductants
N_{ss}	density (concentration) of the surface states on semiconductors
N_V	effective state density of electrons in the valence band of semiconductors
n	concentration of electrons in the conduction band of semiconductors
n_i	concentration of a particle i
n_s	concentration of electrons at semiconductor surfaces (interfaces)
n^*	concentration of electrons in the photostationary state in semiconductors
p	concentration of holes in the valence band of semiconductors
p_s	concentration of holes in the valence band at semiconductor surfaces (interfaces)
p_i	pressure of a particle i
p^*	concentration of holes in photoexcited semiconductors
pH_{iep}	isoelectric point pH
Q_{ss}	charge in the surface state and other surface charges
R	electric resistance of aqueous solution
R_p	polarization resistance of electrodes
R_t	reaction resistance of particle transfers near equilibrium
V_i	vacancy of a particle i in solid crystals
V_{SC}	potential difference across a space charge layer in semiconductors
v	reaction rate
v^+, v^-	rate of forward or backward reactions
v_{eq}	exchange reaction rate in equilibrium
v_T	overall reaction rate of a reaction
$W(\varepsilon)$	probability density of the electron energy state as a function of ε

W_i	probability density of the electron energy state in a redox particle i
w_a	probability for a thermally activated transfer of particles
w_t	probability for a quantum mechanical tunneling of particles
x_{dip}	distance from the metal surface to the closest approach of adsorbed water molecules
x_i	molar fraction of a particle i
x_{im}	distance from the metal surface to the effective image plane
Y_i	sublimation energy of a particle i
z	ionic charge number (ion valence)
α	transfer coefficient (Tafel constant) in electrode reactions
α^+, α^-	anodic or cathodic transfer coefficient in an electrode reaction
α_i	real potential of a particle i
α_i^*	unitary real potential of a particle i
α_i^0	standard real potential of a particle i
β	symmetry factor for an elemental step in a reaction
γ	interfacial tension (surface tension)
γ	activity coefficient
γ_i	activity coefficient of a particle i
Γ	interfacial excess
Γ_i	interfacial excess of a particle i
ε	permittivity (dielectric constant)
ε	energy level
ε^s	electron energy level at surfaces (interface) of semiconductor electrodes
$\varepsilon_C, \varepsilon_V$	electron energy level of the conduction or valence band edges in semiconductor
$\varepsilon_C^s, \varepsilon_V^s$	electron energy level of the conduction or valence band edges at semiconductor electrode interfaces
ε_e	energy of electrons
ε_F	Fermi level
ε_g	band gap energy in semiconductors
$\varepsilon_{F(M)}$	Fermi level of electrons in metals
$\varepsilon_{F(REDOX)}$	Fermi level of redox electrons
$\varepsilon_{F(REDOX)}^0$	standard Fermi level of redox electrons
$\varepsilon_{F(SC)}$	Fermi level of electrons in semiconductors
$\varepsilon_{F(SC)}^i$	Fermi level of electrons in intrinsic semiconductors
$\varepsilon_{F(aq)}^0$	Fermi level of standard redox electrons in hydrated redox particles
$\varepsilon_{F(comp)}^0$	Fermi level of standard redox electrons in complexed redox particles

LIST OF SYMBOLS

$\varepsilon^0_{F(ad)}$	Fermi level of standard redox electrons in adsorbed redox particles
$_n\varepsilon_F, _p\varepsilon_F$	Fermi level of n-type or p-type semiconductor electrodes
$_n\varepsilon_F^*$	quasi-Fermi level of electrons in semiconductor electrodes
$_p\varepsilon_F^*$	quasi-Fermi level of holes in semiconductor electrodes
ε_i	energy of a particle i
ε_{mg}	mobility gap in amorphous semiconductors
$\varepsilon_{OX}, \varepsilon_{RED}$	most probable electron level of oxidant or reductant particles
ε_{STD}	standard gaseous electron level ($\varepsilon_{STD} = 0$)
ε_{ss}	electron level of the surface states on semiconductors
ζ_k	reaction order with respect to a particle k in a reaction
η	rate of increase in electric field across the inner Helmholtz layer
η	overvoltage of an electrode reaction
η_H	overvoltage across a compact layer at electrodes
η_H	interfacial overvoltage at photoexcited semiconductor electrodes
η_{SC}	overvoltage across a space charge layer in semiconductors
η_T	overvoltage for transport of charge carriers in semiconductors
$\eta_{n,SC}$	overvoltage for generation and transport of electrons in semiconductors
$\eta_{p,SC}$	overvoltage for generation and transport of holes in semiconductors
θ	adsorption coverage
θ_i	adsorption coverage of a particle i
κ	relative dielectric constant
λ	a kinetic parameter
λ	wave length of photons
λ,	reorganization energy of redox particles, oxidants or reductants
λ_{ad}	reorganization energy of adsorbed redox particles
λ_i	absolute activity of a particle i
λ_F	Fermi wave length of electrons
$\lambda_{OX}, \lambda_{RED}$	reorganization energy of oxidant or reductant particles
μ	dipole moment
μ_{ad}	adsorption-induced dipole moment
μ_i	chemical potential of a particle i
μ_i^*	unitary chemical potential of a particle i
μ_i^0	standard chemical potential of a particle i
$\bar{\mu}_i$	electrochemical potential of a charged particle i
$\bar{\mu}_i^0$	standard electrochemical potential of a charged particle i
ν_i	stoichiometrical coefficient of a particle i in a reaction
ν_r^*	stoichiometric number of an elemental step, r, in a reaction

LIST OF SYMBOLS

σ	electric charge for unit area
σ_{ad}	electric charge due to surface adsorption for unit area
σ_M	electric charge on the metal side of metal elecxytodes for unit area
σ_S	electric charge on the solution side of electrodes for unit area
σ_{SC}	electric charge in a space charge layer for unit area of semiconductors
σ_{ss}	electric charge of the surface states for unit area
ϕ	electrostatic inner potential
$\Delta\phi_H$	potential difference across a compact layer (Helmholtz layer)
$\Delta\phi_{SC}$	potential difference across a space charge layer in semiconductor
ψ	electrostatic outer potential
χ	surface potential difference (difference in potential across a surface)
Φ	work function

INDEX

A
absolute activity 4
absolute electrode potential 99
acceptor 25
acceptor level 27
accumulation layer 42, 175, 177
acid site 181,189
acidic proton hydration 78
acidic proton level 78, 320
acidic proton transfer 82
activated flow 224
activated state 290
activation energy 73, 222
active state 382
activity 5, 33
activity coefficient 5, 33
adsorbed intermediate 294
adsorbed redox particle 165
adsorption 121, 138, 158
adsorption energy 124
adsorption isotherm 143
adsorption of redox particle 274
adsorption-induced dipole moment 155
adsorption-induced surface state 41
amorphous semiconductor 44
anion 88
anode 201, 213
anodic polarization 219
anodic reaction 213
antibonding band 24, 36

B
band edge level pinning 171, 252
band edge potential 195
band gap 24, 37
band model 19, 24, 32
base site 181, 189
basic proton hydration 79
basic proton level 78, 320
Boltzmann factor 3
Boltzmann function 3
bonding band 24, 36
Bose particle 2

Bose-Einstein function 2
Bose-Einstein level 2

C
capacity of space charge layer 177
cathode 201, 213
cathodic hole injection 347, 355
cathodic polarization 219
cathodic reaction 213
cation 88
cation–anion vacancy pair formation 73
cell diagram 203
cell reaction 202
charge transfer coefficient 155, 167
chemical cell 202
chemical potential 4
chemical potential of electron 8, 22
chemisorption 121
clathrate hydrates 162
clean surface 119
cluster-like network structure 159
coadsorption 122
communal energy 5
compact layer 128, 132, 148, 150, 154, 156, 181, 184
compact layer capacity 148
complexation 274
conduction band 24
conduction band mechanism 251
contact adsorption 140, 142, 153, 155, 163, 278
contact potential 92, 99
corrosion potential 377
covalent adsorption 125
covalent semiconductor 298, 302
cratic energy 5

D
Debye length 130, 132, 169, 176
deep depletion layer 175, 179
degeneracy 44, 174
depletion layer 42, 175, 178
diffuse charge layer 129

D

diffuse layer 130
diffusion overvoltage 246
direct band gap 24, 36
direct electron transfer 281
donor 25
donor level 27

E

effective Debye length 177
effective excess charge plane 146
effective image plane 144
effective state density 27, 28, 251
electric double layer 127, 168
electromotive force 110
electrocapillary curve 139
electrochemical cell 201
electrochemical potential 4
electrochemical potential of electron 8, 20
electrochemical potential of ion 64
electrode 87
electrode potential 96, 99, 100, 104, 105, 108, 109
electrode reaction 213
electrolytic cell 202
electromotive force 110, 205, 367, 369
electron accepting reaction 47
electron donating reaction 47
electron transfer 213, 214, 235, 237, 249
electron–hole pair formation 72
electronic electrode 88
electronic electrode potential 102
emersed electrode 112
energy band 16
equilibrium electrode potential 206
equivalent Fermi level 107, 311
exchange reaction current 240, 254
exponential kinetics 218
extrinsic semiconductor 25

F

faradaic current 90
Fermi energy 2
Fermi function 1
Fermi level 2, 5, 172
Fermi level of intrinsic semiconductor 29
Fermi level of metal 19
Fermi level of n-type semiconductor 31
Fermi level of p-type semiconductor 31
Fermi level of redox electron 239
Fermi level pinning 42, 252
Fermi particle 1
Fermi temperature 3
Fermi wave length 21
fermion 1

flat band potential 185, 192, 337, 344
fluctuation band 54
fluctuation of electron energy level 51
Franck-Condon level splitting 50
free energy 4
free enthalpy 4
Frenkel defect 74
frontier band 19
Frumkin's adsorption isotherm 143

G

Galvani potential difference 92
Galvanic cell 201
gap state 45
gas electrode 108
Gibbs adsorption equation 138

H

Henry's law 5
hole 24, 27
hopping model 32
hydrogen electrode 108
hydrogen-oxygen fuel cell 202
hydroxyl radical 160
hydroxylation 181

I

IHP (inner Helmholtz plane) 141
indirect band gap 24, 36
indirect electron transfer 281
inner Helmholtz layer 141, 156, 157
inner potential 9, 10, 90
inner potential difference 92, 95
inner-sphere electron transfer 216, 278, 283
interfacial charge 92
interfacial complex 182
interfacial dipole 92
interfacial ion pair 182
interfacial lattice transformation 162
interfacial overvoltage 246, 350
interfacial potential difference 95
interfacial redox electron 315
interfacial redox reaction 317
interfacial state 39
intrinsic semiconductor 25, 27
inverse overvoltage 349, 353, 363, 365, 368
inversion layer 175, 179
ion adsorption 314, 317
ion transfer 213, 289, 302, 306, 314, 317
ionic electrode 88
ionic electrode potential 102
ionic semiconductor 306, 309
iso-electric point 183, 321

INDEX

J
jellium model of metal 21

K
Kelvin's method 114
kink site 66, 69

L
Langmuir's adsorption isotherm 143, 296
Lewis acid 158
Lewis base 158
linear kinetics 218
localized electron level 29, 38, 45, 47, 51

M
majority charge carrier 32
mean stoichiometric number 230
metal ion level 63
metal oxide 35
metallic adsorption 125
minority charge carrier 32
mixed electrode 373
mixed electrode potential 377
mobility edge level 45
mobility gap 45
mol fraction 77
molality 76
molar fraction 4
most probable electron level 52, 55, 238
Mott-Schottky plot 178

N
nonfaradaic current 90
nonpolarizable electrode 89
normal hydrogen electrode 55, 110
n-type semiconductor 25, 29

O
occupied proton level 78
OHP (outer Helmholtz plane) 128
operational potential 363
outer Helmholtz layer 141, 156
outer potential 10, 90
outer potential difference 92, 95
outer-sphere electron transfer 215, 235, 249, 283, 285
overvoltage 220, 258, 348
overvoltage for hole generation and transport 349
oxidant 47
oxidation 47
oxidative dissolution 309

P
passivation 382
passivation potential 382
passive state 382
Peltier effect 206
photocurrent 334
photoelectrochemical semiconductor cell 357
photoelectrode reaction 334
photoelectrolytic cell 357, 364
photoelectron 325
photoexcitation 325
photopotential 330
photo-shift of flat band potential 344
photostationary state 325
photovoltaic cell 357, 367
physisorption 121
polarizable electrode 89
polarization 89, 219
polarization curve 218
polarization resistance 292, 379
potential of space charge layer 43
potential of zero charge 130, 135, 139
proton level 62
proton level on electrode surface 319
p-type semiconductor 25, 29

Q
quasi-Fermi level 326, 348
quasi-metallization of semiconductor surface 44, 171

R
Raoult's law 5
rate-determining multiple steps 230
ratio of affinity distribution 231
reaction affinity 216
reaction current 242, 259
reaction order 225
reaction rate 217
reaction resistance 292
real potential 12
real potential of electron 21, 97
real potential of hydrated ion 76
recombination current 268
recombination of photoexcited electron–hole pair 352
redox electron 47, 53, 105
redox electron level 54
redox electron transfer 268
redox particle 48, 88, 235
redox reaction 47
reductant 47
reduction 47
reductive dissolution 311

relative electrode potential 112
reorganization energy 48, 51, 279

S
Schottky barrier 181
Schottky defect 74
self-passivation 388
single electrode 373
single rate-determining step 228
space charge layer 42, 129, 174, 176
specific adsorption 140
spontaneous passivation 388
standard electrochemical potential of electron 33
standard equilibrium potential 208
standard Fermi level of redox electron 238, 255
standard gaseous electron 56
standard gaseous electron level 9, 48
standard ion level 77
standard real potential of ion 77
standard redox electrode potential 105, 207
state density 6, 18
stoichiometric number 220
surface atom ionization 298
surface dangling state 40
surface degeneracy 44
surface Fermi level 42
surface ion level 74
surface ion-induced state 41
surface lattice transformation 120, 162
surface metal ion level 65
surface potential difference 10, 21, 46, 90
surface radical 298
surface reconstruction 119
surface relaxation 119
surface state 39, 188, 272
surface state capacity 191
symmetry factor 223, 239, 290

T
Tafel constant 225, 264, 293
Tafel equation 245
Tafel plot 220
Tafel slope 220
Temkin's adsorption isotherm 296
thermodynamic potential 2
transpassivation 382
transfer coefficient 225
transition metal oxide 36
transpassivation potential 382
transpassive state 382
transport coefficient 247
transport overvoltage 266

tunneling transfer 223

U
undervoltage 349
unitary electrochemical potential of electron 32
unitary energy level 5
unitary ion level 61
unitary real potential of ion 68
unitary surface ion level 63, 65, 67, 71

V
vacant proton level 78
vacuum electron level 8
valence band 24
valence band mechanism 251
Volta potential difference 92

W
work function 21, 23

Z
zero charge interface 93